MIMO Wireless Communications

Multiple-input multiple-output (MIMO) technology constitutes a breakthrough in the design of wireless communication systems, and is already at the core of several wireless standards. Exploiting multi-path scattering, MIMO techniques deliver significant performance enhancements in terms of data transmission rate and interference reduction. This book is a detailed introduction to the analysis and design of MIMO wireless systems. Beginning with an overview of MIMO technology, the authors then examine the fundamental capacity limits of MIMO systems. Transmitter design, including precoding and space–time coding, is then treated in depth, and the book closes with two chapters devoted to receiver design. Written by a team of leading experts, the book blends theoretical analysis with physical insights, and highlights a range of key design challenges. It can be used as a textbook for advanced courses on wireless communications, and will also appeal to researchers and practitioners working on MIMO wireless systems.

Ezio Biglieri is a professor in the Department of Technology at the Universitat Pompeu Fabra, Barcelona.

Robert Calderbank is a professor in the Departments of Electrical Engineering and Mathematics at Princeton University, New Jersey.

Anthony Constantinides is a professor in the Department of Electrical and Electronic Engineering at Imperial College of Science, Technology and Medicine, London.

Andrea Goldsmith is a professor in the Department of Electrical Engineering at Stanford University, California.

Arogyaswami Paulraj is a professor in the Department of Electrical Engineering at Stanford University, California.

H. Vincent Poor is a professor in the Department of Electrical Engineering at Princeton University, New Jersey.

MIMO Wireless Communications

EZIO BIGLIERI
Universitat Pompeu Fabra

ROBERT CALDERBANK
Princeton University

ANTHONY CONSTANTINIDES
Imperial College of Science, Technology and Medicine

ANDREA GOLDSMITH
Stanford University

AROGYASWAMI PAULRAJ
Stanford University

H. VINCENT POOR
Princeton University

CAMBRIDGE UNIVERSITY PRESS
Cambridge, New York, Melbourne, Madrid, Cape Town, Singapore, São Paulo

Cambridge University Press
The Edinburgh Building, Cambridge CB2 2RU, UK

Published in the United States of America by Cambridge University Press, New York

www.cambridge.org
Information on this title: www.cambridge.org/9780521873284

First published 2007

Printed in the United Kingdom at the University Press, Cambridge

A catalog record for this publication is available from the British Library

ISBN-13 978-0-521-87328-4 hardback
ISBN-10 0-521-87328-2 hardback

To our families.

Contents

Contributors

1 Introduction

ROHIT U. NABAR
Marvell Semiconductor, Inc., Santa Clara, California

2 Capacity limits of MIMO systems

SYED ALI JAFAR
University of California, Irvine

NIHAR JINDAL
University of Minnesota, Minneapolis

SRIRAM VISHWANATH
University of Texas at Austin

3 Precoding design

MAI VU
Stanford University, California

4 Space–time coding for wireless communications: principles and applications

NAOFAL AL-DHAHIR
University of Texas at Dallas

SUHAS N. DIGGAVI
École Polytechnique Fédérale de Lausanne (EPFL), Switzerland

5 Fundamentals of receiver design

GIORGIO TARICCO
Politecnico di Torino, Italy

6 Multi-user receiver design

Huaiyu Dai
North Carolina State University, Raleigh

Sudharman Jayaweera
University of New Mexico

Daryl Reynolds
West Virginia University

Xiaodong Wang
Columbia University, New York

Preface

Facies non omnibus una,
Nec diversa tamen
(Ovid, Metamorphoses)

Wireless is one of the most rapidly developing technologies in our time, with dazzling new products and services emerging on an almost daily basis. These developments present enormous challenges for communications engineers, as the demand for increased wireless capacity grows explosively. Indeed, the discipline of wireless communications presents many challenges to designers that arise as a result of the demanding nature of the physical medium and the complexities in the dynamics of the underlying network. The dominant technical issue in wireless communications is that of multipath-induced fading, namely the random fluctuations in the channel gain that arise due to scattering of transmitted signals from intervening objects between the transmitter and the receiver. Multipath scattering is therefore commonly seen as an impairment to wireless communication. However, it can now also be seen as providing an opportunity to significantly improve the capacity and reliability of such systems. By using multiple antennas at the transmitter and receiver in a wireless system, the rich scattering channel can be exploited to create a multiplicity of parallel links over the same radio band, and thereby to either increase the rate of data transmission through multiplexing or to improve system reliability through the increased antenna diversity. Moreover, we need not choose between multiplexing and diversity, but rather we can have both subject to a fundamental tradeoff between the two.

This book addresses multiple-input/multiple-output (MIMO) wireless systems in which transmitters and receivers may have multiple antennas. Since the emergence of several key ideas in this field in the mid-1990s, MIMO systems have been one of the most active areas of research and development in the broad field of wireless communications. An enormous body of work has been created in this area, leading to many immediate applications and to future opportunities. This book provides an entrée into this very active field, aiming at covering the main aspects of analysis and design of MIMO wireless. It is intended for graduate students as well as practicing engineers and researchers with a basic knowledge of digital communications and wireless systems, roughly at the level of [1–4].

The present book gives a unified and comprehensive view of MIMO wireless. After a general overview in Chapter 1, it covers the basic elements of the field in depth, including the fundamental capacity limits of MIMO systems in Chapter 2, transmitter design (including precoding and space–time coding) in Chapters 3 and 4, and receiver design in Chapters 5 and 6. Although the book is designed to be accessible to individual readers, it can also be used as an advanced graduate textbook, either in its entirety, or perhaps in one of two ways: for a course on MIMO Wireless Communication Systems (Chapters 1, 3, 5 and 6) or for a course on Information Theory and Coding in MIMO Wireless (Chapters 1–4).

Barcelona, Spain
London, UK
Princeton, NJ, USA
Stanford, CA, USA

References

[1] S. Benedetto and E. Biglieri, *Principles of Digital Transmission: With Wireless Applications* (New York: Plenum, 1999).
[2] A. Goldsmith, *Wireless Communications* (Cambridge: Cambridge University Press, 2005).
[3] J. Proakis, *Digital Communications,* 4th edn (New York: McGraw-Hill, 2000).
[4] T. Rappaport, *Wireless Communications: Principles and Practice*, 2nd edn (Upper Saddle River, NJ: Prentice-Hall, 2001).

Acknowledgements

This book would not have been possible without the very significant contributions of our contributing co-authors, who worked closely with the principal authors on individual chapters. In particular, we would like to acknowledge the contributions of Rohit Nabar to Chapter 1, Syed Ali Jafar, Nihar Jindal and Sriram Vishwanath to Chapter 2, Mai Vu to Chapter 3, Naofal Al-Dhahir and Suhas Diggavi to Chapter 4, Giorgio Taricco to Chapter 5, and Huaiyu Dai, Sudharman Jayaweera, Daryl Reynolds and Xiaodong Wang to Chapter 6.

The authors also thank the anonymous reviewers who helped shape this book, as well as Hui-Ling Lou, Kedar Shirali, and Peter Loc of Marvell Semiconductor, Inc., for the experimental results provided in Figure 1.16; Holger Boche and Sergio Verdú for their unique insights and helpful suggestions on Chapter 2; Helmut Bölcskei, Robert Heath, Björn Ottersten, Peter Wrycza, and Xi Zhang for their valuable comments on Chapter 3; and Sushanta Das, Sanket Dusad, Christina Fragouli, Anastasios Stamoulis, and Waleed Younis for many stimulating discussions and technical contributions to the results in Chapter 4. Finally, thanks are also due to Phil Meyler of Cambridge University Press for his efficient handing of this project.

Arogyaswami Paulraj and Mai Vu would like to acknowledge the valuable discussions with Professor George Papanicolaou and the members of the Smart Antenna Research Group at Stanford University. This work was supported in part by NSF Contract DMS-0354674-001 and ONR Contract N00014-02-0088. The work of Mai Vu is also supported in part by the Rambus Stanford Graduate Fellowship and the Intel Foundation Ph.D. Fellowship.

The authors further wish to express their gratitude to the following organizations whose support was invaluable in the preparation of this book: the STREP program of the European Commission, the UK Engineering and Physical Sciences Research Council, the US Air Force Research Laboratory, the US Army Research Laboratory, the US Army Research Office, the US Defense Advanced Research Projects Agency, the US National Science Foundation, and the US Office of Naval Research.

Notation

General notation

$\mathbf{X}$	Matrix $\mathbf{X}$ (boldface capital letter)
$\mathbf{x}$	Vector $\mathbf{x}$ (boldface lowercase letter)
$[\mathbf{X}]_{i,j}$	The element on row i and column j of matrix $\mathbf{X}$
$\mathbf{X}^T$	Transpose of $\mathbf{X}$
$\mathbf{X}^*$	Conjugate transpose of $\mathbf{X}$
$\det(\mathbf{X})$	Determinant of $\mathbf{X}$
$\mathrm{tr}(\mathbf{X})$	Trace of $\mathbf{X}$
$\|\|\mathbf{X}\|\|_F$	Frobenius norm of $\mathbf{X}$
$\lambda(\mathbf{X})$	Eigenvalues of $\mathbf{X}$
Λ_X	The diagonal matrix of the eigenvalues of the Hermitian matrix $\mathbf{X}$
Σ_X	The diagonal matrix of the singular values of $\mathbf{X}$
$\mathbf{U}_X$	The eigen- or singular-vector matrix of $\mathbf{X}$
$\mathbf{X} \succcurlyeq \mathbf{0}$	$\mathbf{X}$ is positive semi-definite
$\mathrm{vec}(\mathbf{X})$	Vectorize $\mathbf{X}$ into a vector by concatenating the columns of $\mathbf{X}$
$\otimes$	Kronecker product
$\mathbf{I}$	An identity matrix
$E[\cdot]$	Expected value
$(x)_+$	$= \begin{cases} x & \text{if } x \geq 0, \quad x \in \mathcal{R} \\ 0 & \text{if } x < 0, \quad x \in \mathcal{R} \end{cases}$

Symbols

M_T	The number of transmit antennas
M_R	The number of receive antennas
$\mathbf{H}$	A MIMO flat-fading channel
$\mathbf{H}_w$	A random channel with i.i.d. zero-mean complex Gaussian elements
$\mathbf{H}_m$	The channel mean
$\mathbf{R}$	A covariance of the channel
$\mathbf{R}_t$	Transmit covariance, also called the transmit antenna correlation
$\mathbf{R}_r$	Receive covariance, also called the receive antenna correlation
K	The Ricean K factor
T_c	The channel coherence time
B_c	The channel coherence bandwidth
D_c	The channel coherence distance
(t)	At time or delay t

ρ	Channel temporal correlation function
$\mathbf{F}$	The precoding matrix
p_i	Power loading on beam i
$\mathbf{C}$	A codeword
$\mathbf{Q}$	The codeword covariance matrix
$\mathbf{A}$	The codeword difference product matrix
γ	The signal-to-noise ratio

Abbreviations

APP	*A posteriori* probability
ARQ	Automatic repeat request
AWGN	Additive white Gaussian noise
BC	Broadcast channel
BCJR	Bahl–Cocke–Jelinek–Raviv
BER	Bit-error rate
BLAST	Bell Laboratories space–time
bps	Bits per second
BPSK	Binary phase-shift keying
CCI	Channel covariance information
CDF	Cumulative distribution function
CDI	Channel distribution information
CDIR	Receiver channel distribution information
CDIT	Transmitter channel distribution information
CDMA	Code-division multiple access
CDMA 2000	A CDMA standard
CIR	Channel impulse response
CMI	Channel mean information
CP	Cyclic prefix
CSI	Channel state information
CSIR	Receiver channel state information
CSIT	Transmitter channel state information
dB	Decibels
DDF	Decorrelating decision feedback
DFT	Discrete Fourier transform
DPC	Dirty paper coding
DS	Direct-sequence
DSL	Digital subscriber line
EDGE	Enhanced data rate for GSM evolution
EM	Expectation-maximization
EXIT	Extrinsic information transfer
FDD	Frequency-division duplex
FDE	Frequency domain equalizer
FDMA	Frequency-division multiple access

FER	Frame-error rate
FFT	Fast Fourier transform
FIR	Finite impulse response
GSM	Global system for mobile communications, a second-generation mobile communications standard
IBI	Inter-block interference
IC	Interference cancellation
IEEE	Institute of Electrical and Electronic Engineers
IFC	Interference channel
IFFT	Inverse FFT
iid	Independent, identically distributed
IO	Individually optimal
ISI	Intersymbol interference
JO	Jointly optimal
KKT	Karush–Kuhn–Tucker
LDC	Linear dispersion code
LDPC	Low-density parity check
LLR	Logarithmic likelihood ratio
LMMSE	Linear minimum mean-square error
LMS	Least mean-squares
LOS	Line-of-sight
MAC	Multiple-access channel
MAI	Multiple-access interference
MAP	Maximum *a posteriori* probability
MBWA	Mobile broadband wireless access
MIMO	Multiple-input multiple-output
MISO	Multiple-input single-output
ML	Maximum likelihood
MMSE	Minimum mean-square error
MRC	Maximum ratio combining
MSE	Mean-square error
MU	Multi-user
MUD	Multi-user detection
NAHJ-FST	Noise-averaged Hamilton–Jacobi fast subspace tracking
NUM	Network utility maximization
OFDM	Orthogonal frequency-division multiplexing
OFDMA	Orthogonal frequency-division multiple access
PEP	Pairwise error probability
PRUS	Perfect root of unity sequences
PSD	Positive semi-definite
PSK	Phase shift keying
QAM	Quadrature amplitude modulation

QCI	Quantized channel information
QPSK	Quadrature phase-shift keying
QSTBC	Quasi-orthogonal STBC
RF	Radio frequency
RLS	Recursive least squares
RSC	Recursive systematic convolutional
RV	Random variable
SAGE	Space-alternating generalized EM
SC	Single carrier
SIMO	Single-input, multiple-output
SINR	Signal-to-interference-plus-noise ratio
SISO	Single-input, single-output
SI/SO	Soft-input/soft-output
SNR	Signal-to-noise ratio
SPA	Sum–product algorithm
ST	Space–time
STBC	Space–time block code
STC	Space–time coding/space–time code
STTC	Space–time trellis code
SU	Single user
SVD	Singular-value decomposition
TCP	Transport control protocol
TDD	Time-division duplex
TDMA	Time-division multiple access
36PP	36 Partnership project
TWLK	Tanner–Wieberg–Loeliger–Koetter
UEP	Unequal error protection
V-BLAST	Vertical BLAST
WCDMA	Wideband code-division multiple access
WiMAX	IEEE 802.16 standard
WLAN	Wireless local area network
WMAN	Wireless metropolitan area network
ZF	Zero-forcing
ZMSW	Zero mean spatially white

1 Introduction

1.1 MIMO wireless communication

The use of multiple antennas at the transmitter and receiver in wireless systems, popularly known as MIMO (multiple-input multiple-output) technology, has rapidly gained in popularity over the past decade due to its powerful performance-enhancing capabilities. Communication in wireless channels is impaired predominantly by multi-path fading. Multi-path is the arrival of the transmitted signal at an intended receiver through differing angles and/or differing time delays and/or differing frequency (i.e., Doppler) shifts due to the scattering of electromagnetic waves in the environment. Consequently, the received signal power fluctuates in space (due to angle spread) and/or frequency (due to delay spread) and/or time (due to Doppler spread) through the random superposition of the impinging multi-path components. This random fluctuation in signal level, known as fading, can severely affect the quality and reliability of wireless communication. Additionally, the constraints posed by limited power and scarce frequency bandwidth make the task of designing high data rate, high reliability wireless communication systems extremely challenging.

MIMO technology constitutes a breakthrough in wireless communication system design. The technology offers a number of benefits that help meet the challenges posed by both the impairments in the wireless channel as well as resource constraints. In addition to the time and frequency dimensions that are exploited in conventional single-antenna (single-input single-output) wireless systems, the leverages of MIMO are realized by exploiting the spatial dimension (provided by the multiple antennas at the transmitter and the receiver).

We indicate the kind of performance gains that are expected from the use of MIMO technology by plotting in Figure 1.1 the data rate versus the receive signal-to-noise ratio (SNR) in a 100 kHz channel for an $M \times M$ (i.e., M receive and M transmit antennas) fading link with $M = 1, 2, 4$. The channel response is assumed constant over the bandwidth of interest for this simple example. Assuming a target receive SNR of 25 decibels (dB), a conventional single-input single-output (i.e., $M = 1$) system can deliver a data rate of 0.7 Mbps (where Mbps denotes Mbits per second). With $M = 2$ and 4 we can realize data rates of 1.4 and 2.8 Mbps respectively. This increase in data rate is realized for no additional power or bandwidth expenditure compared to the single-input single-output system. In principle, the single-input single-output system can achieve the data rate of 2.8 Mbps with a receive SNR of 25 dB if the bandwidth is increased to 400 kHz, or

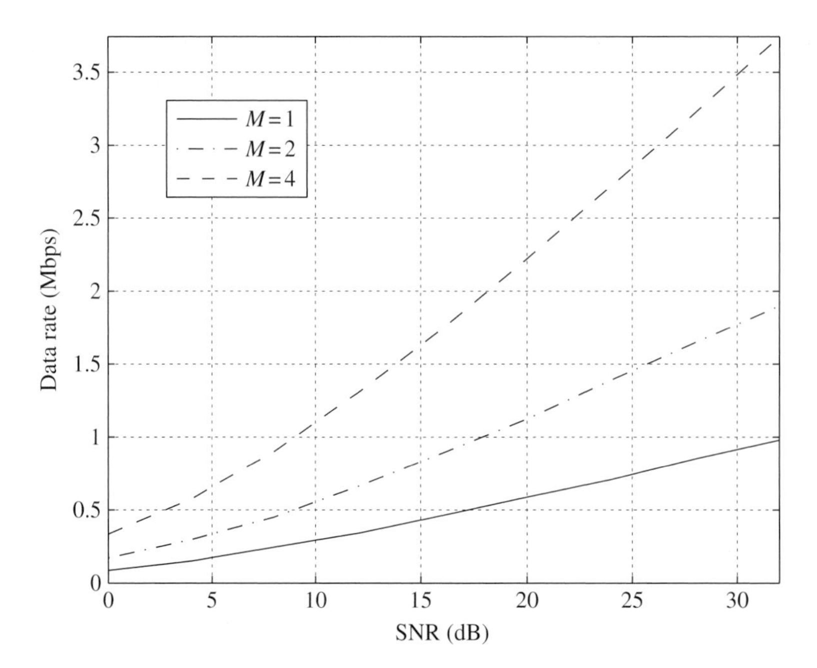

Fig. 1.1. Average data rate versus SNR for different antenna configurations. The channel bandwidth is 100 kHz.

alternatively, with a bandwidth of 100 kHz if the receive SNR is increased to 88 dB! The result presented in this example is based on optimal transceiver design. In practice, the modulation and impairments-constrained data rate delivered will be less but the general trend will still hold.

1.1.1 *Benefits of MIMO technology*

The benefits of MIMO technology that help achieve such significant performance gains are *array gain*, *spatial diversity gain*, *spatial multiplexing gain* and *interference reduction*. These gains are described in brief below.

Array gain

Array gain is the increase in receive SNR that results from a coherent combining effect of the wireless signals at a receiver. The coherent combining may be realized through spatial processing at the receive antenna array and/or spatial pre-processing at the transmit antenna array. Array gain improves resistance to noise, thereby improving the coverage and the range of a wireless network.

Spatial diversity gain

As mentioned earlier, the signal level at a receiver in a wireless system fluctuates or fades. Spatial diversity gain mitigates fading and is realized by providing the receiver with multiple (ideally independent) copies of the transmitted signal in space, frequency or time. With an increasing number of independent copies (the number of copies is often referred to as the diversity order), the probability that at least one of the copies is not experiencing a deep fade increases, thereby improving the quality and reliability of reception. A MIMO channel with M_T transmit antennas and M_R receive antennas potentially offers $M_T M_R$ independently fading links, and hence a spatial diversity order of $M_T M_R$.

Spatial multiplexing gain

MIMO systems offer a linear increase in data rate through spatial multiplexing [5, 9, 22, 35], i.e., transmitting multiple, independent data streams within the bandwidth of operation. Under suitable channel conditions, such as rich scattering in the environment, the receiver can separate the data streams. Furthermore, each data stream experiences at least the same channel quality that would be experienced by a single-input single-output system, effectively enhancing the capacity by a multiplicative factor equal to the number of streams. In general, the number of data streams that can be reliably supported by a MIMO channel equals the minimum of the number of transmit antennas and the number of receive antennas, i.e., $\min\{M_T, M_R\}$. The spatial multiplexing gain increases the capacity of a wireless network.

Interference reduction and avoidance

Interference in wireless networks results from multiple users sharing time and frequency resources. Interference may be mitigated in MIMO systems by exploiting the spatial dimension to increase the separation between users. For instance, in the presence of interference, array gain increases the tolerance to noise as well as the interference power, hence improving the signal-to-noise-plus-interference ratio (SINR). Additionally, the spatial dimension may be leveraged for the purposes of interference avoidance, i.e., directing signal energy towards the intended user and minimizing interference to other users. Interference reduction and avoidance improve the coverage and range of a wireless network.

In general, it may not be possible to exploit simultaneously all the benefits described above due to conflicting demands on the spatial degrees of freedom. However, using some combination of the benefits across a wireless network will result in improved capacity, coverage and reliability.

1.1.2 *Basic building blocks*

Figure 1.2 shows the basic building blocks that comprise a MIMO communication system. The information bits to be transmitted are encoded (using, for instance, a convolutional encoder) and interleaved. The interleaved codeword is mapped to data symbols (such as quadrature amplitude modulation or QAM symbols) by the symbol mapper. These data symbols are input to a space–time encoder that outputs one or more spatial data streams. The spatial data streams are mapped to the transmit antennas by the space–time precoding block. The signals launched from the transmit antennas propagate through the channel and arrive at the receive antenna array. The receiver collects the signals at the

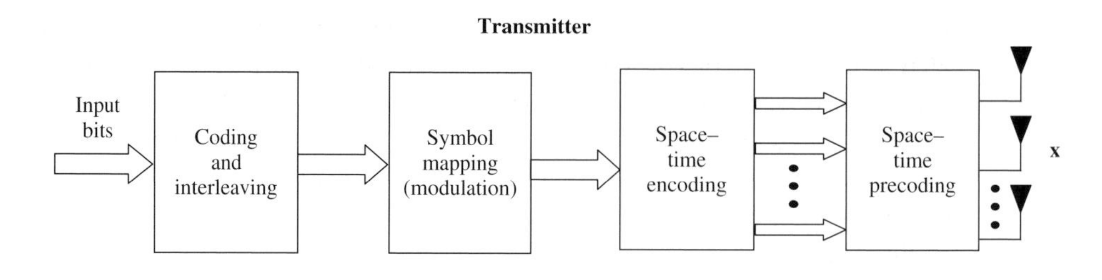

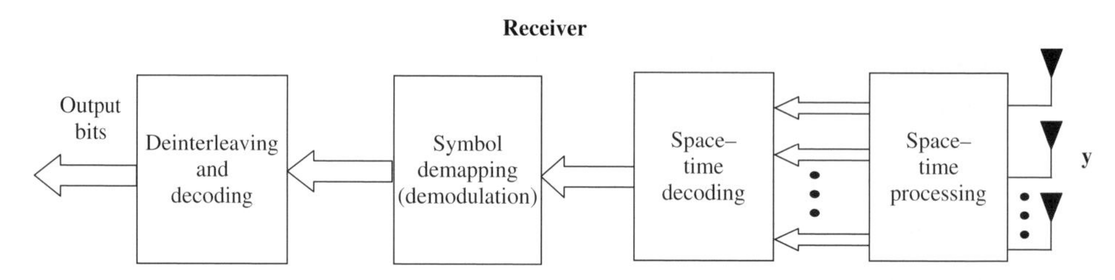

Fig. 1.2. Diagram of a complex equivalent baseband MIMO communication system. **x** and **y** stand for the transmitted and received signal vectors respectively.

output of each receive antenna element and reverses the transmitter operations in order to decode the data: receive space–time processing, followed by space–time decoding, symbol demapping, deinterleaving and decoding. Each of the building blocks offers the opportunity for significant design challenges and complexity–performance trade-offs. Furthermore, a number of variations can exist in the relative placement of the blocks, the functionality and the interactions between the blocks.

This book addresses key concepts and challenges in designing and understanding the performance limits of a MIMO communication system.

1.2 MIMO channel and signal model

In order to design efficient communication algorithms for MIMO systems and to understand the performance limits, it is important to understand the nature of the MIMO channel. For a system with M_T transmit antennas and M_R receive antennas, assuming frequency-flat fading[1] over the bandwidth of interest, the MIMO channel at a given time instant may be represented as an $M_R \times M_T$ matrix

$$\mathbf{H} = \begin{bmatrix} H_{1,1} & H_{1,2} & \cdots & H_{1,M_T} \\ H_{2,1} & H_{2,2} & \cdots & H_{2,M_T} \\ \vdots & \vdots & \ddots & \vdots \\ H_{M_R,1} & H_{M_R,2} & \cdots & H_{M_R,M_T} \end{bmatrix}, \tag{1.1}$$

[1] The delay spread in the channel is negligible compared to the inverse bandwidth.

where $H_{m,n}$ is the (single-input single-output) channel gain between the mth receive and nth transmit antenna pair. The nth column of $\mathbf{H}$ is often referred to as the spatial signature of the nth transmit antenna across the receive antenna array. The relative geometry of the M_T spatial signatures determines the distinguishability of the signals launched from the transmit antennas at a receiver. This is particularly important when independent data streams are launched from the transmit antennas, as is done in the case of spatial multiplexing.

As for the case of single-input single-output channels, the individual channel gains comprising the MIMO channel are commonly modeled as zero-mean circularly symmetric complex Gaussian random variables. Consequently, the amplitudes $|H_{m,n}|$ are Rayleigh-distributed random variables and the corresponding powers $|H_{m,n}|^2$ are exponentially distributed.

1.2.1 Classical independent, identically distributed (i.i.d.) Rayleigh fading channel model

The degree of correlation between the individual $M_T M_R$ channel gains comprising the MIMO channel is a complicated function of the scattering in the environment and antenna spacing at the transmitter and the receiver. Consider an extreme condition were all antenna elements at the transmitter are collocated and likewise at the receiver. In this case, all the elements of $\mathbf{H}$ will be fully correlated (in fact identical) and the spatial diversity order of the channel is one.

Decorrelation between the channel elements will increase with antenna spacing. However, antenna spacing alone is not sufficient to ensure decorrelation. Rich (i.e., omni-directional and isotropic) scattering in the environment in combination with adequate antenna spacing ensures decorrelation of the MIMO channel elements. With rich scattering, the typical antenna spacing required for decorrelation is approximately $\lambda/2$, where λ is the wavelength corresponding to the frequency of operation. Under ideal conditions, when the channel elements are perfectly decorrelated, we have[2] $H_{m,n}(m = 1, 2, \ldots, M_R, n = 1, 2, \ldots, M_T) \sim$ i.i.d. $\mathcal{CN}(0, 1)$. Summarizing, we get $\mathbf{H} = \mathbf{H}_w$, the classical i.i.d. frequency-flat Rayleigh fading MIMO channel. The spatial diversity order of $\mathbf{H}_w$ is $M_T M_R$.

1.2.2 Frequency-selective and time-selective fading

The channel model above assumes that the product of the bandwidth and the delay spread is very small. With increasing bandwidth and/or delay spread, this product is no longer negligible (0.1 is often considered the threshold for voice communication [11]), resulting in channel realizations that are frequency-dependent, i.e., $\mathbf{H}(f)$. While fading at a given frequency may be decorrelated in the spatial domain (resulting in $\mathbf{H}_w(f)$), correlation may exist across channel elements in the frequency domain. The correlation properties in the frequency domain are a function of the power delay profile. The coherence bandwidth B_c

[2] A complex-valued random variable $Z = X + jY$ is $\mathcal{CN}(0, 1)$ if X and Y are independent and normally distributed with zero mean and variance $\frac{1}{2}$.

is defined as the minimum separation in bandwidth required to achieve decorrelation. For two frequencies f_1 and f_2 with $|f_1 - f_2| > B_c$, we have $E[\text{vec}(\mathbf{H}(f_1))\text{vec}^H(\mathbf{H}(f_2))] = 0$. The coherence bandwidth is inversely proportional to the delay spread of the channel.

Furthermore, due to the motion of scatterers in the environment or of the transmitter or receiver, the channel realizations will vary with time. As with the case of frequency-selective fading, we can define a coherence time T_c, defined as the minimum separation in time required for decorrelation of the time-varying channel realizations. For two time instances t_1 and t_2 with $|t_1 - t_2| > T_c$, we have $E[\text{vec}(\mathbf{H}(t_1))\text{vec}^H(\mathbf{H}(t_2))] = 0$. The coherence time is inversely proportional to the Doppler spread of the channel.

1.2.3 *Real-world MIMO channels*

In practice, the behavior of $\mathbf{H}$ can significantly deviate from $\mathbf{H}_w$ due to a combination of inadequate antenna spacing and/or inadequate scattering leading to spatial fading correlation. Furthermore, the presence of a fixed (possibly line-of-sight or LOS) component in the channel will result in Ricean fading. Extensive measurements of real-world MIMO channels [3, 8, 15, 17, 30, 32] have been carried out by researchers across the world to develop accurate models [5, 10, 24]. Figure 1.3 shows the measured

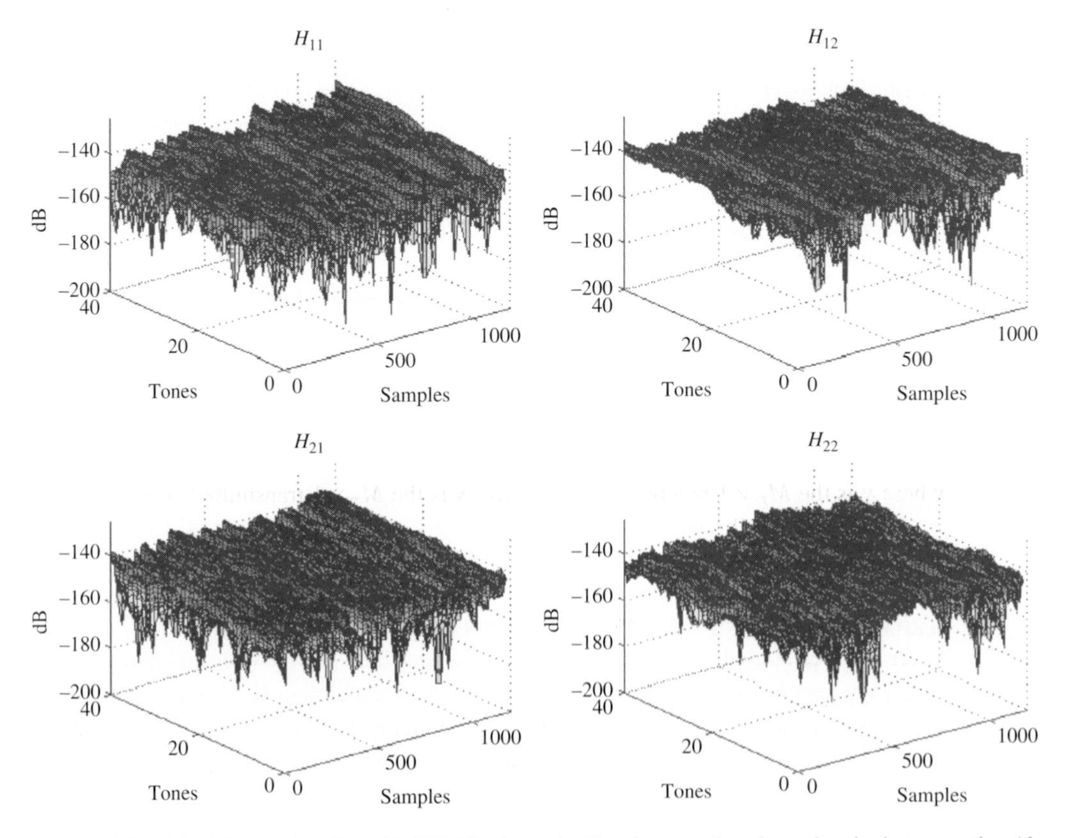

Fig. 1.3. Measured real-world MIMO channel. $H_{i,j}$ denotes the channel gain between the jth transmit antenna and ith receive antenna.

time–frequency response for an $M_T = M_R = 2$ MIMO channel in a fixed broadband wireless access system at 2.5 GHz. It is clear from the figure that real-world MIMO channels are triply selective, i.e., they exhibit fading across space, time and frequency.

In the presence of an LOS component between the transmitter and the receiver, the MIMO channel may be modeled as the sum of a fixed component and a fading component

$$\mathbf{H} = \sqrt{\frac{K}{1+K}}\overline{\mathbf{H}} + \sqrt{\frac{1}{1+K}}\mathbf{H}_w, \tag{1.2}$$

where $\sqrt{\frac{K}{1+K}}\overline{\mathbf{H}} = E[\mathbf{H}]$ is the LOS component of the channel and $\sqrt{\frac{1}{1+K}}\mathbf{H}_w$ is the fading component, assuming uncorrelated fading. $K \geq 0$ in (1.2) is the Ricean K-factor of the channel and is defined as the ratio of the power in the LOS component of the channel to the power in the fading component. When $K = 0$ we have pure Rayleigh fading. At the other extreme $K = \infty$ corresponds to a non-fading channel.

In general, real-world MIMO channels will exhibit some combination of Ricean fading and spatial fading correlation. Spatial correlation models will be discussed in Chapter 2. Furthermore, the use of polarized antennas will necessitate additional modifications to the channel model. These factors collectively will impact (probably adversely) the performance of a given MIMO signaling scheme. With appropriate knowledge of the MIMO channel at the transmitter, the signaling strategy can be appropriately adapted to meet performance requirements. The channel state information could be complete (i.e., the precise channel realization) or partial (i.e., knowledge of the spatial correlation, K-factor, etc.). Space–time precoding techniques to exploit channel knowledge at the transmitter are detailed in Chapter 3.

1.2.4 *Discrete-time signal model*

For a frequency-flat fading MIMO channel, the commonly used discrete-time input–output relation over a symbol period is given by

$$\mathbf{y} = \sqrt{\frac{E_x}{M_T}}\mathbf{H}\mathbf{x} + \mathbf{n}, \tag{1.3}$$

where $\mathbf{y}$ is the $M_R \times 1$ received signal vector, $\mathbf{x}$ is the $M_T \times 1$ transmitted signal vector, $\mathbf{n}$ is additive temporally white complex Gaussian noise with $E[\mathbf{n}\mathbf{n}^H] = N_o\mathbf{I}_{M_R}$ and E_x is the total average energy available at the transmitter over a symbol period having removed losses due to propagation and shadowing. We constrain the total average transmitted power over a symbol period by assuming that the covariance matrix of $\mathbf{x}$, $\mathbf{R}_{\mathbf{xx}} = E[\mathbf{x}\mathbf{x}^H]$, satisfies $\text{Tr}(\mathbf{R}_{\mathbf{xx}}) = M_T$. The ratio $\rho = E_x/N_o$ equals the SNR per receive antenna (simply referred to as SNR henceforth). Furthermore, it is commonly assumed that the channel is block fading [4], i.e., the channel remains constant over N consecutive symbol periods (determined by the coherence time) and then changes in an independent fashion to a new realization. Frequency-selective fading can be incorporated into the channel model by using a matrix tapped-delay line. The appropriate changes to the channel model will be described when required in subsequent chapters.

1.3 A fundamental trade-off

Two key performance metrics associated with any communication system are the transmission rate and the frame-error rate (FER). In the following, the transmission rate, R, is defined as the data rate transmitted per unit bandwidth. The FER, $P_e(\rho, R)$, is defined as the probability with which the transmitted frame (i.e., packet) is incorrectly decoded at the receiver, and is a function of the SNR and transmission rate.

Intuitively, for a fixed transmission rate an increase in SNR will result in reduced FER. Similarly, at a fixed target FER, an increase in SNR may be leveraged to increase the transmission rate. Hence, a fundamental trade-off exists in any communication system between the transmission rate and FER. In the context of MIMO systems, this trade-off is often referred to as the diversity–multiplexing trade-off [41] with diversity signifying the FER reduction and multiplexing signifying an increase in transmission rate. The diversity–multiplexing trade-off is central to MIMO communication theory and is described in brief in this section.

1.3.1 Outage capacity

The capacity of a communication channel is the maximum, asymptotic (in block length) error-free transmission rate that can be achieved. The capacity of a MIMO channel is a complicated function of the channel conditions and transmit/receive processing constraints [4, 9, 13, 19, 21, 35, 40]. A detailed discussion on the capacity of MIMO channels is provided in Chapter 2. The development below focuses on the outage capacity of MIMO channels, which gains operational significance when the fading channel holds constant over the entire duration of the transmitted frame.

The p percentage outage capacity at SNR ρ, $C_{out,p}(\rho)$, is defined as the transmission rate that can be supported by $(100-p)\%$ of the fading realizations of the channel [35]. Hence at SNR ρ, if a frame is transmitted with rate $C_{out,p}(\rho)$, the probability that the frame will be decoded correctly is $(100-p)\%$. Equivalently the FER associated with transmission rate $C_{out,p}(\rho)$ is $p\%$, i.e.,

$$P_e(\rho, C_{out,p}(\rho)) = p\%. \tag{1.4}$$

1.3.2 Multiplexing gain

The maximum multiplexing gain r_{max} that can be achieved over a MIMO channel is given by the asymptotic (in SNR) slope of the outage capacity (for fixed FER) plotted as a function of the SNR on a linear–log scale, i.e.,

$$r_{max} = \lim_{\rho \to \infty} \frac{C_{out,p}(\rho)}{\log_2 \rho}. \tag{1.5}$$

For the $\mathbf{H}_w$ MIMO channel with optimal transceiver design (i.e., Gaussian code books, asymptotically large frame length, maximum-likelihood detection, etc.) $r_{max} = \min\{M_R, M_T\}$ indicating that for a fixed FER, the transmission rate may be increased by $\min\{M_R, M_T\}$bps/Hz for every 3 dB increase in SNR.

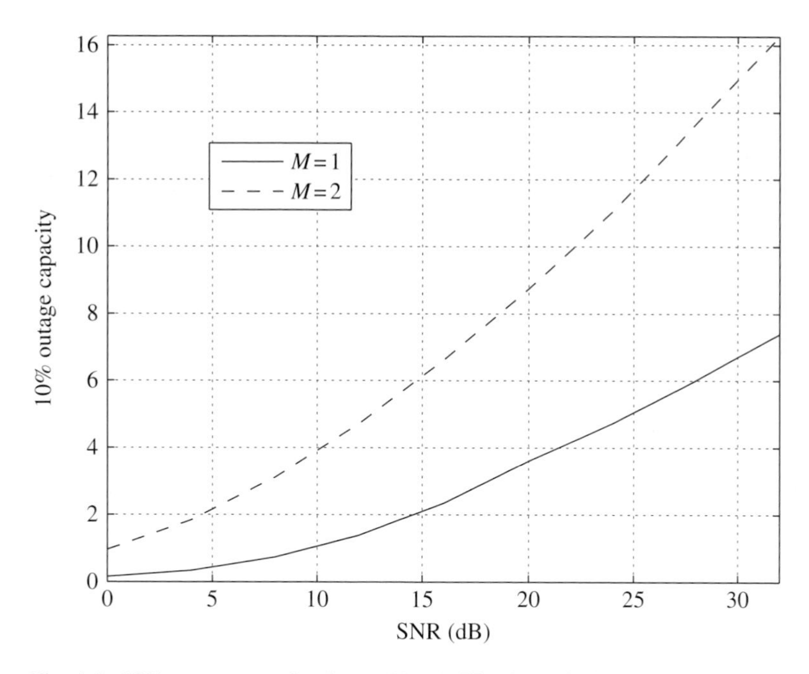

Fig. 1.4. 10% outage capacity for an $M \times M$ $\mathbf{H}_w$ channel plotted as a function of SNR.

Figure 1.4 shows a comparison of the 10% outage capacity of a 2×2 $\mathbf{H}_w$ MIMO channel to the 10% outage capacity of a single-input single-output channel, plotted as a function of SNR. At high SNR, the outage capacity for the MIMO channel grows with 2 bps/Hz/3 dB slope compared to 1 bps/Hz/3 dB slope for the single-input single-output channel.

1.3.3 Diversity gain

The maximum diversity gain d_{max} that can be achieved over a MIMO channel is given by the negative of the asymptotic (in SNR) slope of FER for a fixed transmission rate, plotted as a function of SNR on a log–log scale, i.e.,

$$d_{max} = -\lim_{\rho \to \infty} \frac{\log_2 P_e(\rho, R)}{\log_2 \rho}. \tag{1.6}$$

For the $\mathbf{H}_w$ MIMO channel with optimal transceiver design (i.e., Gaussian code books, asymptotically large frame length, maximum-likelihood detection, etc.) $d = M_R M_T$, indicating that for a fixed transmission rate, with every 3 dB increase in SNR, the FER decreases by a factor of $2^{-M_R M_T}$.

For $R = 2$ bps/Hz, Figure 1.5 compares the FER in a 2×2 $\mathbf{H}_w$ MIMO channel to the FER in a single-input single-output channel, plotted as a function of SNR. The fourth-order diversity provided by the MIMO channel is clearly reflected by the slope of the FER curve – at high SNR, the FER decreases by a factor of 2^{-4} in the MIMO channel (compared to 2^{-1} in the single-input single-output channel) for a 3 dB increase in SNR.

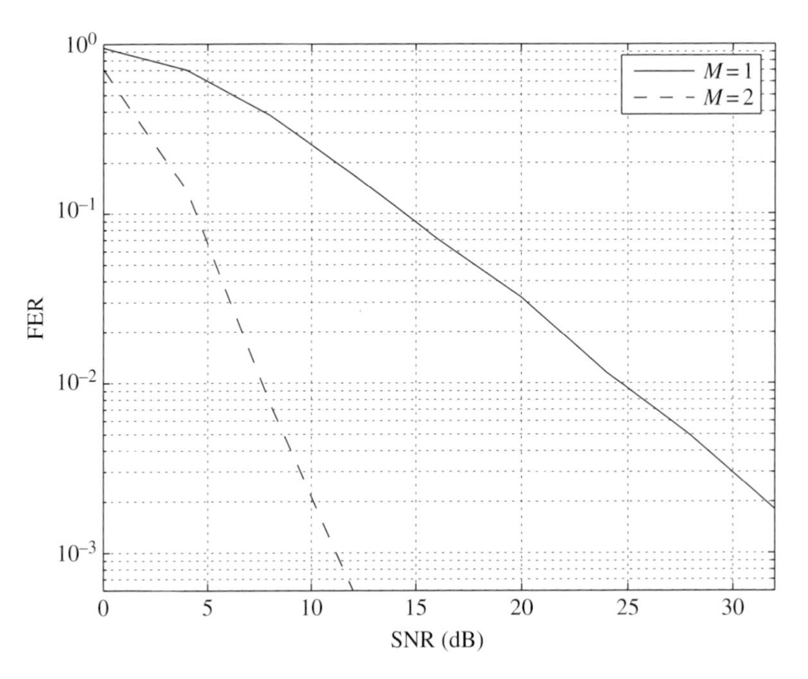

Fig. 1.5. FER at $R = 2$ bps/Hz for an $M \times M$ $\mathbf{H}_w$ channel plotted as a function of SNR.

1.3.4 *Flexible trade-off*

Individually, (1.5) and (1.6) represent the extremities of the diversity–multiplexing trade-off for MIMO channels. In (1.5) an increase in SNR is completely utilized to linearly (in $\min\{M_R, M_T\}$) increase the transmission rate, keeping the FER fixed. At the other extreme, in (1.6), an increase in SNR yields an exponential (the exponent is $-M_R M_T$) reduction in FER at a fixed transmission rate. Furthermore, (1.5) and (1.6) represent the gains of MIMO communication over single-input single-output systems as demonstrated in Figures 1.5 and 1.4.

In certain scenarios, we may desire to utilize an increase in SNR for some combination of transmission rate increase and FER reduction. It has been shown that a flexible trade-off between diversity and multiplexing can be achieved – the optimal trade-off curve for the $\mathbf{H}_w$ MIMO channel, $d(r)$, is piecewise linear (see Figure 1.6) connecting $(r, d(r))$, $r = 0, 1, \ldots, r_{max}$, where

$$d(r) = (M_R - r)(M_T - r). \tag{1.7}$$

The trade-off curve implies that if the transmission rate is increased by r bps/Hz over a 3 dB increase in SNR, the corresponding reduction in FER will equal $2^{-d(r)}$. Hence it is not possible to increase the transmission rate and decrease the FER simultaneously to the fullest extent (represented by r_{max} and d_{max} respectively) possible.

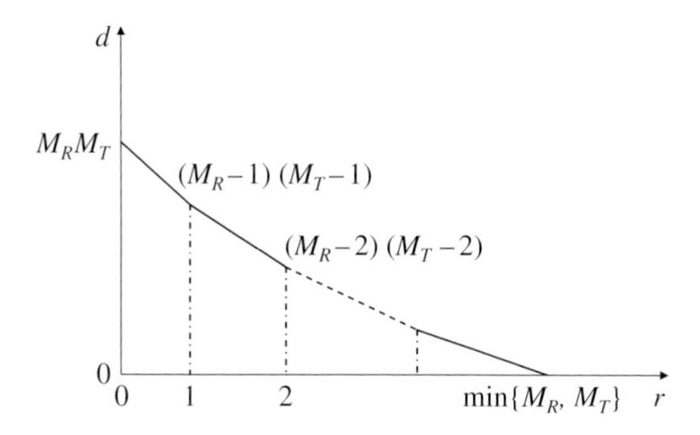

Fig. 1.6. The optimal diversity–multiplexing trade-off curve in an $\mathbf{H}_w$ MIMO channel.

1.4 MIMO transceiver design

Transceiver algorithms for MIMO systems may be broadly classified into two categories, i.e., those designed to increase the transmission rate and those designed to increase reliability. The former are often collectively referred to as spatial multiplexing and the latter as transmit diversity. Spatial multiplexing and transmit diversity techniques achieve either one of the two extremities in the diversity–multiplexing trade-off curve. Spatial multiplexing provides maximum multiplexing gain at fixed FER, while transmit diversity techniques provide maximum diversity gain for fixed transmission rate. Two representative transceiver algorithms, one from each of these classifications, are described below.

1.4.1 *Alamouti scheme*

The Alamouti scheme [2] is a simple transmit diversity technique that may be applied in systems with $M_T = 2$ and any number of receive antennas. The transmission strategy for the Alamouti scheme is shown in the schematic in Figure 1.7. Assuming two data symbols x_0 and x_1 to be transmitted, the transmitter launches x_0 and x_1 from the first and second transmit antenna, respectively, during the first symbol period, followed by x_1^* and $-x_0^*$ from the first and second transmit antenna, respectively, during the second symbol period. Hence, effectively, only one data symbol is transmitted per symbol period. Furthermore, through appropriate processing at the receiver, the matrix channel is collapsed into a scalar channel for either data symbol with the effective input–output relation

$$\widetilde{y}_i = \sqrt{\frac{E_x}{2}}\|\mathbf{H}\|_F x_i + \widetilde{n}_i, \qquad i = 0, 1, \tag{1.8}$$

where $\widetilde{y}_i$ is the processed (scalar) received signal and $\widetilde{n}_i \sim \mathcal{CN}(0, N_o)$ is temporally white processed noise.

The Alamouti scheme realizes a diversity gain of $2M_R$ at a fixed transmission rate. For transmission rate $R = 2$ bps/Hz, Figure 1.8 compares the FER for the Alamouti scheme in a 2×2 $\mathbf{H}_w$ MIMO channel with the FER in a single-input single-output channel. The Alamouti scheme clearly achieves fourth-order diversity.

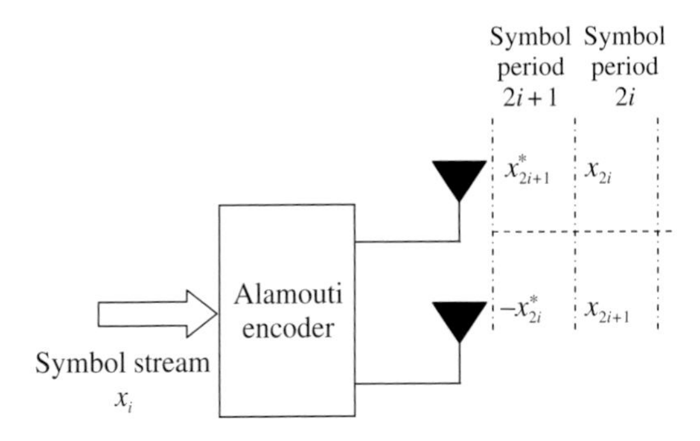

Fig. 1.7. Schematic of the Alamouti scheme.

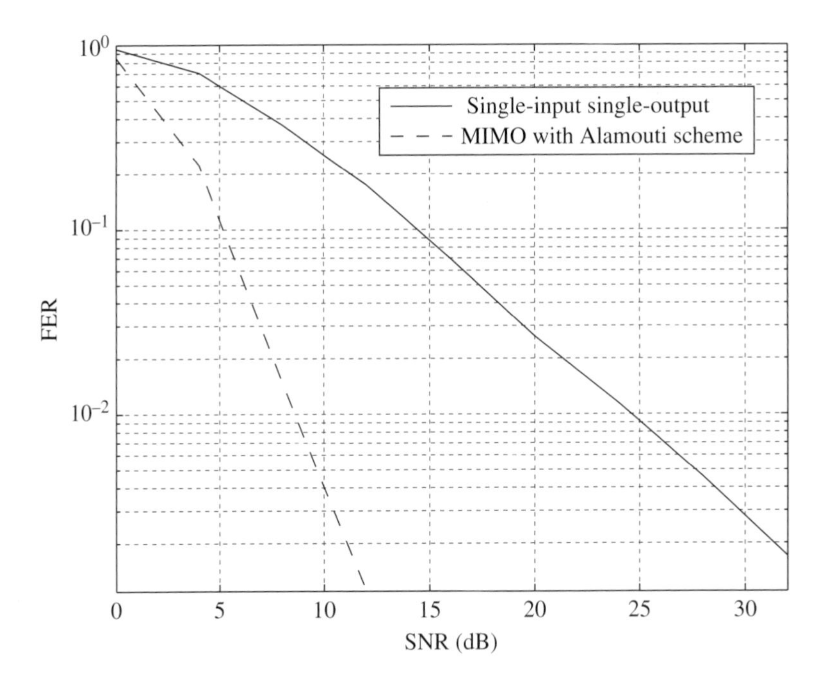

Fig. 1.8. FER at $R = 2$ bps/Hz for a 2×2 $\mathbf{H}_w$ MIMO channel with the Alamouti scheme.

Figure 1.9 compares the 10% outage capacity of the Alamouti scheme in a 2×2 $\mathbf{H}_w$ MIMO channel to the 10% outage capacity of a single-input single-output channel. The figure shows that the multiplexing gain provided by the Alamouti scheme at fixed FER is one, equal to the multiplexing gain in the single-input single-output channel.

Hence, while transmit diversity techniques are capable of realizing full diversity gain, they fall considerably short when a sharp increase in data rate is desired. Transmit diversity techniques are of value for lower data rate, high reliability communication. The Alamouti scheme may be extended to channels with more than two transmit antennas. The techniques governing signal construction for transmit diversity fall within the realm of space–time coding [2, 12, 16, 27, 29, 33, 34], which is described in Chapter 4.

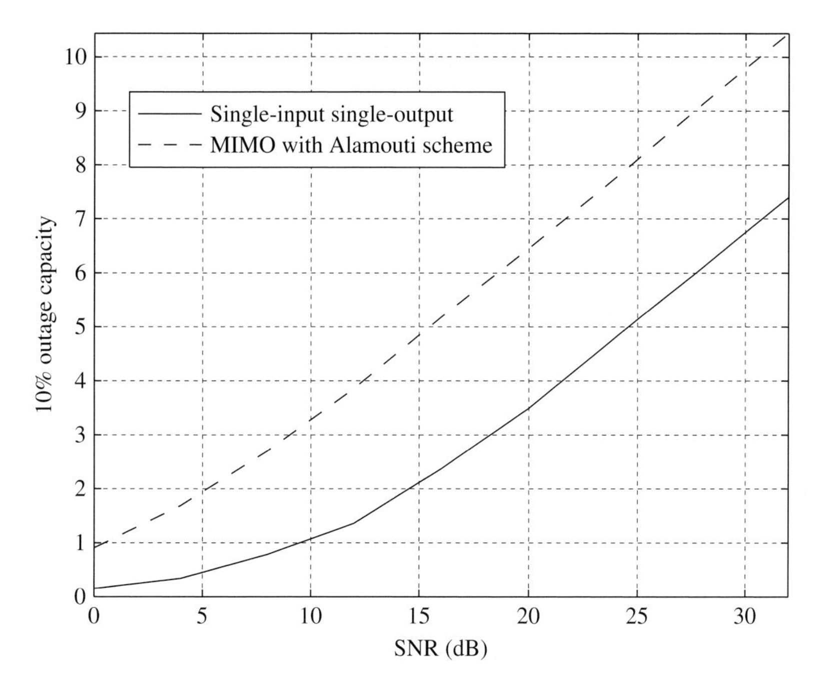

Fig. 1.9. 10% outage capacity for a 2×2 $\mathbf{H}_w$ MIMO channel with the Alamouti scheme.

1.4.2 *Spatial multiplexing*

Spatial multiplexing gain can be realized by transmitting independent data streams from each of the transmit antennas. Figure 1.10 is the schematic of spatial multiplexing for a transmitter with $M_T = 2$.

Unlike the Alamouti scheme, the multiple transmitted data streams interfere with one another at the receiver. In order to reliably separate the received data streams we require $M_R \geq M_T$. For this simple example, assuming $M_R = M_T = 2$, a zero-forcing (ZF) receiver [38] may be applied that completely eliminates the multi-stream interference. More complex receiver architectures and the associated trade-offs are the subject of

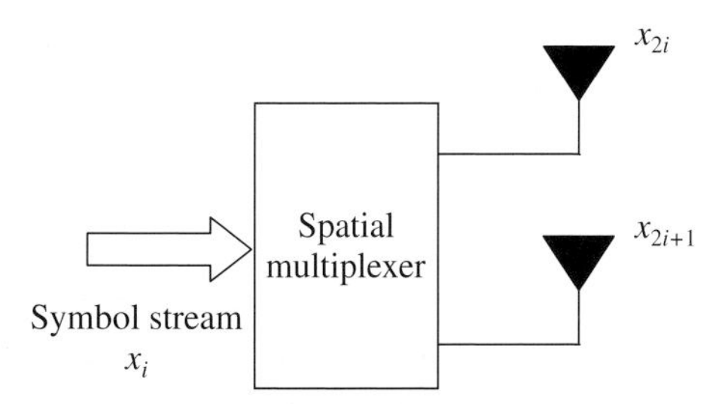

Fig. 1.10. Schematic of spatial multiplexing for a transmitter with two antennas.

Chapter 5. With a zero-forcing receiver the effective input–output relation for the spatially multiplexed MIMO channel is

$$\widetilde{\mathbf{y}} = \mathbf{x} + \widetilde{\mathbf{n}},$$

where $\widetilde{\mathbf{y}}$ is the processed (zero-forced) received $M_T \times 1$ signal vector, $\mathbf{x}$ is the $M_T \times 1$ transmitted signal vector and $\widetilde{\mathbf{n}}$ is the temporally white zero-mean processed noise vector with covariance matrix $\mathbf{R}_{\widetilde{\mathbf{n}}\widetilde{\mathbf{n}}} = \frac{M_T}{\rho}(\mathbf{H}^H\mathbf{H})^{-1}$. Hence, with a ZF receiver, the spatially multiplexed MIMO channel is decomposed into parallel, scalar sub-channels with correlated noise across the sub-channels. Furthermore, it can be shown that for the $\mathbf{H}_w$ MIMO channel with $M_R = M_T$, the average receive SNR in a sub-channel is equal to ρ/M_T, with Rayleigh fading across the sub-channels. Ignoring correlation across the sub-channels allows independent encoding/decoding of the multiplexed streams. The composite transmitted frame is then in error if any one of the (sub-)frames transmitted over the sub-streams is in error.

Clearly, with a 3 dB increase in SNR, the transmission rate in each of the parallel sub-channels may be increased by 1 bps/Hz at fixed FER, giving an overall increase in transmission rate of M_T bps/Hz, equal to the maximum multiplexing gain of the MIMO channel. However, since each of the scalar sub-channels is single-input single-output Rayleigh fading, we can only expect a maximum diversity gain of one.

Figure 1.11 shows the 10% outage capacity of a spatially multiplexed 2×2 $\mathbf{H}_w$ MIMO channel with a ZF receiver. The figure clearly shows the superiority of spatial multiplexing

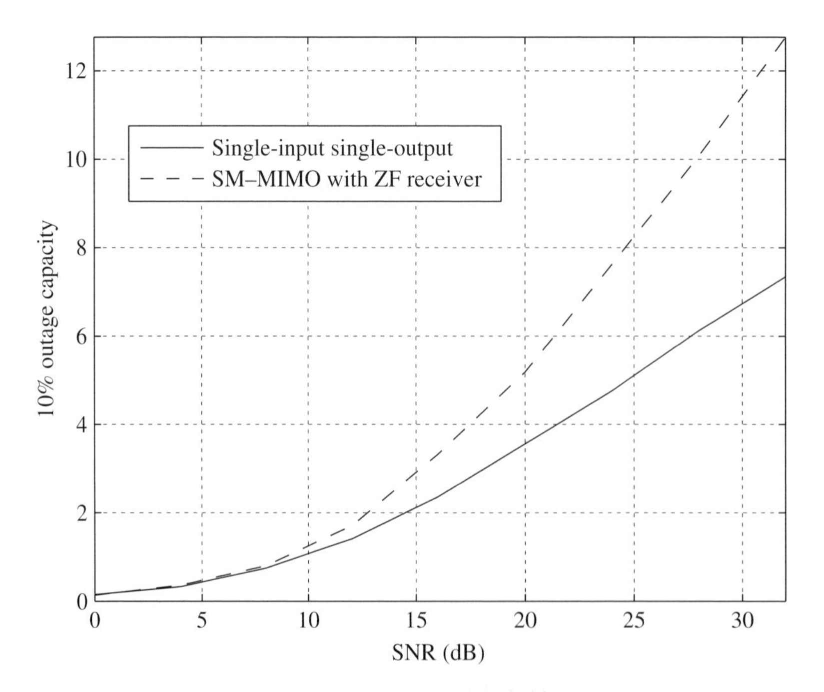

Fig. 1.11. 10% outage capacity of a spatially multiplexed (SM) 2×2 $\mathbf{H}_w$ MIMO channel with a ZF receiver.

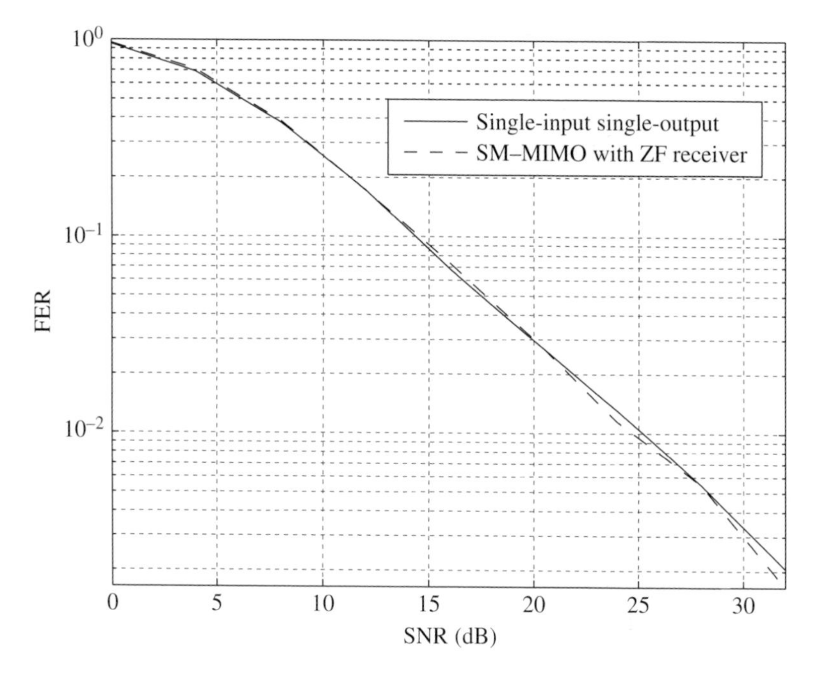

Fig. 1.12. FER at $R = 2$ bps/Hz for a spatially multiplexed 2×2 $\mathbf{H}_w$ MIMO channel with a ZF receiver.

in providing high data rate. Figure 1.12 shows the FER for the same MIMO channel with spatial multiplexing and ZF reception at a transmission rate of $R = 2$ bps/Hz. Spatial multiplexing with ZF reception does not offer a diversity advantage over the single-input single-output channel. Spatial multiplexing is therefore advantageous when we require a dramatic increase in spectral efficiency, but not in reliability.

Each of the MIMO transceiver techniques described above achieves one of the two extremities in the diversity–multiplexing trade-off curve, and is clearly suboptimal at the other extremity. Techniques that achieve a flexible diversity–multiplexing trade-off form an important topic of current research.

1.5 MIMO in wireless networks

Wireless networks may be broadly classified as cellular or ad hoc networks. A cellular network is characterized by centralized communication – multiple users within a cell communicate with a base-station that controls all transmission/reception and forwards data to the users. On the contrary, in an ad hoc network, all terminals are on an equal footing – any terminal can act as a sender or receiver of data or as a relay for other transmissions. This section briefly reviews the use of MIMO technology in each of these networks and also discusses a new form of the technology, known as distributed MIMO.

1.5.1 MIMO in cellular networks

In a cellular wireless communication network, multiple users may communicate at the same time and/or frequency. The more aggressive the reuse of time and frequency resources, the higher the network capacity will be, provided that transmitted signals can be detected reliably. Multiple users may be separated in time (time-division) or frequency (frequency-division) or code (code-division). The spatial dimension in MIMO channels, provides an extra dimension to separate users, allowing more aggressive reuse of time and frequency resources, thereby increasing the network capacity.

Figure 1.13 is the schematic of a cell in a MIMO cellular network. A base-station equipped with L antennas communicates with P users, each equipped with M antennas. The channel from the base-station to the users (the downlink) is a broadcast channel (BC) while the channel from the users to the base-station (the uplink) is a multiple-access channel (MAC). The set of rate-tuples $(R_1, R_2, \ldots, R_P)$ that can be reliably supported on the downlink or uplink constitutes the capacity rate region for that link. Recently, an important duality has been discovered between the rate regions for the downlink and uplink channels [7, 31, 37, 39]. This result along with other capacity results for multi-user MIMO systems will be discussed in Chapter 2.

In order to understand the possible gains from MIMO technology in a multi-user environment, consider the uplink of a cellular MIMO system where all the users simultaneously transmit independent data streams from each of their transmit antennas, i.e., each user signals with spatial multiplexing. To the base-station, the users combined, appear as a multi-antenna transmitter with PM antennas. Thus the effective uplink channel has a dimension of $L \times PM$. This effective channel will have a considerably different structure from the $\mathbf{H}_w$ MIMO single user channel due to path-loss and shadowing

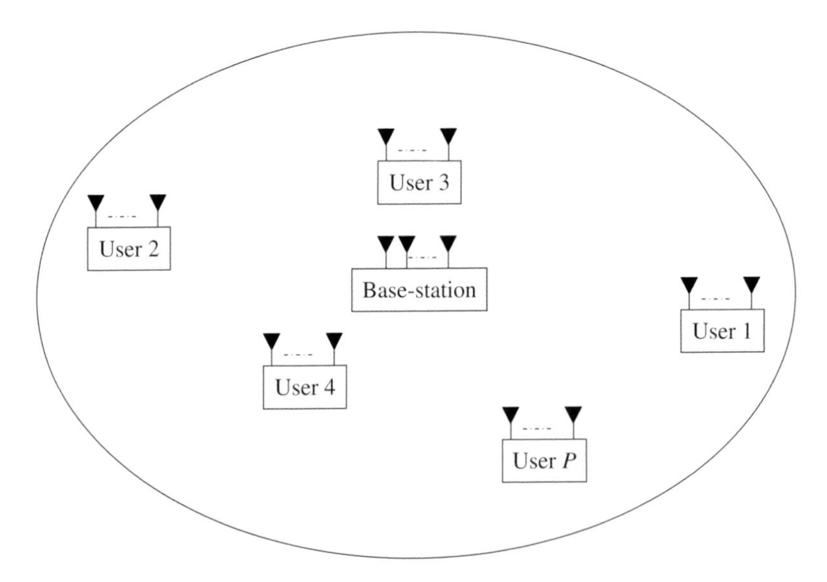

Fig. 1.13. MIMO cellular system. A base-station with L antennas communicates with P users, each equipped with M antennas.

differences between users. However, with rich scattering and $L \geq PM$, we can expect that the spatial signatures of the users are well separated to allow reliable detection. Using a multi-user ZF receiver will allow perfect separation of all the data streams at the base-station, yielding a multi-user multiplexing gain of PM. The use of more complex receivers for multi-user detection and the associated performance trade-offs are the subject of Chapter 6. A similar thought experiment can be applied for the downlink, where the base-station exploits the spatial dimension to beam information intended for a particular user towards that user and steers nulls in the directions of the other users, thus completely eliminating interference.

1.5.2 *MIMO in ad hoc networks*

Figure 1.14 shows a wireless ad hoc network. At a given instant of time, a subset of terminals will be sources of data and another subset the intended destinations. The terminals in the network that are neither sources nor destinations may act as relays to assist data transmission in the network. Thus the number of operating modes in an ad hoc network is very large and will, in general, comprise of combinations of multiple-access, broadcast, relay and interference channels. While the ultimate performance limit of an ad hoc network is unknown, it is clear that leveraging the spatial dimension through the use of MIMO technology in each of the building blocks (i.e., the constituent multiple-access, broadcast, relay and interference channels) will increase the overall network capacity. A discussion on the capacity benefits of MIMO in ad hoc networks is provided in Chapter 2.

Distributed MIMO

While MIMO technology provides substantial performance gains, the cost of deploying multiple antennas at terminals in a network can be prohibitive, at least for the immediate future. Distributed MIMO is a means of realizing the gains of MIMO with single-antenna terminals in a network, allowing a gradual migration to a true MIMO network. The approach requires some level of cooperation between network terminals. This can be

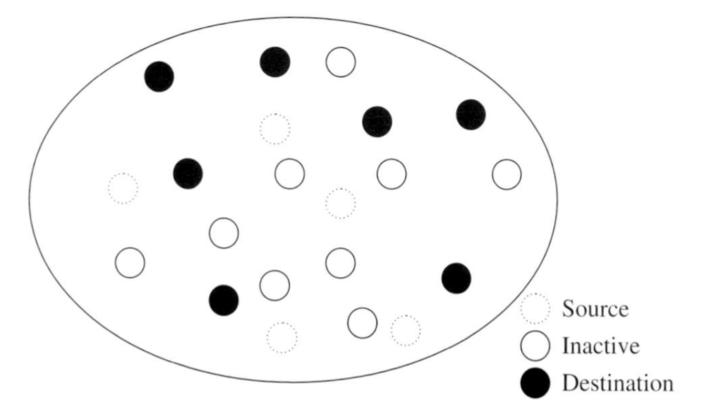

Fig. 1.14. An ad hoc wireless network.

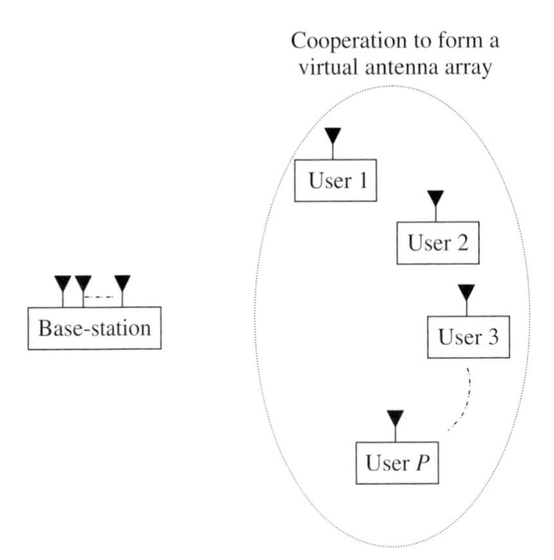

Fig. 1.15. Distributed MIMO: multiple users cooperate to form a virtual antenna array that realizes the gains of MIMO in a distributed fashion.

accomplished through suitably designed protocols [14, 18, 25, 26, 28]. The cooperating terminals form a virtual antenna array (see Figure 1.15) that leverages the gains of MIMO in a distributed fashion. Substantial performance gains can be realized through this approach. The concept may be applied to both cellular as well as ad hoc wireless networks.

1.6 MIMO in wireless standards

With the advent of the Internet and rapid proliferation of computational and communication devices, the demand for higher data rates is ever growing. In many circumstances, the wireless medium is an effective means of delivering a high data rate at a cost lower than that of wireline techniques (such as DSL or cable modem). Limited bandwidth and power makes MIMO technology indispensable in meeting the increasing demand for data.

MIMO technology is now at the core of many existing and emerging wireless standards such as IEEE 802.11 (for wireless local area networks or WLAN), IEEE 802.16 (for wireless metropolitan area networks or WMAN) and IEEE 802.20 (for mobile broadband wireless access or MBWA). While MIMO is compatible with any modulation scheme, the preferred choice for next-generation networks, to tackle the increased bandwidth delay spread product, is orthogonal frequency-division multiplexing (OFDM). OFDM significantly reduces the computational complexity of equalization at the receiver by dividing an otherwise frequency-selective channel into narrower frequency-flat fading sub-channels called tones. The associated multi-user format is orthogonal frequency-division multiple access (OFDMA). In OFDMA multiple users are allocated tones

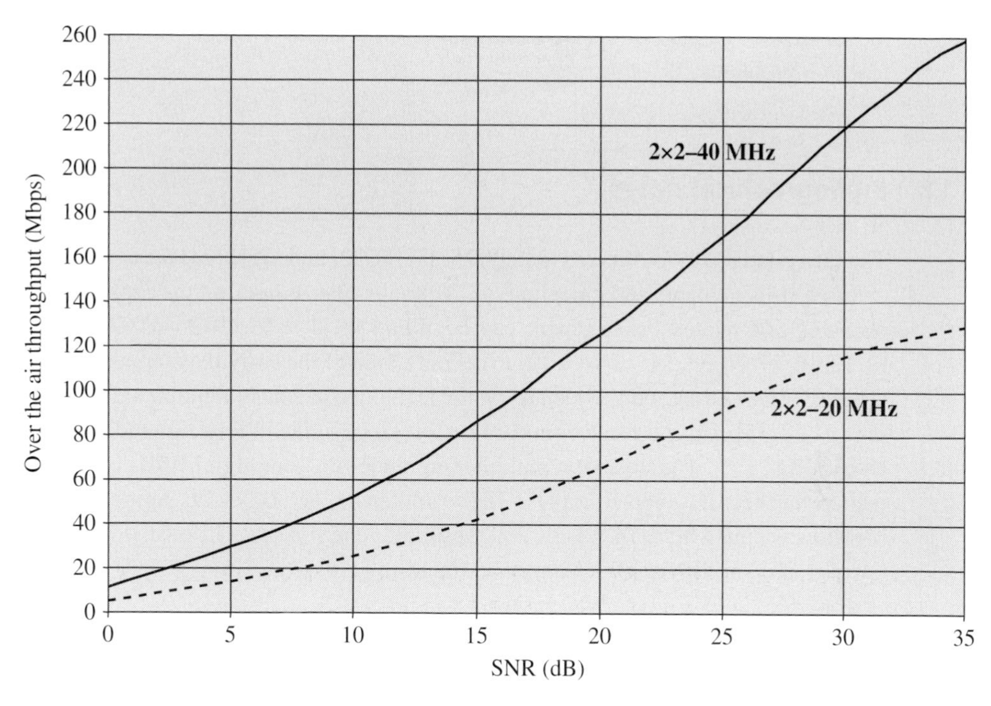

Fig. 1.16. Experimental throughput achieved in a WLAN system based on the IEEE 802.11n specification at Marvell Semiconductor, Inc.

depending on individual throughput and delay constraints. As an example of the data rates that will be realized, the IEEE 802.11n specification promises a throughput in excess of 200 Mbps. Figure 1.16 shows the experimental throughput for the IEEE 802.11n specification with a 2×2 MIMO configuration in 20 and 40 MHz channels.

1.7 Organization of the book and future challenges

This book addresses some of the key challenges in designing MIMO communication systems and understanding the ultimate performance limits. Chapter 2 provides a detailed discussion of the capacity limits of single- and multi-user MIMO systems. Chapter 3 discusses how knowledge of the MIMO channel at the transmitter may be leveraged to improve performance through space–time precoding. Space–time code construction is discussed in Chapter 4. Chapter 5 describes a unified framework for designing MIMO receivers. The final chapter, Chapter 6, details multi-user detection techniques for MIMO systems.

MIMO communication systems have been attracting considerable research attention from both academia and industry. Topics of research include channel modeling, capacity limits, coding, modulation, receiver design and multi-user communication. From an implementation viewpoint practical MIMO systems present a plethora of challenges in such areas as synchronization, channel estimation, training, power consumption, complexity reduction and efficiency. A few select topics have been chosen for inclusion in this book. The area of MIMO communication presents a variety of other challenging research problems.

1.8 Bibliographical notes

The interested reader is directed to [3, 8, 15, 17, 30, 32] and [5, 10, 24] for details on MIMO channel measurements and modeling, respectively. Discussions on the capacity of MIMO channels and related development can be found in [4, 9, 13, 19–21, 35, 40]. The works [2, 12, 16, 27, 29, 33, 34] serve as an excellent guide to the early developments in the area of space–time coding. Further details on the technique of spatial multiplexing can be found in [5, 9, 22, 35]. The ZF receiver and its performance limits in Rayleigh fading channels is discussed in [38]. The diversity–multiplexing trade-off, central to MIMO communication theory, has been developed in [41]. The results in [7, 31, 36, 37, 39] pertain to multi-user MIMO systems while [14, 18, 25, 26, 28] discuss the recent field of distributed MIMO. Early results on MIMO-OFDM may be found in [1, 6, 23].

References

[1] D. Agarwal, V. Tarokh, A. Naguib, and N. Seshadri, "Space–time coded OFDM for high data rate wireless communication over wideband channels," in *Proc. IEEE VTC*, vol. 3, pp. 2232–2236, May 1998.

[2] S. M. Alamouti, "A simple transmit diversity technique for wireless communications," *IEEE J. Select. Areas Commun.*, vol. 16, no. 8, pp. 1451–1458, Oct. 1998.

[3] D. S. Baum, D. Gore, R. Nabar, S. Panchanathan, K. V. S. Hari, V. Erceg, and A. J. Paulraj, "Measurement and characterization of broadband MIMO fixed wireless channels at 2.5 GHz," in *Proc. IEEE ICPWC*, Hyderabad, pp. 203–206, Dec. 2000.

[4] E. Biglieri, J. Proakis, and S. Shamai, "Fading channels: information-theoretic and communications aspects," *IEEE Trans. Inform. Theory*, vol. 44, no. 6, pp. 2619–2692, Oct. 1998.

[5] H. Bölcskei, D. Gesbert, and A. J. Paulraj, "On the capacity of OFDM-based spatial multiplexing systems," *IEEE Trans. Commun.*, vol. 50, no. 2, pp. 225–234, Feb. 2002.

[6] H. Bölcskei and A. J. Paulraj, "Space–frequency coded broadband OFDM systems," in *Proc. IEEE WCNC*, Chicago, IL, vol. 1, pp. 1–6, Sept. 2000.

[7] G. Caire and S. Shamai, "On the achievable throughput of a multiantenna gaussian broadcast channel," *IEEE Trans. Inform. Theory*, vol. 49, no. 7, pp. 1691–1706, July 2003.

[8] D. Chizhik, J. Ling, P. W. Wolniansky, R. A. Valenzuela, N. Costa, and K. Huber, "Multiple-input–multiple-output measurements and modeling in Manhattan," *IEEE J. Select Areas Commun.*, vol. 23, no. 3, pp. 321–331, Apr. 2003.

[9] G. J. Foschini, "Layered space–time architecture for wireless communication in a fading environment when using multi-element antennas," *Bell Labs Tech. J.*, pp. 41–59, 1996.

[10] D. Gesbert, H. Bölcskei, D. A. Gore, and A. J. Paulraj, "Outdoor MIMO wireless channels: models and performance prediction," *IEEE Trans. Commun.*, vol. 50, no. 12, pp. 1926–1934, Dec. 2002.

[11] A. Goldsmith, *Wireless Communications*, Cambridge: Cambridge University Press, 2005.

[12] J. Guey, M. P. Fitz, M. R. Bell, and W. Kuo, "Signal design for transmitter diversity wireless communication systems over Rayleigh fading channels," in *Proc. IEEE VTC*, Atlanta, GA, vol. 1, pp. 136–140, Apr./May 1996.

[13] B. M. Hochwald and T. L. Marzetta, "Unitary space–time modulation for multiple antenna communications in Rayleigh fading," *IEEE Trans. Inform. Theory*, vol. 46, pp. 543–564, March 2000.

[14] T. E. Hunter and A. Nosratinia, "Cooperative diversity through coding," in *Proc. IEEE ISIT*, Lausanne, Switzerland, June 2002, p. 220.

[15] J. P. Kermoal, L. Schumacher, P. E. Mogensen, and K. I. Pedersen, "Experimental investigation of correlation properties of MIMO radio channels for indoor picocell scenarios," in *Proc. IEEE VTC*, vol. 1, pp. 14–21, Sept. 2000.

[16] W. Kuo and M. P. Fitz, "Design and analysis of transmitter diversity using intentional frequency offset for wireless communications," *IEEE Trans. Vehicular. Tech.*, vol. 46, no. 4, pp. 871–881, Nov. 1997.

[17] P. Kyritsi, *Capacity of multiple input–multiple output wireless systems in an indoor environment*, Ph.D. thesis, Stanford University, Jan. 2002.

[18] J. N. Laneman and G. W. Wornell, "Distributed space–time-coded protocols for exploiting cooperative diversity in wireless networks," *IEEE Trans. Inform. Theory*, vol. 49, no. 10, pp. 2415–2425, Oct. 2003.

[19] T. L. Marzetta and B. M. Hochwald, "Capacity of a mobile multiple-antenna communication link in Rayleigh flat fading," *IEEE Trans. Inform. Theory*, vol. 45, no. 1, pp. 139–157, Jan. 1999.

[20] Ö. Oyman, R. U. Nabar, H. Bölcskei, and A. J. Paulraj, "Characterizing the statistical properties of mutual information in MIMO channels," *IEEE Trans. Sig. Proc.*, vol. 51, no. 11, pp. 2784–2795, Nov. 2003.

[21] L. H. Ozarow, S. Shamai, and A. D. Wyner, "Information theoretic considerations for cellular mobile radio," *IEEE Trans. Vehicular Technol.*, vol. 43, no. 2, pp. 359–378, May 1994.

[22] A. J. Paulraj and T. Kailath, "Increasing capacity in wireless broadcast systems using distributed transmission/directional reception," *U.S. Patent*, 1994, no. 5,345,599.

[23] G. G. Raleigh and J. M. Cioffi, "Spatio-temporal coding for wireless communication," *IEEE Trans. Commun.*, vol. 46, no. 3, pp. 357–366, March 1998.

[24] F. Rashid-Farrokhi, A. Lozano, G. J. Foschini, and R. A. Valenzuela, "Spectral efficiency of wireless systems with multiple transmit and receive antennas," in *Proc. IEEE Intl. Symp. on PIMRC*, London, vol. 1, pp. 373–377, Sept. 2000.

[25] A. Sendonaris, E. Erkip, and B. Aazhang, "User cooperation diversity – Part I: System description," *IEEE Trans. Commun.*, vol. 51, no. 11, pp. 1927–1938, Nov. 2003.

[26] A. Sendonaris, E. Erkip, and B. Aazhang, "User cooperation diversity – Part II: Implementation aspects and performance analysis," *IEEE Trans. Commun.*, vol. 51, no. 11, pp. 1939–1948, Nov. 2003.

[27] N. Seshadri and J. H. Winters, "Two signaling schemes for improving the error performance of frequency-division-duplex (FDD) transmission systems using transmitter antenna diversity," *Intl. J. Wireless Information Networks*, vol. 1, pp. 49–60, Jan. 1994.

[28] A. Stefanov and E. Erkip, "Cooperative coding for wireless networks," in *Proc. Intl. Workshop on Mobile and Wireless Commun. Networks*, Stockholm, Sweden, pp. 273–277, Sept. 2002.

[29] P. Stoica and E. Lindskog, "Space–time block coding for channels with intersymbol interference," in *Proc. Asilomar Conf. on Signals, Systems and Computers*, Pacific Grove, CA, vol. 1, pp. 252–256, Nov. 2001.

[30] R. Stridh, B. Ottersten, and P. Karlsson, "MIMO channel capacity of a measured indoor radio channel at 5.8 GHz," in *Proc. Asilomar Conf. on Signals, Systems and Computers*, vol. 1, pp. 733–737, Nov. 2000.

[31] B. Suard, G. Xu, and T. Kailath, "Uplink channel capacity of space-division-multiple-access schemes," *IEEE Trans. Inform. Theory*, vol. 44, no. 4, pp. 1468–1476, July 1998.

[32] A. L. Swindlehurst, G. German, J. Wallace, and M. Jensen, "Experimental measurements of capacity for MIMO indoor wireless channels," in *Proc. IEEE Signal Proc. Workshop on Signal Processing Advances in Wireless Communications*, Taoyuan, Taiwan, pp. 30–33, March 2001.

[33] V. Tarokh, H. Jafarkhani, and A. R. Calderbank, "Space–time block codes from orthogonal designs," *IEEE Trans. Inform. Theory*, vol. 45, no. 5, pp. 1456–1467, July 1999.

[34] V. Tarokh, N. Seshadri, and A. R. Calderbank, "Space–time codes for high data rate wireless communication: performance criterion and code construction," *IEEE Trans. Inform. Theory*, vol. 44, no. 2, pp. 744–765, March 1998.

[35] I. E. Telatar, "Capacity of multi-antenna Gaussian channels," *European Trans. Tel.*, vol. 10, no. 6, pp. 585–595, Nov./Dec. 1999.

[36] P. Viswanath, D. N. C. Tse, and V. Anantharam, "Asymptotically optimal water-filling in vector multiple-access channels," *IEEE Trans. Inform. Theory*, vol. 47, no. 1, pp. 241–267, Jan 2001.

[37] S. Vishwanath, N. Jindal, and A. Goldsmith, "Duality, achievable rates, and sum-rate capacity of Gaussian MIMO broadcast channels," *IEEE Trans. Inform. Theory*, vol. 49, no. 10, pp. 2658–2668, Oct. 2003.

[38] J. H. Winters, J. Salz, and R. D. Gitlin, "The impact of antenna diversity on the capacity of wireless communications systems," *IEEE Trans. Commun.*, vol. 42, no. 2, pp. 1740–1751, Feb. 1994.

[39] W. Yu and J. M. Cioffi, "Trellis precoding for the broadcast channel," in *Proc. IEEE GLOBECOM*, San Antonio, TX, vol. 2, pp. 1338–1344, Nov. 2001.

[40] L. Zheng and D. N. C. Tse, "Communicating on the Grassmann manifold: a geometric approach to the non-coherent multiple antenna channel," *IEEE Trans. Inform. Theory*, vol. 48, no. 2, pp. 359–383, Feb. 2002.

[41] L. Zheng and D. N. C. Tse, "Diversity and multiplexing: a fundamental trade-off in multiple-antenna channels," *IEEE Trans. Inform. Theory*, vol. 49, no. 5, pp. 1073–1096, May 2003.

2 Capacity limits of MIMO systems

2.1 Introduction

Chapter 1 introduced the basic concepts behind multiple-input multiple-output (MIMO) communications along with their performance advantages. In particular, we saw that MIMO systems provide tremendous capacity gains, which has spurred significant activity to develop transmitter and receiver techniques that realize these capacity benefits and exploit diversity–multiplexing trade-offs. In this chapter we will explore in more detail the Shannon capacity limits of single- and multi-user MIMO systems. These fundamental limits dictate the maximum data rates that can be transmitted over the MIMO channel to one or more users (not in outage) with asymptotically small error probability, assuming no constraints on the delay or the complexity of the encoder and decoder. Much of the initial excitement about MIMO systems was due to pioneering work by Foschini [32] and Telatar [121] predicting remarkable capacity growth for wireless systems with multiple antennas when the channel exhibits rich scattering and its variations can be accurately tracked. This promise of exceptional spectral efficiency almost "for free," also studied in earlier work by Winters [142], resulted in an explosion of research and commercial activity to characterize the theoretical and practical issues associated with MIMO systems. However, these predictions are based on somewhat unrealistic assumptions about the underlying time-varying channel model and how well it can be tracked at the receiver as well as at the transmitter. More realistic assumptions can dramatically impact the potential capacity gains of MIMO techniques. This chapter provides a comprehensive summary of MIMO Shannon capacity for both single- and multi-user systems with and without fading under different assumptions about what is known at the transmitter(s) and receiver(s).

We first provide some background on Shannon capacity and mutual information, and then apply these ideas to the single-user additive white Gaussian noise (AWGN) MIMO channel. We next consider MIMO fading channels, describe the different capacity definitions that arise when the channel is time-varying, and present the MIMO capacity under these different definitions. These results indicate that the capacity gain obtained from multiple antennas heavily depends on the available channel information at either the receiver or the transmitter, the channel signal-to-noise ratio (SNR), and the correlation between the channel gains on each antenna element. We then focus attention on the capacity region of multi-user channels: in particular, the multiple access (many-to-one, or uplink) and broadcast (one-to-many, or downlink) channels. In contrast to single-user MIMO channels, capacity results for these multi-user MIMO channels can be quite difficult to obtain, especially when the channel is not known perfectly at all transmitters and receivers. We will show that with perfect channel knowledge, the capacity region of the MIMO multiple access and broadcast channels are intimately related via a duality transformation, which can greatly simplify capacity analysis. This transformation facilitates finding the transmission strategies that achieve a point on the boundary of the MIMO multiple access channel (MAC) capacity region in terms of the transmission strategies of the MIMO broadcast channel (BC) capacity region, and vice versa. We then consider MIMO cellular systems with frequency reuse, where the base-stations cooperate. With cooperation the base-stations act as a spatially distributed antenna array, and transmission strategies that exploit this structure exhibit significant capacity gains. The chapter concludes with a discussion of fundamental capacity results of wireless ad hoc networks where nodes either have multiple antennas or cooperate to form multiple antenna transmitters and/or receivers. Open problems in this field abound and are discussed throughout the chapter.

2.2 Mutual information and Shannon capacity

Channel capacity was pioneered by Claude Shannon in the late 1940s, using a mathematical theory of communication [111–113]. The capacity of a channel, denoted by C, is the maximum rate at which reliable communication can be performed, without any constraints on transmitter and receiver complexity. Shannon showed that for any rate $R < C$, there exist rate R channel codes with arbitrarily small block (or symbol) error probabilities. Thus, for any rate $R < C$ and *any* desired non-zero probability of error P_e, there exists a rate R code that achieves P_e. However, such a code may have a very long block length, and the encoding and decoding complexity may also be extremely large. In fact, the required block length may increase as the desired P_e is decreased and/or the rate R is increased towards C. In addition, Shannon showed that codes operating at rates $R > C$ cannot achieve an arbitrarily small error rate, and thus the error probability of a code operating at a rate above capacity is bounded away from zero. Therefore, the channel capacity is truly the fundamental limit to communication.

Although it is theoretically possible to communicate at any rate below capacity, it is actually a very difficult problem to design practical channel codes (or codes with reasonable block length and encoding/decoding complexity) at rates close to capacity. Tremendous progress has been made in code design over the past few decades, and practical codes at rates very close to capacity do exist for certain channels, such as single-antenna Gaussian channels. However, these codes generally cannot be directly used for MIMO channels, as codes for MIMO channels must also utilize the spatial dimension. Practical space–time coding and decoding techniques for MIMO channels are described in Chapter 4, and are shown to achieve near-capacity limits in some scenarios. The capacity limits of MIMO channels provide a benchmark against which performance of space–time codes and general MIMO transmission and reception strategies can be compared. In addition, the study of MIMO channel capacity often yields insights into near-capacity achieving transmission strategies, receiver structures, and codes.

The following sections give the precise mathematical definition of channel capacity as well as discuss the capacity of time-varying and multi-user channels.

2.2.1 Mathematical definition of capacity

Shannon's pioneering work showed that the capacity of a channel, defined to be the maximum rate at which reliable communication is possible, can be simply characterized in terms of the mutual information between the input and the output of the channel. The basic channel model consists of a random input X, a random output Y, and a probabilistic relationship between X and Y which is generally characterized by the conditional distribution of Y given X, or $f(y|x)$. The *mutual information* of a single-user channel with random input X and random output Y is defined as

$$I(X;Y) = \int_{S_x,S_y} f(x,y) \log\left(\frac{f(x,y)}{f(x)f(y)}\right) dx\, dy, \tag{2.1}$$

where the integral is taken over the supports S_x, S_y of the random variables X and Y, respectively, and $f(x)$, $f(y)$, and $f(x,y)$ denote the probability distribution functions of the random variables. The log function is typically with respect to base 2, in which case the units of mutual information are bits per channel use. Mutual information can also be written in terms of the *differential entropy* of the channel output and conditional output as $I(X;Y) = h(Y) - h(Y|X)$, where $h(Y) = -\int_{S_y} f(y) \log f(y) dy$ and $h(Y|X) = -\int_{S_x,S_y} f(x,y) \log f(y|x)\, dx\, dy$.

Shannon proved that the channel capacity of most channels is equal to the mutual information of the channel maximized over all possible input distributions:

$$C = \max_{f(x)} I(X;Y) = \max_{f(x)} \int_{S_x,S_y} f(x,y) \log\left(\frac{f(x,y)}{f(x)f(y)}\right). \tag{2.2}$$

For a time-invariant AWGN channel with bandwidth B and received SNR γ, the maximizing input distribution is Gaussian, which results in the channel capacity

$$C = B \log_2(1+\gamma) \text{ bps}. \tag{2.3}$$

The definition of entropy and mutual information is the same when the channel input and output are vectors instead of scalars, as in the MIMO channel. Referring back to Figure 1.2, the channel input is the vector $\mathbf{x}$ sent from the transmit antennas and the channel output is the vector $\mathbf{y}$ obtained at the receive antennas. Thus, the Shannon capacity of the MIMO AWGN channel is based on the maximum mutual information between its input and output vectors, as described in Section 2.3.2.

2.2.2 Time-varying channels

When the channel is time-varying the channel capacity has multiple definitions, depending on what is known about the channel state or its distribution at the transmitter and/or receiver, as well as the time scale of the underlying channel fading process. These definitions have different operational meanings. Specifically, when the instantaneous channel gains, also called the *channel state information* (CSI), are known perfectly at both the transmitter and the receiver, the transmitter can adapt its transmission strategy (rate and/or power) relative to the instantaneous channel state. In this case the Shannon (ergodic) capacity is the maximum mutual information averaged over all channel states. Recall that ergodic essentially means that a reasonably long time sample of channel (fading) realizations has a distribution similar to the statistical distribution of the channel. The opposite of an ergodic channel is, for example, a quasi-static channel in which the channel realization is chosen at random initially but does not subsequently change (or changes extremely slowly). Ergodic capacity is an appropriate capacity metric for channels that vary quickly, or where the channel is ergodic over the time period of interest. With CSI at the transmitter (CSIT), ergodic capacity can be achieved using an adaptive transmission policy where the power and data rate vary relative to the channel state variations [40]. If an adaptive rate and power policy are used, the rate varies as a function of the instantaneous channel state and the ergodic capacity refers to the maximum possible long-term average of the instantaneous rates. Thus it is not necessary for the channel to change rapidly during a single codeword, as ergodic capacity is meaningful even in relatively slow fading environments where the long-term average rates are of interest. Ergodic capacity can also be achieved using a fixed rate code with varying power [12]. If such a code is used, it is necessary for the channel to change rapidly over the duration of a single codeword. Therefore, these codes must either be very long or the fading must change on a very fast time scale.

An alternate capacity definition for time-varying channels with perfect transmitter and receiver CSI is outage capacity. Outage capacity requires a fixed data rate in all non-outage channel states, which is needed for applications with delay-constrained data where the data rate cannot depend on channel variations (except in outage states, where no data are transmitted). The average rate associated with outage capacity is typically smaller than the ergodic capacity due to the additional constraint associated with this definition. Outage capacity is the appropriate capacity metric in slowly varying channels, where the channel coherence time exceeds the duration of a codeword. In this case each codeword experiences only one channel state: if the channel state is not good enough to support the desired rate then an outage is declared and no data are transmitted, since the transmitter knows that the channel is in outage. Note that ergodic capacity is not a relevant metric for slowly varying channels since each codeword is affected by a single channel realization. Similarly, outage capacity is not an appropriate capacity metric for channels that vary quickly: since the channel experiences all possible channel states over the duration of a codeword, there is no notion of poor states where an outage must be declared. Outage capacity under perfect CSI at the transmitter and the receiver (CSIR) has been studied for single-antenna channels [14, 46, 83], but this work has yet to be extended to MIMO channels. A more common assumption for studying capacity of time-varying MIMO channels is perfect CSIR but no CSIT. This assumption leads to a different notion of outage capacity, as described next.

When only the channel distribution is known at the transmitter (receiver), the transmission (reception) strategy is based on the channel distribution instead of the instantaneous channel state. The channel coefficients are typically assumed to be jointly Gaussian, so the channel distribution is specified by the channel mean and covariance matrices. We will refer to knowledge of the channel distribution as *channel distribution information* (CDI): CDI at the transmitter is abbreviated as CDIT and CDI at the receiver is denoted by CDIR. We assume throughout the chapter that CDI is always perfect, so there is no mismatch between the CDI at the transmitter or receiver and the true channel distribution. When only the receiver has perfect CSI the transmitter must maintain a fixed-rate transmission strategy optimized with respect to its CDI. In this case the ergodic capacity defines the rate that can be achieved based on averaging over all channel states [121], and this metric is relevant for channels that vary quickly so that codeword transmissions are affected by all possible channel states. Note that this is similar to the situation where ergodic capacity is achieved using constant rate transmission in channels with perfect CSI. Alternatively, the transmitter can send at a rate that cannot be supported by all channel states: in these poor channel states the receiver declares an outage and the transmitted data are lost. For this scenario, as described earlier in Section 1.3.1, we define the percentage outage capacity p to be the transmission rate that can be supported $(100-p)\%$ of the time. The outage probability, defined as the probability that the transmission rate cannot be supported and hence the transmitted data are received in error, is $p/100$. Intuitively, an outage occurs whenever the channels enter a deep fade that makes it impossible to communicate reliably. In this outage scenario each transmission rate has an outage probability associated with it so capacity is parameterized by the outage probability[1] (capacity versus outage or capacity CDF) [32]. An excellent tutorial on fading channel capacity for single-antenna channels can be found in [4]. This chapter extends these results to MIMO systems.

2.2.3 *Multi-user channels*

In multi-user channels, the capacity is a K-dimensional region instead of a single number. The capacity region is defined to be the set of all rate vectors $(R_1, \ldots, R_K)$ simultaneously achievable (with arbitrarily small probability of error) by all K users. The multiple capacity definitions for time-varying channels under different transmitter and receiver CSI and CDI assumptions extend to the capacity region of the MAC and BC in the obvious way, as we will describe in Section 2.4. However, these MIMO multi-user capacity regions are very difficult to find even for time-invariant channels. Few capacity results exist for time-varying multi-user MIMO channels under the realistic assumption that the transmitter(s) and/or receiver(s) have CDI only. The results become even more sparse for more complex systems such as cellular and ad hoc wireless networks, as will be seen in Sections 2.5 and 2.6. Thus, there are many open problems in these areas, as will be highlighted in the related sections.

[1] Note that an outage under perfect CSI at the receiver is only different from an outage when both the transmitter and the receiver have perfect CSI. Under receiver CSI, an outage occurs when the transmitted data cannot be reliably decoded at the receiver, so that data are lost. When both the transmitter and the receiver have perfect CSI the channel is not used during outage (no service), so no data are lost.

2.3 Single-user MIMO

In this section we focus on the capacity of single-user MIMO channels. While most wireless systems today support multiple users, single-user results are still of much interest for the insight they provide and their application to channelized systems where users are allocated orthogonal resources (time, frequency bands, etc.). MIMO channel capacity is also much easier to derive for single-users than for multiple users. Indeed, single-user MIMO capacity results are known for many cases where the corresponding multi-user problems remain unsolved. In particular, very little is known about multi-user capacity without the assumption of perfect channel state information at the transmitter (CSIT) and at the receiver (CSIR). While there remain many open problems in obtaining the single-user capacity under general assumptions of CSI and CDI, for several interesting cases the solution is known. This section will discuss fundamental capacity limits for single-user MIMO channels with a particular focus on special cases of CDI at the transmitter as well as the receiver. We begin with a description of the channel model and the different CSI and CDI models we consider, along with their motivation.

2.3.1 *Channel and side information model*

Consider a transmitter with M_T transmit antennas and a receiver with M_R receive antennas. The channel can be represented by the $M_R \times M_T$ matrix $\mathbf{H}$ of channel gains h_{ij} representing the gain from transmit antenna j to receive antenna i. The $M_R \times 1$ received signal $\mathbf{y}$ is equal to

$$\mathbf{y} = \mathbf{H}\mathbf{x} + \mathbf{n}, \tag{2.4}$$

where $\mathbf{x}$ is the $M_T \times 1$ transmitted vector and $\mathbf{n}$ is the $M_R \times 1$ additive white circularly symmetric complex Gaussian noise vector,[2] normalized so that its covariance matrix is the identity matrix. The normalization of any non-singular noise covariance matrix $\mathbf{K}_w$ to fit the above model is as straightforward as multiplying the received vector $\mathbf{y}$ by $\mathbf{K}_w^{-1/2}$ to yield the effective channel $\mathbf{K}_w^{-1/2}\mathbf{H}$ and a white noise vector. The channel state information is the channel matrix $\mathbf{H}$ and/or its distribution.

The transmitter is assumed to be subject to an average power constraint of P across all transmit antennas, i.e. $E[\mathbf{x}^*\mathbf{x}] \le P$. Since the noise power is normalized to unity, we commonly refer to the power constraint P as the SNR.

Perfect CSIR and/or CSIT

With perfect CSIT or CSIR, the channel matrix $\mathbf{H}$ is assumed to be known perfectly and instantaneously at the transmitter or receiver, respectively. When the transmitter or receiver knows the channel state perfectly, we also assume that it knows the distribution of this state perfectly, since the distribution can be obtained from the state observations.

[2] A complex Gaussian vector $\mathbf{x}$ is circularly symmetric if for any $\theta \in [0, 2\pi]$, the distribution of $\mathbf{x}$ is the same as the distribution of $e^{j\theta}\mathbf{x}$.

Perfect CSIR and CDIT

The perfect CSIR and CDIT model is motivated by the scenario where the channel state can be accurately tracked at the receiver and the statistical channel model at the transmitter is based on channel distribution information fed back from the receiver. This distribution model is typically based on receiver estimates of the channel state and the uncertainty and delay associated with these estimates. Figure 2.1 illustrates the underlying communication model in this scenario, where $\tilde{\mathcal{N}}$ denotes the complex Gaussian distribution.

The salient features of the model are as follows. The channel distribution is defined by the parameter θ and, conditioned on this parameter, the channel realizations $\mathbf{H}$ at different time instants are independently and identically distributed (i.i.d.). Since the channel statistics will change over time due to mobility of the transmitter, receiver, and the scattering environment, we assume that θ is time-varying. Note that the statistical model depends on the time scale of interest. For example, in the short-term the channel coefficients may have a non-zero-mean and one set of correlations reflecting the geometry of the particular propagation environment. However, over a long-term the channel coefficients may be described as being zero-mean and uncorrelated due to the averaging over several propagation environments. For this reason, uncorrelated, zero-mean channel coefficients are commonly assumed for the channel distribution in the absence of distribution feedback or when it is not possible to adapt to the short-term channel statistics. However, if the transmitter receives frequent updates of θ and it can adapt to these time-varying short-term channel statistics then capacity is increased relative to the transmission strategy associated with just the long-term channel statistics. In other words, adapting the transmission strategy to the short-term channel statistics increases capacity. In the literature adaptation to the short-term channel statistics (the feedback model of Figure 2.1) is referred to by many names including mean and covariance feedback, quantized feedback, imperfect feedback, and partial CSI [57, 61, 63, 72, 73, 93, 117, 134]. The feedback channel is assumed to be free from noise. This makes the CDIT a deterministic function of the CDIR and allows optimal codes to be constructed directly over the input alphabet [12]. We assume a power constraint such that for each realization of θ, the conditional average transmit power is constrained as $\mathbb{E}\left[||\mathbf{x}||^2|\Theta = \theta\right] \leq P$.

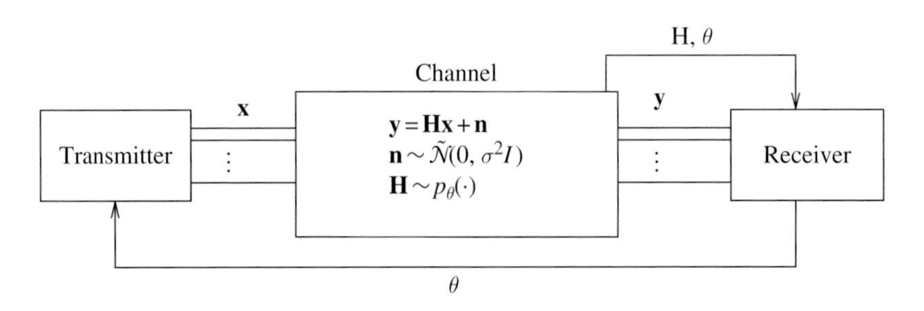

Fig. 2.1. MIMO channel with perfect CSIR and distribution feedback.

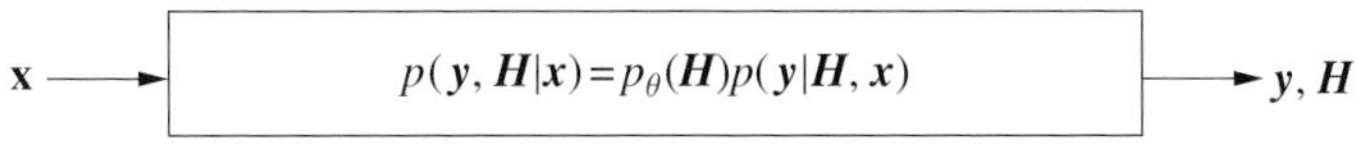

$$x \longrightarrow \boxed{p(y, H|x) = p_\theta(H)p(y|H, x)} \longrightarrow y, H$$

Fig. 2.2. MIMO channel with perfect CSIR and CDIT (θ fixed).

The ergodic capacity C of the system in Figure 2.1 is the capacity $C(\theta)$ averaged over the different θ realizations:

$$C = E_\theta[C(\theta)],$$

where $C(\theta)$ is the ergodic capacity of the channel shown in Figure 2.2. This figure represents a MIMO channel with perfect CSI at the receiver and only CDI about the *constant* distribution θ at the transmitter. Channel capacity calculations generally implicitly assume CDI at both the transmitter and the receiver except for special channel classes, such as the compound channel or an arbitrarily varying channel [5, 26, 27]. This implicit knowledge of θ is justified by the fact that the channel coefficients are typically modeled based on their long-term average distribution. Alternatively, θ can be obtained by the feedback model of Figure 2.1. The distribution feedback model of Figure 2.1 along with the system model of Figure 2.2 lead to various capacity results under different distribution (θ) models. The availability of CDI at either the transmitter or the receiver is explicitly indicated to contrast with the case where CSI is also available.

Computation of $C(\theta)$ for general $p_\theta(\cdot)$ is a hard problem. With the exception of a quantized channel information model, almost all research in this area has focused on three special cases for this distribution: zero-mean spatially white channels, spatially white channels with a non-zero-mean, and zero-mean channels with non-white channel covariance. In all three cases the channel coefficients are modeled as complex jointly Gaussian random variables. Under the zero-mean spatially white (ZMSW) model, the channel mean is zero and the channel covariance is modeled as being white, i.e. the channel elements are assumed to be i.i.d. random variables. This model typically captures the long-term average distribution of the channel coefficients averaged over multiple propagation environments. Under the channel mean information (CMI) model, the mean of the channel distribution is non-zero while the covariance is modeled as being white with a constant scale factor. This model is motivated by a system where the delay in the feedback leads to an imperfect estimate at the transmitter, so the CMI reflects the outdated channel measurement and the constant factor reflects the estimation error. Under the channel covariance information (CCI) model, the channel is assumed to be varying too rapidly to track its mean, so the mean is set to zero and the information regarding the relative geometry of the propagation paths is captured by a non-white covariance matrix. Based on the underlying system model shown in Figure 2.1, in the literature the CMI model is also called mean feedback and the CCI model is also called covariance feedback. Mathematically, the three distribution models for $\mathbf{H}$ can be described as follows:

$$\begin{aligned}
\mathbf{ZMSW}: E[\mathbf{H}] &= \mathbf{0}, & \mathbf{H} &= \mathbf{H}_w \\
\mathbf{CMI}: E[\mathbf{H}] &= \mathbf{H}_m, & \mathbf{H} &= \mathbf{H}_m + \sqrt{\alpha}\mathbf{H}_w \\
\mathbf{CCI}: E[\mathbf{H}] &= \mathbf{0}, & \mathbf{H} &= (\mathbf{R}_r)^{1/2}\mathbf{H}_w(\mathbf{R}_t)^{1/2}.
\end{aligned}$$

Here $\mathbf{H}_w$ is an $M_R \times M_T$ matrix of i.i.d. zero-mean, unit variance complex circularly symmetric Gaussian random variables. In the CMI model, the channel mean $\mathbf{H}_m$ and α are constants that may be interpreted as the channel estimate based on the feedback, and the variance of the estimation error, respectively. In the CCI model, $\mathbf{R}_r$ and $\mathbf{R}_t$ are referred to as the receive and transmit antenna correlation matrices, respectively. Although not completely general, this simple correlation model has been validated through field measurements as a sufficiently accurate representation of the fade correlations seen in actual cellular systems [19]. Under CMI the channel mean $\mathbf{H}_m$ and the variance of the estimation error α are assumed to be known when there is CDI, and under CCI the transmit and receive covariance matrices $\mathbf{R}_r$ and $\mathbf{R}_t$ are assumed to be known when there is CDI.

In addition to the CMI, CCI, and ZMSW models of CDIT, research has explored the effects of quantized channel state information (QCI) at the transmitter based on a finite bit rate feedback channel. In this model, the receiver is assumed to have perfect CSI and feeds back a B-bit quantization of the channel instantiation to the transmitter. This model is most applicable to relatively slow fading scenarios, where the receiver feeds back quantized CSI at the beginning of each block, thereby allowing the transmitter to adapt. Note that this is a very practical model, as many wireless systems have a low-rate feedback link from the receiver to the transmitter. With B bits of QCI, a predetermined set of $N = 2^B$ quantization vectors is used to represent the channel at the transmitter. Finding the best quantization vectors is equivalent to the Grassmannian packing of subspaces within a vector space to maximize the minimum-distance between them [85, 94]. Conditioned on the QCI, the channel distribution assumed at the transmitter is constrained within the Voronoi region of the quantization vector used to represent the channel.

CDIT and CDIR

In highly mobile channels the assumption of perfect CSI at the receiver can be unrealistic. This motivates system models where both the transmitter and the receiver only have information about the channel distribution. Even for a rapidly fluctuating channel where reliable channel estimation is not possible, it might be possible for the receiver to track the short-term distribution of the channel fades, as the channel distribution changes much more slowly than the channel itself. The estimated distribution can be made available to the transmitter through a feedback channel. Figure 2.3 illustrates the underlying communication model.

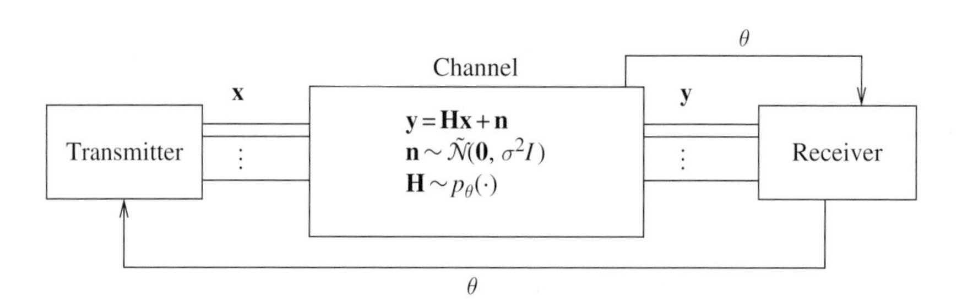

Fig. 2.3. MIMO channel with CDIR and distribution feedback.

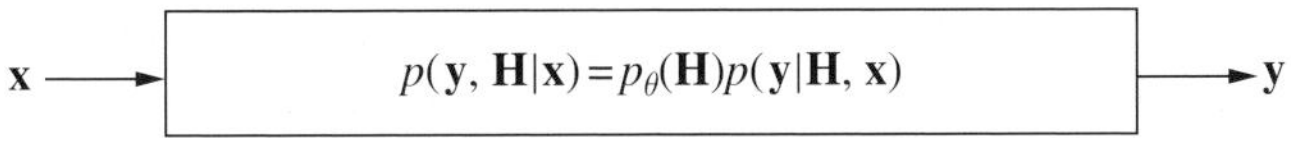

$\mathbf{x} \longrightarrow$ $p(\mathbf{y}, \mathbf{H}|\mathbf{x}) = p_\theta(\mathbf{H})p(\mathbf{y}|\mathbf{H}, \mathbf{x})$ $\longrightarrow \mathbf{y}$

Fig. 2.4. MIMO channel with CDIT and CDIR (θ fixed).

Note that the estimation of the channel statistics at the receiver is captured in the model as a genie that provides the receiver with the correct channel distribution. The feedback channel represents the same information being made available to the transmitter simultaneously. This model is slightly optimistic because in practice the receiver estimates θ only from the received signal $\mathbf{v}$ and therefore will not have a perfect estimate.

As in the previous subsection the ergodic capacity turns out to be the expected value (expectation over θ) of the ergodic capacity $C(\theta)$, where $C(\theta)$ is the ergodic capacity of the channel in Figure 2.4. In this figure θ is constant and known at both the transmitter and receiver (CDIT and CDIR). As in the previous section, the computation of $C(\theta)$ is difficult for general θ and the capacity investigations are limited mainly to the same channel distribution models described in the previous subsection: the ZMSW, CMI, CCI and QCI models.

Next, we summarize single-user MIMO capacity results under various assumptions on CSI and CDI.

2.3.2 *Constant MIMO channel capacity*

When the channel is constant and known perfectly at the transmitter and the receiver, the capacity (maximum mutual information) is

$$C = \max_{\mathbf{Q}:\text{tr}(\mathbf{Q})=P} \log\det\left(\mathbf{I}_N + \mathbf{H}\mathbf{Q}\mathbf{H}^\star\right) \tag{2.5}$$

where the optimization is over the input covariance matrix $\mathbf{Q}$, which is $M \times M$ and must be positive semi-definite by definition. Using the singular value decomposition (SVD) of the $M_R \times M_T$ matrix $\mathbf{H}$, this channel can be converted into $\min(M_T, M_R)$ parallel, non-interfering single-input/single-output channels [41, 121]. The SVD allows us to rewrite $\mathbf{H}$ as $\mathbf{H} = \mathbf{U}\mathbf{\Sigma}\mathbf{V}^\star$, where $\mathbf{U}$ is $M_R \times M_R$ and unitary, $\mathbf{V}$ is $M_T \times M_T$ and unitary, and $\mathbf{\Sigma}$ is $M_T \times M_R$ and diagonal with non-negative entries. The diagonal elements of the matrix $\mathbf{\Sigma}$, denoted by σ_i, are the singular values of $\mathbf{H}$ and are assumed to be in descending order (i.e. $\sigma_1 \geq \sigma_2 \cdots \geq \sigma_{\min(M_T, M_R)}$). The matrix $\mathbf{H}$ has exactly R_H positive singular values, where R_H is the rank of $\mathbf{H}$, which by basic principles satisfies $R_H \leq \min(M, N)$.

The MIMO channel is converted into parallel, non-interfering channels by pre-multiplying the input by the matrix $\mathbf{V}$ (i.e. transmit precoding) and post-multiplying the output by the matrix $\mathbf{U}^\star$. This conversion is illustrated in Figure 2.5. Note that transmit

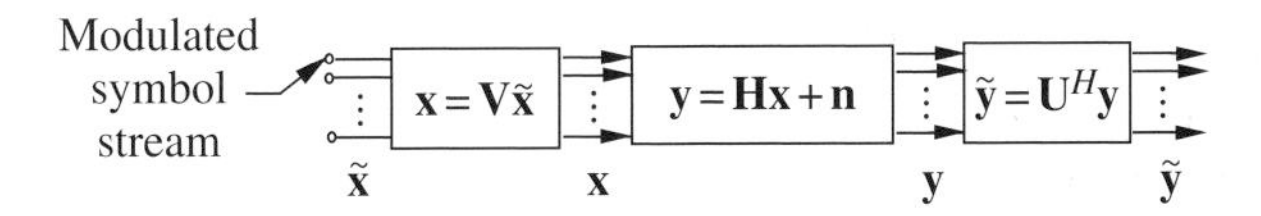

Fig. 2.5. MIMO channel decomposition.

precoding is widely used in practical systems and provides significant performance benefits, as discussed in Chapter 3. The optimal MIMO receiver requires joint maximum-likelihood detection across all receive antennas, but practical techniques of lower complexity can achieve very good performance, as described in Chapter 5. The input $\mathbf{x}$ to the channel is generated by multiplying the data stream $\tilde{\mathbf{x}}$ by the matrix $\mathbf{V}$, i.e. $\mathbf{x} = \mathbf{V}\tilde{\mathbf{x}}$. Since $\mathbf{V}$ is a unitary matrix, this is a power-preserving linear transformation, i.e. $E[||\mathbf{x}||^2] = E[||\tilde{\mathbf{x}}||^2]$. The vector $\mathbf{x}$ is fed into the channel and the output $\mathbf{y}$ is multiplied by the matrix $\mathbf{U}^\star$, resulting in $\tilde{\mathbf{y}} = \mathbf{U}^\star\mathbf{y}$, which can be expanded as

$$\begin{aligned}\tilde{\mathbf{y}} &= \mathbf{U}^\star(\mathbf{H}\mathbf{x} + \mathbf{n}) \\ &= \mathbf{U}^\star(\mathbf{U}\boldsymbol{\Sigma}\mathbf{V}^\star(\mathbf{V}\tilde{\mathbf{x}}) + \mathbf{n}) \\ &= \boldsymbol{\Sigma}\tilde{\mathbf{x}} + \mathbf{U}^\star\mathbf{n} \\ &= \boldsymbol{\Sigma}\tilde{\mathbf{x}} + \tilde{\mathbf{n}},\end{aligned}$$

where $\tilde{\mathbf{n}} = \mathbf{U}^\star\mathbf{n}$. Since $\mathbf{U}$ is unitary and $\mathbf{n}$ is a spatially white complex Gaussian, $\tilde{\mathbf{n}}$ and $\mathbf{n}$ have the same distribution. Using the fact that $\boldsymbol{\Sigma}$ is diagonal, we have $\tilde{\mathbf{y}}_i = \sigma_i\tilde{\mathbf{x}}_i + \tilde{\mathbf{n}}_i$ for $i = 1, \ldots, \min(M_T, M_R)$. Since R_H of the singular values σ_i are strictly positive, the result is R_H parallel, non-interfering channels. The parallel channels are commonly referred to as the *eigenmodes* of the channel, because the singular values of $\mathbf{H}$ are equal to the square root of the eigenvalues of the matrix $\mathbf{H}\mathbf{H}^\star$.

Because the parallel channels are of different quality, the water-filling algorithm can be used to optimally allocate power over the parallel channels, leading to the following allocation:

$$P_i = \left(\mu - \frac{1}{\sigma_i^2}\right)^+, \qquad 1 \le i \le R_H, \tag{2.6}$$

where P_i is the power of $\tilde{\mathbf{x}}_i$, x^+ is defined as $\max(x, 0)$, and the waterfill level μ is chosen such that $\sum_{i=1}^{R_H} P_i = P$. Capacity is therefore achieved by choosing each component $\tilde{\mathbf{x}}_i$ according to an independent Gaussian distribution with power P_i. The covariance which achieves the maximum in (2.5) (i.e. the covariance of the capacity-achieving input) is $\mathbf{Q} = \mathbf{V}\mathbf{P}\mathbf{V}^\star$, where the $M_T \times M_T$ matrix $\mathbf{P}$ is defined as $\mathbf{P} = \text{diag}(P_1, \ldots, P_{R_H}, 0, \ldots, 0)$. The resulting capacity is given by

$$C = \sum_i^{R_H} (\log(\mu\sigma_i^2))^+. \tag{2.7}$$

At low SNRs, the water-filling algorithm allocates all power to the strongest of the R_H parallel channels (i.e. $P_1 = P$ and $P_i = 0$ for $i \neq 1$). At high SNR, the water-filling algorithm allocates approximately equal power to each of the R_H, and a first-order approximation of the capacity at high SNR is $C \approx R_H \log_2(P) + O(1)$, where the constant term depends on the singular values of $\mathbf{H}$. From this approximation, one can see that every increase of 3 dB in transmission power leads to an increase of approximately R_H bps/Hz in spectral efficiency; this contrasts with single-antenna systems, where every 3 dB of power only leads to an additional bit of spectral efficiency. The pre-log term R_H is often referred to as the MIMO spatial *multiplexing gain*, described in Section 1.1.1, as it dictates the multiplicative capacity increase of MIMO relative to single-input single-output (SISO) systems.

There are a number of intuitive observations that can be made regarding the capacity-achieving transmission scheme. First, note that transmit precoding by the matrix $\mathbf{V}$ "aligns" the inputs with the eigenmodes of the channel. By aligning the inputs with the eigenmodes, simple post-multiplication of the output signal results in independent (noisy) observations of each of the inputs, i.e. a complete decoupling of the different data streams. If the transmitter did not perform this alignment process, e.g. transmitted independent inputs on each of the M_T transmit antennas, then the receiver would not be able to completely decouple the multiple data streams, which leads to an effective loss in SNR of each of the streams and thus is not capacity achieving. In addition, the transmitter performs water-filling across the different eigenmodes of the channel in order to take advantage of channels of different quality. As expected, the channels with the highest SNR are loaded with the most power and highest rate.

A key advantage of the decomposition of the MIMO channel into parallel non-interfering channels is that the decoding complexity is only linear in R_H, the rank of the channel. When it is not possible to perform such a decomposition (e.g. when the transmitter does not have perfect knowledge of the matrix $\mathbf{H}$), the maximum-likelihood decoding complexity is typically exponential in R_H. Chapter 5 discusses practical receiver structures that reduce this complexity in exchange for some performance penalty.

Although the constant channel model is relatively easy to analyze, wireless channels in practice are not fixed or constant. Instead, due to the changing propagation environment wireless channels vary over time, assuming values over a continuum. The capacity of fading channels is investigated next.

2.3.3 Fading MIMO channel capacity

With slow fading, the channel may remain approximately constant long enough to allow reliable estimation of the channel state at the receiver (perfect CSIR) and timely feedback of this state information to the transmitter (perfect CSIT). However, in systems with moderate to high user mobility, the system designer is inevitably faced with channels that change rapidly. Fading models where only the channel distribution is available to the receiver (CDIR) and/or transmitter (CDIT) are more applicable to such systems. Capacity results under various assumptions regarding CSI and CDI are summarized in this section.

Capacity with perfect CSIT and perfect CSIR

Perfect CSIT and perfect CSIR model a fading channel that changes slowly enough to be reliably measured by the receiver and fed back to the transmitter without significant delay. The ergodic capacity of a flat-fading channel with perfect CSIT and CSIR is simply the average of the capacities achieved with each channel realization. The capacity for each channel realization is given by the constant channel capacity expression in the previous section. Thus the fading MIMO channel capacity assuming perfect channel knowledge at both the transmitter and the receiver is

$$C = E_{\mathbf{H}}\left[\max_{\mathbf{Q}(\mathbf{H}):\mathrm{tr}(\mathbf{Q}(\mathbf{H}))=P} \log\det\left(\mathbf{I}_{M_R} + \mathbf{H}\mathbf{Q}(\mathbf{H})\mathbf{H}^{\star}\right)\right]. \tag{2.8}$$

The covariance of the input is written as $\mathbf{Q}(\mathbf{H})$ to emphasize the fact that the covariance can be changed as a function of the channel realization. In fact, the covariance for each channel realization is chosen using the water-filling procedure described in Section 2.3.2. Thus, each MIMO channel realization is decomposed into parallel channels, and water-filling is performed over both space and time, i.e. over the $\min(M_T, M_R)$ eigenmodes in each state and across the fading distribution. Some results on the computation of the water-filling level and the corresponding capacity are given in [64]. Note that the capacity expression in (2.8) is valid for *any* fading distribution.[3]

Obtaining CSIT can be rather difficult in time-varying channels, as it generally requires either high-rate feedback from the receiver, or time-division duplex (TDD) operation on a sufficiently fast scale. A detailed discussion of obtaining CSIT in practical systems is given in Section 3.1. However, there are both capacity and implementation benefits relative to having CDIT. In the next section we study the capacity with CSIR and CDIT, and briefly compare this scenario to CSIT/CSIR for the ZMSW model.

Capacity with perfect CSIR and CDIT: ZMSW model

For the case of perfect CSIR and a ZMSW channel distribution at the transmitter, the channel matrix $\mathbf{H}$ is assumed to have i.i.d. complex Gaussian entries (i.e. $\mathbf{H} \sim \mathbf{H}_{\mathrm{w}}$). As described in Section 2.2, the two relevant capacity definitions in this case are capacity versus outage (capacity cumulative distribution function, CDF) and ergodic capacity. For any given input covariance matrix the input distribution that achieves the ergodic capacity is shown in [33] and [121] to be a complex vector Gaussian, mainly because the vector Gaussian distribution maximizes the entropy for any given covariance matrix. This leads to the transmitter optimization problem, i.e. finding the optimum input covariance matrix to maximize ergodic capacity subject to a transmission power (the trace of the input covariance matrix) constraint. Mathematically, the problem is to characterize the optimum $\mathbf{Q}$ to maximize

$$C = \max_{\mathbf{Q}:\mathrm{tr}(\mathbf{Q})=P} C(\mathbf{Q}), \tag{2.9}$$

where

$$C(\mathbf{Q}) \triangleq E_{\mathbf{H}}\left[\log\det\left(\mathbf{I}_{M_R} + \mathbf{H}\mathbf{Q}\mathbf{H}^{\star}\right)\right] \tag{2.10}$$

[3] Additionally, this capacity only depends on the stationary distribution of the fading process and is thus independent of any memory in the fading process. We investigate only memoryless fading processes in this chapter, but this assumption does not affect the capacity if perfect CSI is available at both the transmitter and the receiver.

is the maximum achievable rate when using the input covariance matrix $E[\mathbf{x}\mathbf{x}^\star] = Q$ and the expectation is with respect to the channel matrix $\mathbf{H}$. The rate $C(\mathbf{Q})$ is achieved by transmitting independent complex circular Gaussian symbols along the eigenvectors of $\mathbf{Q}$. The powers allocated to the eigenvectors are given by the eigenvalues of $\mathbf{Q}$.

When there is CSIR and CSIT, as discussed in the previous section, the transmitter can use its instantaneous knowledge of $\mathbf{H}$ to align its transmission (i.e. the input covariance matrix) with the eigenmodes of the channel $\mathbf{H}$. When the transmitter does not know the instantaneous channel realization and only has knowledge of the fading distribution, it is not possible to align the input (the covariance of which is fixed for all time) with every possible realization of the channel. In addition to not being able to identify the eigenmodes of the channel, the transmitter is also not able to identify the directions in which the channel is stronger in the sense of delivering more power. Therefore, one might intuitively expect the optimal strategy to involve transmitting power in all spatial directions, without any form of power control. In fact, it is shown in [121] and [33] that the optimum transmit strategy is to indeed transmit power in all spatial directions with equal power allocated to each direction. More specifically, the optimum input covariance matrix that maximizes ergodic capacity is the scaled identity matrix, i.e. $\mathbf{Q} = \frac{P}{M_T}\mathbf{I}_{M_T}$, and thus the transmit power is divided equally among all the transmit antennas. Thus the ergodic capacity is given by

$$C = E_{\mathbf{H}}\left[\log\det\left(\mathbf{I}_{M_R} + \frac{P}{M_T}\mathbf{H}\mathbf{H}^\star\right)\right]. \tag{2.11}$$

An integral form of this expectation involving Laguerre polynomials is derived in [121]. Explicit capacity expressions are also obtained in [115].

Results on the asymptotic behavior of ergodic capacity as either the SNR or the number of antennas are taken to infinity are useful for gaining intuition. Though these results are asymptotic, they illustrate many features that are applicable for moderate SNR values and relatively small antenna arrays. If M_T and M_R are fixed and the SNR (P) is taken to infinity, the capacity grows approximately as $C \approx \min(M_T, M_R)\log_2 P + O(1)$. Thus the ergodic capacity has a multiplexing gain of $\min(M_T, M_R)$, i.e. each 3 dB of SNR leads to an increase of $\min(M_T, M_R)$ bps/Hz in spectral efficiency.[4] Plots of ergodic capacity for 1×1, 4×4, and 4×10 systems are shown in Figure 2.6. Notice that the linear growth of capacity with respect to SNR begins around 10 dB, which is quite reasonable. The 1×1 system has a slope of 1 bit/3 dB, whereas the 4×4 and 4×10 systems both have a slope of 4 bits/3 dB. Notice also that the 4×4 and 4×10 systems have the same slope but different constant terms, i.e. the 4×10 system has a power gain relative to the 4×4 system. Expressions for the constant term (for the ZMSW model as well as the CCI and CMI model) are available in the literature on the high SNR behavior of MIMO channels [89].

It is also possible to study the asymptotic behavior as the number of transmit and/or receive antennas is taken to infinity while the transmit power is kept fixed. If the number of transmit antennas (M_T) is taken to infinity while keeping M_R fixed, the capacity is

[4] This is related to the fact that the rank of $\mathbf{H}$ is $\min(M_T, M_R)$ with probability one for the ZMSW model.

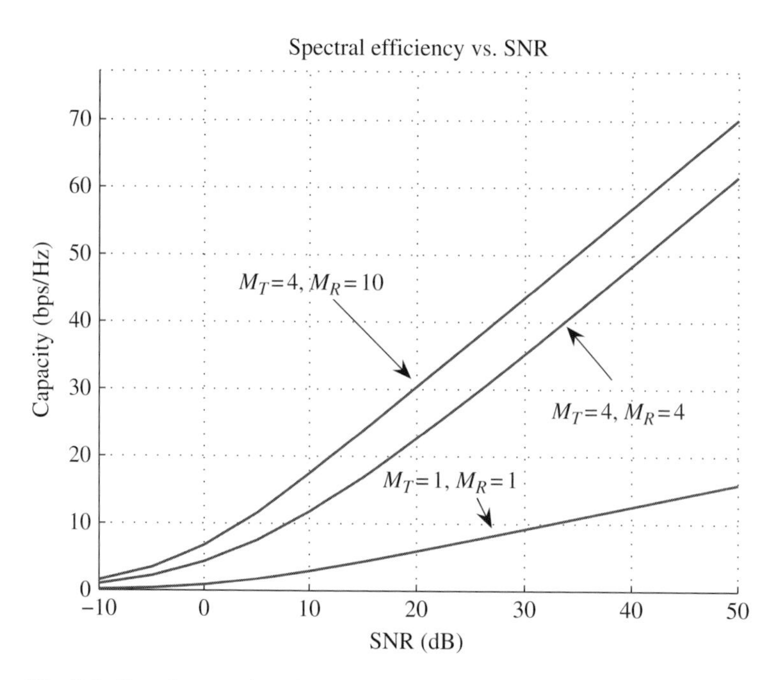

Fig. 2.6. Ergodic capacity with CSIR/CDIT versus SNR.

bounded in M_T and converges to $M_R \log(1+P)$ [121]. This is due to the fact that a fixed amount of transmitted power is equally divided between more and more antennas. If the number of receiving antennas M_R is taken to infinity while keeping M_T fixed, the capacity does indeed go to infinity approximately as $\log(M_R)$. The key difference is that adding receive antennas increases the amount of received power, while adding transmit antennas does not do so because the power is split between all transmit antennas. If M_T and M_R are simultaneously taken to infinity, the capacity is seen to grow *linearly* with $\min(M_T, M_R)$, i.e. $C \approx \min(M_T, M_R) \cdot c$, where c is a constant depending on the ratio of M_T and M_R, and on the SNR. Expressions for the growth rate constant can be found in [50, 121]. In summary, increasing the number of receive antennas leads to a logarithmic growth in capacity, while simultaneously increasing the number of receive and transmit antennas leads to linear growth. Increasing the number of transmit antennas, on the other hand, provides only a bounded increase in capacity. These points are illustrated in Figure 2.7, where capacity is plotted versus the number of antennas. In the linear curve, the number of transmit and receive antennas are set equal to the x-axis parameter r. In the second curve, only the number of receive antennas is increased (i.e. $M_R = r$) while M_T is kept at one, leading to logarithmic growth. In the final curve, only the number of transmit antennas is increased (i.e. $M_T = r$) while M_R is kept at one, leading to only bounded growth. One crucial point to take away from this section is that the capacity of an $r \times r$ MIMO system supports approximately r times the capacity of a SISO system at *any SNR*. This approximation has a maximum error of around 10%, and thus is extremely accurate.

In general, vector codebooks are needed to achieve capacity on a channel with multiple inputs. The decoding complexity for vector codebooks can increase exponentially with

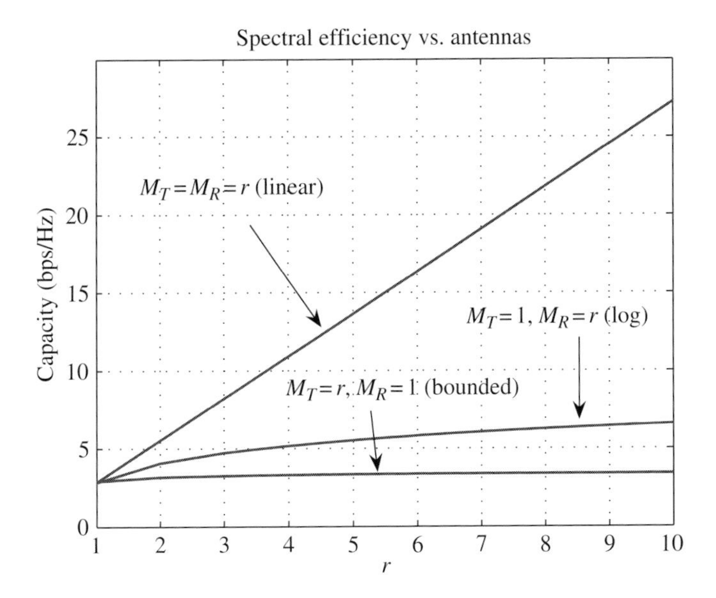

Fig. 2.7. Ergodic capacity with CSIR/CDIT versus antennas.

the number of inputs. Therefore capacity-achieving schemes that use scalar codes are of much practical interest. For the two transmit, one receive antenna ZMSW case, the Alamouti coding scheme (described in Chapters 1 and 4) is an extremely simple method to achieve capacity. Such a scheme, however, cannot be generalized to an arbitrary number of transmit and receive antennas. BLAST (Bell labs layered space–time) is a well-known layered architecture which can be used to achieve (or come close to) capacity for an arbitrary number of transmit and receive antennas, and practical implementations have even been shown to provide enormous capacity gains over single-antenna systems. For example, at 1% outage, 12 dB SNR, and with 12 antennas, the spectral efficiency is shown to be 32 bps/Hz as opposed to the spectral efficiencies of around 1 bps/Hz achieved in present-day single-antenna systems. While initial results on BLAST assumed uncorrelated and frequency-flat fading, practical channels exhibit both correlated fading as well as frequency selectivity. The need to estimate the capacity gains of BLAST for practical systems in the presence of channel-fade correlations and frequency-selective fading sparked off the measurement campaigns reported in [37, 92]. The measured capacities are found to be about 30% smaller than would be anticipated from an idealized model. However, the capacity gains over single-antenna systems are still overwhelming. Different low complexity receiver structures are analyzed in detail in Chapter 5.

In the previous section an expression for the capacity with perfect CSIR and CSIT is given. Of course, capacity with CSIR/CSIT must be larger than with CSIR/CDIT for the ZMSW model. Note that the multiplexing gain is $\min(M_T, M_R)$ with either CSIT or CDIT; thus, transmitter CSI can only provide a power or rate gain relative to CDIT. In general, CSIT provides the most benefit relative to CDIT at low SNRs (for any number of antennas), and at all SNRs when the number of transmit antennas is strictly larger than the number of receive antennas. This comparison is discussed in more detail in Section 3.2.1.

It is conjectured in [121] that the optimal input covariance matrix that maximizes capacity versus outage is a diagonal matrix with the power equally distributed among a *subset* of the transmit antennas. The principal observation is that as the capacity CDF becomes steeper, capacity versus outage increases for low outage probabilities and decreases for high outage probabilities. This is reflected in the fact that the higher the outage probability, the smaller the number of transmit antennas that should be used. As the transmit power is shared equally between more antennas the expectation of C increases (so the ergodic capacity increases) but the tails of its distribution decay faster. While this improves capacity versus outage for low outage probabilities, the capacity versus outage for high outages is decreased. Usually we are interested in low outage probabilities[5] and therefore the usual intuition for outage capacity is that it increases as the diversity order of the channel increases, i.e. as the capacity CDF becomes steeper. Telatar's conjecture is proven to hold for the MISO case in [6].

Capacity with perfect CSIR and CDIT: CMI, CCI and QCI models

For MIMO channels the capacity improvement resulting from some knowledge of the short-term channel statistics at the transmitter has been shown to be substantial, igniting much interest in the capacity of MIMO channels with perfect CSIR and CDIT under general distribution models. In this section we focus on the cases of CMI, CCI, and QCI channel distributions, corresponding to distribution feedback of the channel mean, covariance matrix, or quantized information of the instantaneous channel state, respectively. Key results on the capacity of such channels can be found in [57, 61, 63, 72, 73, 85, 93, 94, 96, 97, 117, 120, 124, 134].

Mathematically the problem is defined by (2.9) and (2.10), with the distribution on $\mathbf{H}$ being determined by the CMI, CCI, or QCI. The optimum input covariance matrix, in general, can be a full-rank matrix which implies either vector coding across the antenna array or transmission of several scalar codes in parallel with successive interference cancellation at the receiver. Limiting the rank of the input covariance matrix to unity, called *beamforming*, essentially leads to a scalar coded system which has a significantly lower complexity for typical array sizes.

The complexity versus capacity trade-off is an interesting aspect of capacity results under CDIT. The ability to use scalar codes to achieve capacity under CDIT for different channel distribution models, also called optimality of beamforming, captures this trade-off and has been the topic of much research in itself. Note that vector coding refers to fully unconstrained signaling schemes for the memoryless MIMO Gaussian channel. Every symbol period, a channel use corresponds to the transmission of a vector symbol comprised of the inputs to each transmit antenna. Ideally, while decoding vector codewords the

[5] The capacity for high outage probabilities becomes relevant for schemes that transmit only to the best user. For such schemes, it is shown in [10] that increasing the number of transmit antennas reduces the average sum capacity.

receiver needs to take into account the dependences in both space and time dimensions, and therefore the complexity of vector decoding grows exponentially in the number of transmit antennas. A lower complexity implementation of the vector coding strategy is also possible in the form of several scalar codewords being transmitted in parallel. It is shown in [57] that without loss of capacity, any input covariance matrix, regardless of its rank, can be treated as several scalar codewords encoded independently at the transmitter and decoded successively at the receiver by subtracting out the contribution from previously decoded codewords at each stage. However, well-known problems associated with successive decoding and interference subtraction, e.g. error propagation, render this approach unsuitable for use in practical systems. It is in this context that the question of optimality of beamforming becomes important. Beamforming transforms the MIMO channel into a SISO channel. Thus, well-established scalar codec technology can be used to approach capacity and since there is only one beam, interference cancellation is not needed. In the summary given below we include the results on both the transmitter optimization problem as well as the optimality of beamforming. We first discuss multiple-input single-output channels, followed by MIMO channels. Notice that if there is perfect CSIR, a single-input multiple-output channel can be converted into a SISO channel by use of maximal-ratio combining at the receiver, and therefore we need not consider such channels.

Multiple-input single-output channels

We first consider systems that use a single receive antenna and multiple transmit antennas. The channel matrix is rank one. With perfect CSIT and CSIR, for every channel matrix realization it is possible to identify the only non-zero eigenmode of the channel accurately and beamform along that mode. On the other hand, with perfect CSIR and CDIT under the ZMSW model, the optimal input covariance matrix is a multiple of the identity matrix. Thus, the inability of the transmitter to identify the non-zero channel eigenmodes forces a strategy where the power is equally distributed in all directions.

For a system using a single receive antenna and multiple transmit antennas, the transmitter optimization problem under CSIR and CDIT is solved for the distribution models of CMI and CCI. For the CMI model ($\mathbf{H} \sim \tilde{\mathcal{N}}(\mathbf{H}_m, \alpha\mathbf{I})$) the principal eigenvector of the optimal input covariance matrix $\mathbf{Q}^o$ is along the channel mean vector and the eigenvalues corresponding to the remaining eigenvectors are equal [134]. When beamforming is optimal, all power is allocated to the principal eigenvector. For the CCI model ($\mathbf{H} \sim \tilde{\mathcal{N}}(\mathbf{0}, \mathbf{R}_t)$) the eigenvectors of the optimal input covariance matrix $\mathbf{Q}^o$ are along the eigenvectors of the transmit fade covariance matrix and the eigenvalues are of the same order as the corresponding eigenvalues of the transmit fade covariance matrix [134]. A general condition that is both necessary and sufficient for optimality of beamforming can be obtained for both the CMI and CCI models by simply taking the derivative of the capacity expression [61].[6]

The optimality conditions are plotted in Figure 2.8. For the CCI model the optimality of beamforming depends on the two largest eigenvalues λ_1, λ_2 of the transmit fade

[6] For special considerations at low SNR see [87].

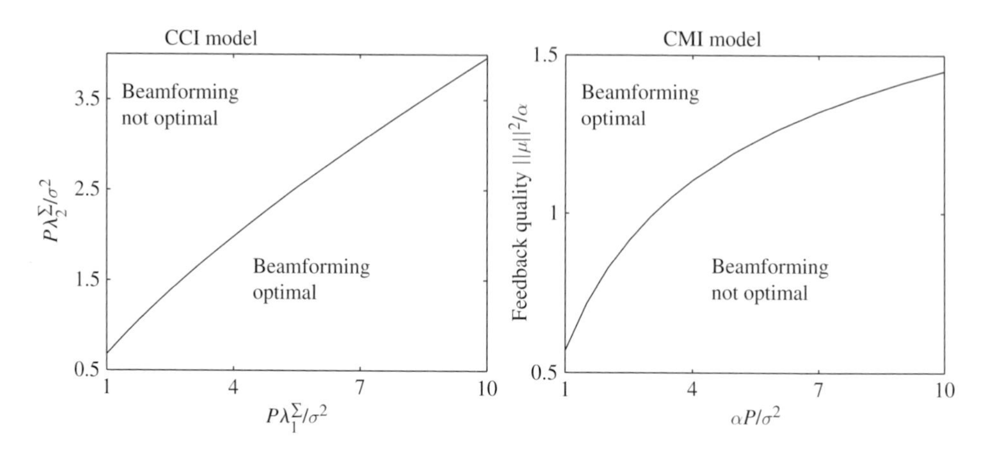

Fig. 2.8. Conditions for optimality of beamforming.

covariance matrix and the transmit power P. Beamforming is found to be optimal when the two largest eigenvalues of the transmit covariance matrix are sufficiently disparate or the transmit power P is sufficiently low. Since beamforming corresponds to using the principal eigenmode alone, this is reminiscent of water-pouring solutions where only the deepest level gets all the water when it is sufficiently deeper than the next deepest level and when the quantity of water is small enough. For the CMI model the optimality of beamforming is found to depend on the transmit power P and the *quality of feedback* associated with the mean information, which is defined mathematically as the ratio $||\mu||^2/\alpha$. Here, $\mu = ||\mathbf{H}_m||$ is the norm of the channel mean vector. As the transmit power P is decreased or the quality of feedback improves beamforming becomes optimal. As mentioned earlier, for perfect CSIT (the uncertainty α goes to zero so the quality of feedback goes to infinity) the optimal input strategy is beamforming, while in the absence of mean feedback (quality of feedback goes to zero so the CMI model becomes the ZMSW model), the optimal input covariance matrix has full rank, i.e. beamforming is necessarily sub-optimal.

Most research on quantized channel state information has assumed a beamforming transmit strategy where the transmitter forms a beam along the quantized channel vector. Beamforming is known to achieve ergodic capacity when the number of transmit antennas is equal to the number of quantization vectors. This is also known as the antenna selection scenario. In general, for symmetric quantization regions, e.g. if the quantization region consists of all channel vectors that make a maximum angle of $\theta_{\max}$ or less with the quantization vector then for $\theta_{\max} \leq 45°$, beamforming is optimal not only for ergodic capacity but for outage capacity as well, regardless of the number of quantization vectors, and the number of transmit antennas. The outage capacity with quantized beamforming approaches the perfect CSIT case as $(t-1)2^{-B/(t-1)}$, where B is the number of feedback bits and t is the number of transmit antennas. In general, the optimality of beamforming for both ergodic capacity as well as outage capacity depends on the number of quantization vectors as well as the symmetry of the Voronoi regions [120].

Next, we summarize the analogous capacity results for MIMO channels.

Multiple-input multiple-output channels

With multiple transmit and receive antennas, capacity with CSIR and CDIT under the CCI model with spatially white fading at the receiver ($\mathbf{R}_r = I$) shows similar behavior to the single receive antenna case. The capacity-achieving input covariance matrix has the same eigenvectors as the transmit fade covariance matrix and the eigenvalues are in the same order as the corresponding eigenvalues of the transmit fade covariance matrix [57, 73]. A direct differentiation of the capacity expression yields a necessary and sufficient condition for optimality of beamforming in this case as well. While the receive fade correlation matrix does not affect the eigenvectors of the optimal input covariance matrix, it does affect the eigenvalues as well as the corresponding capacity. The general condition for optimality of beamforming depends upon the two largest eigenvalues of the transmit covariance matrix and all the eigenvalues of the receive covariance matrix.

Transmitter optimization under the CMI model with multiple transmit and receive antennas is also similar to the single receive antenna case. The eigenvectors of the capacity achieving input covariance matrix coincide with the eigenvectors of $\mathbf{H}_m{}^*\mathbf{H}_m$, where $\mathbf{H}_m$ is the mean value of the random channel matrix [52, 57, 128]. The capacity for the CMI model has been shown to be monotonic in the singular values of $\mathbf{H}_m$ [52].

These results summarize our discussion of channel capacity with CDIT and perfect CSIR under different channel distribution models. From these results we notice that the benefits of adapting to distribution information regarding CMI or CCI fed back from the receiver to the transmitter are two fold. Not only does the capacity increase with more information about the channel distribution, but this feedback also allows the transmitter to identify the stronger channel modes and achieve this higher capacity with simple scalar codewords.

We conclude this subsection with a discussion on the growth of capacity with the number of antennas. With perfect CSIR and CDIT under the ZMSW channel distribution, it was shown in [33, 121] that the channel capacity grows linearly with $\min(M_T, M_R)$. This linear increase occurs whether the transmitter knows the channel perfectly (perfect CSIT) or only knows its distribution (CDIT). The proportionality constant of this linear increase, called the rate of growth, has also been characterized in [20, 49, 118, 121]. With perfect CSIR and CSIT, the rate of growth of capacity with $\min(M_T, M_R)$ is reduced by channel fading correlations at high SNR but is increased at low SNR. The mutual information under CSIR increases linearly with $\min(M_T, M_R)$ even when a spatially white transmission strategy is used on a correlated fading channel, although the slope is reduced relative to the uncorrelated fading channel. As we will see in the next section, the assumption of perfect CSIR is crucial for the linear growth behavior of capacity with the number of antennas. Interestingly, it has been shown in [18] that the effect of correlation is not significant when the maximum correlation between pairs of antenna elements is less than 0.5.

In the next subsection we explore the capacity when only CDI is available at the transmitter and the receiver.

Capacity with CDIT and CDIR: ZMSW model

We saw in the last subsection that with perfect CSIR, channel capacity grows linearly with the minimum of the number of transmit and receive antennas. However, reliable channel estimation may not be possible for a mobile receiver that experiences rapid fluctuations of the channel coefficients. Since user mobility is the principal driving force for wireless communication systems, the capacity behavior with CDIT and CDIR under the ZMSW distribution model (i.e. $\mathbf{H}$ is distributed as $\mathbf{H}_w$ with no knowledge of $\mathbf{H}$ at either the receiver or the transmitter) is of particular interest. In this section we summarize some MIMO capacity results in this area.

With CDIR and CDIT, and the ZMSW model, in a block fading scenario the channel matrix components are modeled as i.i.d. complex Gaussian random variables that remain constant for a coherence interval of T_c symbol periods after which they change to another independent realization. In the block fading model, the effective input to the channel is the inputs over the duration of the length T_c block. Capacity is achieved when the $T_c \times M_T$ transmitted signal matrix is equal to the product of two statistically independent matrices: a $T_c \times T_c$ isotropically distributed unitary matrix times a certain $T_c \times M_T$ random matrix that is diagonal, real, and non-negative [90]. This result enables the computation of capacity for many interesting cases. Not unexpectedly, for a fixed number of antennas, as the length of the coherence interval T_c increases, the capacity approaches the capacity obtained as if the receiver knew the propagation coefficients. However, there is a surprising result obtained for this channel model: in contrast to the linear growth of capacity with $\min(M_T, M_R)$ under the perfect CSIR assumption, in the absence of CSIT and CSIR, capacity does not increase at all as the number of transmit antennas is increased beyond the length of the coherence interval T_c. At high SNRs, capacity is achieved using no more than $M^\star = \min(M_T, M_R, \lfloor T_c/2 \rfloor)$ transmit antennas [155]. In particular, having more transmit antennas than receive antennas does not provide any capacity increase at high SNR. For each 3 dB SNR increase, the capacity gain is $M^\star(1 - 1/T_c)$.

Crucial to these results is the assumption of a block fading model, i.e. the channel fade coefficients are assumed to be constant for a block of T_c symbol durations. With a continuous fading model, within each independent T_c-symbol block, the fading coefficients have an arbitrary time correlation. If the correlation vanishes beyond some lag τ, called the *correlation time* of the fading, then it is shown in [91] that increasing the number of transmit antennas beyond $\min(\tau, T_c)$ antennas does not increase the capacity. However, the results obtained without the block fading assumption can be quite different. It is shown in [80] that without the block fading distribution for the CDIT/CDIR model with the ZMSW distribution, the capacity at high SNR grows only double-logarithmically in SNR. This result is shown to hold under very general conditions, even allowing for memory and partial receiver side information.

Capacity with CDIT and CDIR: CCI model

The results described in the previous section assume a somewhat pessimistic model for the channel distribution. That is because most channels when averaged over a relatively small area have either a non-zero mean or a non-white covariance. Thus, if these distribution parameters can be tracked, the channel distribution corresponds to either the CMI or the CCI model.

With CDIT and CDIR under the CCI distribution model, the channel matrix components are modeled as spatially correlated complex Gaussian random variables that remain constant for a coherence interval of T_c symbol-periods after which they change to another independent realization based on the spatial correlation model. The channel correlations are assumed to be known at the transmitter and the receiver. As in the case of spatially white fading (ZMSW model), with the CCI model the capacity is achieved when the $T_c \times M_T$ transmitted signal matrix is equal to the product of a $T_c \times T_c$ isotropically distributed unitary matrix, a statistically independent $T_c \times M_T$ random matrix that is diagonal, real, and non-negative, and the matrix of the eigenvectors of the transmit fade covariance matrix $\mathbf{R}_t$ [59]. The channel capacity is independent of the smallest $(M_T - T_c)^+$ eigenvalues of the transmit fade covariance matrix as well as the eigenvectors of the transmit and receive fade covariance matrices $\mathbf{R}_t$ and $\mathbf{R}_r$. Also, in contrast to the results for the spatially white fading model where adding more transmit antennas beyond the coherence interval length ($M_T > T_c$) does not increase capacity, additional transmit antennas always increase capacity as long as their channel fading coefficients are spatially correlated. Thus, in contrast to the results in favor of independent fades with perfect CSIR, these results indicate that with CCI at the transmitter and the receiver, transmit fade correlations can be beneficial, making the case for minimizing the spacing between transmit antennas when dealing with highly mobile, fast fading channels that cannot be accurately measured. Mathematically, for fast fading channels ($T_c = 1$), capacity is a Schur-concave function of the vector of eigenvalues of the transmit fade correlation matrix. The maximum possible capacity gain due to transmitter fade correlations is shown to be $10 \log_{10} M_T$ dB in terms of power.

Capacity with correlated fading

The impact of channel correlations on the capacity of a MIMO channel is of interest because the channels encountered in practice invariably exhibit non-zero correlations in time and space. Temporal correlations are those that exist between the channel matrix realizations at different time instants. Spatial correlations are those that exist between the elements of the channel matrix for each realization.

First, we discuss the impact of temporal correlations. In general, if perfect CSIR is not assumed, a single letter capacity characterization for a temporally correlated channel is hard to obtain because the correlations introduce memory in the channel. Typically if channel memory is limited to a block of length τ then we need to consider the τ symbol extension of the channel. Because the τ symbol extension of the channel has an input alphabet size that is exponentially larger (exponential in the memory τ), the complexity of the input optimization problem for channels with memory can increase exponentially with τ. While a complete characterization of the impact of temporal correlations is not known, it is easy to see that temporal correlations will increase the capacity when no CSIR is available. This is because the channel correlations allow some amount of channel estimation that is not possible in a memoryless channel. Another intuitive argument to show that the capacity with temporally correlated channels cannot be smaller than the capacity with temporally uncorrelated channels is that by interleaving codewords it is often possible to transform the correlated channel into an uncorrelated one.

Capacity characterization with temporal correlations is significantly simpler when the channel realizations are assumed to be known perfectly to the receiver (perfect CSIR). In this case, temporal channel correlations do not affect ergodic capacity. This is because, conditioned on the channel knowledge available at the receiver, the channel randomness is only due to the additive noise which is memoryless from one symbol to the next. Therefore, with perfect CSIR, a single letter characterization of the capacity is possible and the ergodic capacity depends only on the marginal distribution of a single realization of the channel matrix. Note that while the capacity with perfect CSIR is independent of the temporal correlations, the performance of practical coding schemes is affected by the temporal correlations. On the one hand, strong temporal correlations signify a slowly varying channel that would require longer codewords to realize the ergodic capacity. On the other hand, many low complexity coding schemes (such as the orthogonal space–time codes discussed in Section 3.4.1) rely on the channel remaining constant over several symbols, and therefore may perform better for slowly varying (high temporal correlation) channels.

Next, we discuss the impact of spatial correlations. Spatial correlations are a function of the scattering environment and the antenna spacing. Roughly speaking, the correlation between fades experienced by different antennas decreases as the density of scatterers in the vicinity increases or as the spacing between the antennas increases. For example, elevated base-stations located in relatively unobstructed surroundings have a larger decorrelating distance between antennas than an indoor mobile device surrounded by scatterers. Since a mobile unit is more size constrained than a base-station the two factors offset each other.

Completely general models for spatial correlations are often analytically intractable and remain an active area of research. The most commonly used model is the Kronecker product form

$$\mathbf{H} = \mathbf{R}_r^{1/2}\mathbf{H}_w\mathbf{R}_t^{1/2} \tag{2.12}$$

where $\mathbf{H}_w$ is an i.i.d. Rayleigh fading channel, and $\mathbf{R}_t$, $\mathbf{R}_r$ are the transmit and receive correlation matrices, respectively. This model is attractive for its analytical tractability and also has been shown to be reasonably accurate through field measurements. The impact of transmitter correlations has been explored in some detail while the impact of receiver correlations is not as easily characterized. Here, using the Kronecker product model as our reference, we discuss the intuition behind the impact of transmitter side spatial correlations on the capacity followed by the references that theoretically validate this intuition. First, let us consider the case of perfect CSIR. In particular, to develop some intuition, let us consider the cases of a perfectly correlated channel ($\text{rank}(\mathbf{R}_t) = 1$) and a perfectly uncorrelated channel ($\mathbf{R}_t = \mathbf{I}_{M_T}$), and the two extremes of high and low SNR. Notice that the perfectly correlated channel has unit rank. Therefore, correlation can make the channel matrix rank-deficient. This is an important limitation at high SNR where the rank of the channel matrix determines the capacity (multiplexing gain). On the other hand, consider a MIMO system at low SNR. At low SNR, if the transmitter can identify the strongest eigenmode of the channel, the optimal transmit strategy is to beamform to the strongest channel mode. Clearly at low SNR the channel capacity is limited by the strength of the strongest channel mode and the ability of the transmitter to beamform in that direction. The rank of the channel matrix does not play an important part at low SNR. Correlation enhances the principal eigenmode of the channel at the cost of the remaining eigenmodes. In other words, correlation reduces the multiplexing gain but supports single beamforming. In light of these observations, one would expect that if both the transmitter and the receiver know the channel state (perfect CSIT, perfect CSIR) then transmitter side correlation will increase the capacity at low SNR where beamforming is the optimal policy and will reduce the capacity at high SNR where the multiplexing gain is the limiting factor. Now suppose the transmitter does not have any CSIT and is forced to split its power uniformly among all transmit antennas. In this case, correlation does not help at low SNR either because the transmitter is not able to identify the strongest channel mode along which to direct its transmit power. At high SNR however, uniform power allocation is nearly optimal (for exceptions to this rule see [87]) and the impact of correlation would be the same as with perfect CSIT. Thus, with uniform power allocation, transmitter side correlation would reduce the capacity both at high SNR as well as at low SNR. Finally, consider the case where the transmitter possesses no CSIT but knows the correlation structure through CDIT. This allows the transmitter to identify the strong channel modes and beamform to them at low SNR. Thus, correlation would be helpful at low SNR. The high SNR case is more complicated. On the one hand, knowledge of the channel correlation structure allows the transmitter to identify the strong channel modes. In fact, for a perfectly correlated channel, knowledge of the channel correlations is as good as perfect CSIT, since there is only one non-zero eigenmode (the principal eigenvector of $\mathbf{R}_t$). However, this is offset by the loss in the multiplexing gain of the channel. Because these two factors work in opposition, the impact of transmitter side correlations is not as easily characterized for a MIMO channel. However, suppose there is only one receive antenna. In that case, the spatial multiplexing gain is unity whether the channel is correlated or uncorrelated. And indeed in this case, correlation increases the capacity because the transmitter is able to identify the strong channel eigenmodes. Analytical results verifying these intuitive explanations are provided in [20, 74, 125].

Frequency-selective fading channels

While flat-fading is a realistic assumption for narrowband systems where the signal bandwidth is smaller than the channel coherence bandwidth, broadband communications involve channels that experience frequency-selective fading. Research on the capacity of MIMO systems with frequency-selective fading typically takes the approach of dividing the channel bandwidth into parallel flat-fading channels, and constructing an overall block-diagonal channel matrix with the diagonal blocks being given by the channel matrices corresponding to each of these subchannels. Under perfect CSIR and CSIT, the total power constraint then leads to the usual closed-form water-filling solution. Note that the water-fill is done simultaneously over both space and frequency. Even SISO frequency-selective fading channels can be represented by the MIMO system model (2.4) in this manner [102]. For MIMO systems, the matrix channel model is derived in [8] based on an analysis of the capacity behavior of OFDM-based MIMO channels in broadband fading environments. Under the assumption of perfect CSIR and CDIT for the ZMSW model, it is shown that in the MIMO case, unlike the SISO case, frequency-selective fading channels may provide advantages over flat-fading channels not only in terms of ergodic capacity but also in terms of capacity versus outage. In other words, MIMO frequency-selective fading channels are shown to provide both higher diversity gain and higher multiplexing gain than MIMO flat-fading channels. The measurements in [92] show that frequency selectivity makes the CDF of the capacity steeper and, thus, increases the capacity for a given outage as compared with the flat-frequency case, but the influence on the ergodic capacity is small.

Training for multiple antenna systems

The results summarized in the previous sections indicate that CSI plays a crucial role in the capacity of MIMO systems. In particular, the capacity results in the absence of CSIR are strikingly different and often quite pessimistic compared to those that assume perfect CSIR. To recapitulate, with perfect CSIR and CDIT the MIMO channel capacity is known to increase linearly with $\min(M_T, M_R)$ when the CDIT assumes the ZMSW or CCI distribution models. However, in fast fading when the channel changes so rapidly that it cannot be estimated reliably at the receiver (CDIR only), the capacity does not increase with the number of transmit antennas at all for $M_T > T_c$, where T_c is the channel decorrelation time. Also at high SNR under the ZMSW distribution model, capacity with perfect CSIR and CDIT increases logarithmically with SNR, while the capacity with CDIR and CDIT increases only double-logarithmically with SNR. Thus, CSIR is critical to obtain the high capacity benefits of multiple-antenna wireless links. CSIR is often obtained by sending known training symbols to the receiver. However, with too little training the channel estimates are poor, whereas with too much training there is no time for data transmission before the channel changes. So the key question to ask is how much training is needed in multiple-antenna wireless links [48]. It turns out that when the training and data powers are allowed to vary, the optimal number of training symbols is equal to the number of transmit antennas, which is also the smallest training interval length that guarantees meaningful estimates of the channel matrix. When the training and data powers are instead required to be equal, the optimal training duration may be longer than the number of antennas. Interestingly, while training-based schemes can be optimal at high SNR, they are sub-optimal at low SNR.

Application to matrix channels

Note that the MIMO capacity results described in the prior subsections are applicable to any channel described by a matrix. Matrix channels describe not only multi-antenna systems but also channels with crosstalk [150] and wideband channels [130]. Code-division multiple-access (CDMA) systems are another prime example of matrix channels [129]. While the focus of this chapter is on flat and frequency-selective fading channels with multiple antennas, the same capacity analysis can be applied to obtain the capacity of these matrix channels as well.

2.3.4 Open problems in single-user MIMO

The results summarized in this section form the basis of our understanding of single-user MIMO channel capacity under different CSI and CDI assumptions. These results serve as useful indicators for the benefits of incorporating training and feedback schemes in a MIMO wireless link to obtain CSIR/CDIT and CSIT/CDIT respectively. However, our knowledge of MIMO capacity with CDI only is still far from complete, even for single-user systems. We conclude this section by pointing out some of the many open problems.

1. Combined CCI and CMI: capacity under CDIT and perfect CSIR is unsolved under a combined CCI and CMI distribution model even with a single receive antenna.
2. CCI: with perfect CSIR and CDIT capacity is not known under the CCI model for completely general (i.e. non-separable) spatial correlations.
3. CDIR: capacity for almost all cases with only CDIR are open problems.
4. Outage capacity: most results for CDI only at either the transmitter or receiver are for ergodic capacity. Capacity versus outage has proven to be less analytically tractable than ergodic capacity and contains an abundance of open problems.

2.4 Multi-user MIMO

In this section we give capacity results for the two basic multi-user MIMO channel models: the MIMO multiple-access channel (MAC or uplink) and the MIMO broadcast channel (BC or downlink). The MIMO MAC consists of many multiple-antenna transmitters sending to a single multiple-antenna receiver and the MIMO BC consists of one multiple-antenna transmitter sending to many multiple-antenna receivers. In cellular-type architectures (e.g. cellular networks or wireless local-area networks), the MAC models the channel from mobile devices to the base-station, and the BC models the channel from the base-station to mobile devices. The uplink and downlink channels are illustrated in Figure 2.9. As discussed in Section 1.4 multiple antennas are becoming increasingly common in such systems (e.g. in IEEE 802.11n or IEEE 802.16), and thus it is important to understand the fundamental limits of such channels. Multi-user MIMO receivers are significantly more complex than single-user MIMO systems, since the signals from all users must be detected simultaneously. Practical techniques for multi-user MIMO detection are described in Chapter 6.

The channel capacity of a point-to-point MIMO channel is a real number which is the fundamental limit on reliable communication: any rate strictly smaller than the capacity

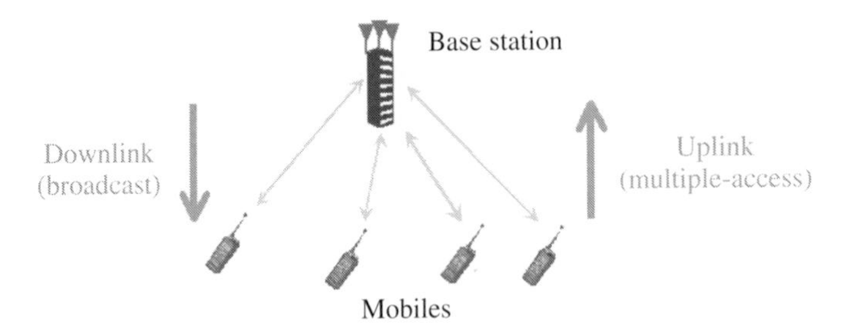

Fig. 2.9. Uplink/downlink channels in cellular systems.

is achievable, while all rates strictly larger than the capacity are not achievable. For multi-user channels, the channel capacity has a similar definition, but the capacity is a region (i.e. a set in K-dimensional space) instead of a single number because different rates are associated with the multiple users. In the MAC, each transmitter is assumed to have an independent message for the base-station, and thus a different rate is associated with each transmitter. In the BC, the transmitter is assumed to have a different (and independent) message for each of the receivers,[7] and similarly a different rate is associated with each transmission. The capacity region is therefore defined as the set of rates that can *simultaneously* be achieved with an arbitrarily small probability of error. It is important to note that multiple messages are sent simultaneously; the rates achieved with schemes such as time-division multiple access (TDMA), where only a single mobile communicates with the base-station or access point on either the uplink or the downlink, are contained in the capacity region, but are generally strictly sub-optimal. In fact, contrary to the method in which most systems are currently designed, tremendous capacity benefits can be achieved by simultaneously transmitting to multiple users (on the downlink) or having multiple users simultaneously transmit (on the uplink) without any separation in the time, frequency, or code domain.

The capacity benefits of multi-user MIMO can be even greater than in the single-user setting discussed in Section 2.3. In single-user systems, multiple antennas are required at both the transmitter and the receiver in order to realize linear capacity gains (i.e. the capacity scales as $\min(M_T, M_R)$, where M_T and M_R are the number of transmit and receive antennas, respectively). In the uplink and downlink channels, however, it is sufficient to deploy multiple antennas at only the access point in order to achieve a similar linear increase in capacity. In this scenario, if we let M_T denote the number of access point antennas, M_R the number of antennas at each mobile, and K the number of mobiles, then the sum rate capacity (the maximum throughput or the point in the capacity region that maximizes the sum of all rates) increases linearly with $\min(M_T, M_R K)$ as the number of antennas and users is increased. Thus, having a large number of mobiles can make up for deploying a small number of antennas at each mobile. This is a key point for space-limited mobile devices.

[7] In networking terms, this is referred to as unicast. In the multicast scenario, which is not discussed here, there is a single common message which all receivers wish to receive [66, 74, 76, 97, 116].

Similar to Section 2.3, we give results on the capacity of the MIMO MAC and MIMO BC for different assumptions on the amount of channel state information available at transmitters and receivers. Multi-user detection or successive interference cancellation can be used to achieve the capacity of the MIMO MAC. This channel is well understood from an information-theoretic point of view, and thus many results are available for the MIMO MAC. In addition, the coding complexity required to achieve the MIMO MAC capacity is essentially the same as the complexity of point-to-point MIMO systems. The broadcast channel, on the other hand, remains one of the key open problems in information theory. However, there has been a great deal of progress on the MIMO BC, and the capacity region is known in some scenarios. A clever pre-processing technique called *dirty paper coding* achieves the capacity of the MIMO BC when the channel is fixed. Though these results have shed a great deal of light on the capacity of the MIMO BC, the problem of finding practical coding schemes for this channel is even more difficult than designing codes for single-user MIMO channels, and thus remains an open research area. Interestingly, the MIMO MAC and MIMO BC have been shown to be duals, as we will discuss in Section 2.4.3.

2.4.1 System model

To describe the MAC and BC models, we consider a cellular-type system in which the base-station has M_T antennas and each of the K mobiles has M_R antennas. The downlink (or forward channel) of this system is a MIMO BC and the uplink (or reverse channel) is a MIMO MAC. We will use $\mathbf{H}_i$ to denote the *downlink* channel matrix from the base-station to user i. Assuming that the same channel is used on the uplink and downlink (i.e. a TDD system), the *uplink* matrix of user i is $\mathbf{H}_i^*$. Note that we assume the base-station has M_T antennas in both the downlink channel (in which case they are transmit antennas) as well as the uplink channel (in which case they are receive antennas). The same is true for M_R, which is the number of antennas per mobile on the downlink (receive antennas) and uplink (transmit antennas). A picture of the system model is shown in Figure 2.10.[8]

In the MAC, let $\mathbf{x_k} \in \mathbb{C}^{M_R \times 1}$ be the transmitted signal of user (i.e. mobile) k. Let $\mathbf{y}_{MAC} \in \mathbb{C}^{M_T \times 1}$ denote the received signal and $\mathbf{n} \in \mathbb{C}^{M_T \times 1}$ the noise vector, where $\mathbf{n} \sim \tilde{\mathcal{N}}(0, \mathbf{I}_{M_T})$ is a circularly symmetric complex Gaussian with an identity covariance matrix. The received signal at the base-station is then equal to

$$\begin{aligned}\mathbf{y}_{MAC} &= \mathbf{H}_1^*\mathbf{x}_1 + \cdots + \mathbf{H}_K^*\mathbf{x}_K + \mathbf{n} \\ &= \mathbf{H}^* \begin{bmatrix} \mathbf{x}_1 \\ \vdots \\ \mathbf{x}_K \end{bmatrix} + \mathbf{n} \text{ where } \mathbf{H}^* = [\mathbf{H}_1^* \ldots \mathbf{H}_K^*].\end{aligned}$$

[8] The system is assumed to be time-division duplex (TDD) for mathematical simplicity, and in order to introduce the concept of MAC–BC duality. However, this assumption need not be true in order for the results of this section to hold true. In addition, we consider the dual uplink channel to be the conjugate transpose of the downlink channel. The true channels should only be related through the transpose operation, but we add the conjugate (which does not affect capacity) for mathematical simplicity.

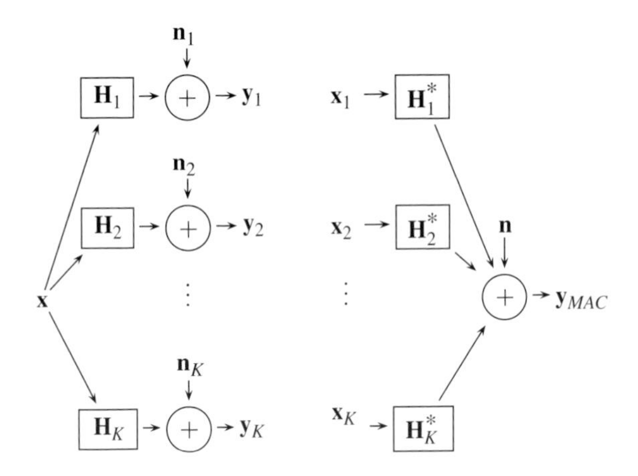

Fig. 2.10. System models of the MIMO BC (left) and the MIMO MAC (right) channels.

In the MAC, each user (i.e. mobile) is subject to an individual power constraint of P_k. The transmit covariance matrix of user k is defined to be $\mathbf{Q}_k \triangleq E[\mathbf{x}_k\mathbf{x}_k^*]$. The power constraint implies $\text{tr}(\mathbf{Q}_k) \leq P_k$ for $k = 1, \ldots, K$.

In the BC, let $\mathbf{x} \in \mathbb{C}^{M_T \times 1}$ denote the transmitted vector signal (from the base-station) and let $\mathbf{y}_k \in \mathbb{C}^{M_R \times 1}$ be the received signal at receiver (i.e. mobile) k. The noise at receiver k is represented by $\mathbf{n}_k \in \mathbb{C}^{M_R \times 1}$ and is assumed to be circularly symmetric complex Gaussian noise ($\mathbf{n_k} \sim \tilde{N}(0, \mathbf{I}_{M_R})$). The received signal of user k is equal to

$$\mathbf{y}_k = \mathbf{H}_k\mathbf{x} + \mathbf{n}_k. \tag{2.13}$$

The transmit covariance matrix of the input signal is $\mathbf{\Sigma}_x \triangleq E[\mathbf{x}\mathbf{x}^*]$. The base-station is subject to an average power constraint P, which implies $\text{tr}(\mathbf{\Sigma}_x) \leq P$.

In the MAC, each transmitter is assumed to have an independent data stream (i.e. a sequence of messages) for the receiver. Thus, a different data rate is associated with each transmitter and the capacity region is a K-dimensional region. Similarly, in the BC the transmitter has an independent message for each of the receivers and the capacity region is therefore also a K-dimensional region.

Though we consider only multiple-antenna channels here, it is important to note that the multiple-antenna MAC with $M_T > 1$ and $M_R = 1$ can be used to model single-antenna CDMA systems. The channel vectors (on either the downlink or uplink) represent the spreading codes of the users, with M_T being the length of each code. A number of papers focusing on the uplink study this connection further (cf. [137]).

2.4.2 *MIMO multiple-access channel*

In this section we summarize capacity results on the multiple-antenna MAC. We first provide general results on the capacity region of multiple-access channels to provide some background. We then analyze the constant-channel MIMO MAC, followed by the fading channel. Since the capacity region of a general MAC is known, the expressions for the capacity of a constant MAC are quite straightforward. For the fading case, one must consider different assumptions about the CSI and CDI available at the transmitter and receiver.

Multiple-access channel capacity

The capacity region of the general multiple-access channel was first derived in the 1970s [2, 84]. The capacity region of the two-transmitter single-antenna AWGN MAC (i.e. $M_T = M_R = 1$) is given by [25]

$$\begin{aligned} R_1 &\leq \log(1+|h_1|^2 P_1) \\ R_2 &\leq \log(1+|h_2|^2 P_2) \\ R_1 + R_2 &\leq \log(1+|h_1|^2 P_1 + |h_2|^2 P_2). \end{aligned}$$

We provide the two-user capacity region for simplicity; this formula can easily be extended to a MAC with an arbitrary number of transmitters.

In order to achieve the capacity region, each transmitter uses a Gaussian codebook (as in a point-to-point AWGN channel) and the receiver employs either multi-user detection or successive interference cancellation to decode the codes of every transmitter. The capacity region clearly corresponds to a pentagonal region. Consider the corner point $(R_1 = \log(1+|h_1|^2 P_1), R_2 = \log(1+\frac{|h_2|^2 P_2}{|h_1|^2 P_1 + 1}))$. In order to achieve this point using interference cancellation, the receiver first decodes the message from transmitter 2 while treating the codeword from transmitter 1 (which is normally distributed with power $|h_1|^2 P_1$) as an extra source of additive Gaussian noise. The receiver subtracts the decoded message from the received signal, and then decodes the message from transmitter 1. During this second decoding operation, note that there is no interference, i.e. transmitter 1 communicates over a "clean" channel to the receiver. The other corner point of the capacity region $(R_1 = \log(1+\frac{|h_1|^2 P_1}{|h_2|^2 P_2 + 1}), R_2 = \log(1+|h_2|^2 P_2))$ can similarly be achieved by first decoding transmitter 1's signal, followed by the codeword of transmitter 2. The same strategy is also optimal for the MIMO channel, which we discuss next.

Constant channel

For any set of power constraints $\mathbf{P} = (P_1, \ldots, P_K)$, the capacity of the MIMO MAC (denoted by $\mathcal{C}_{MAC}(\mathbf{P}; \mathbf{H}^*)$) is given by

$$\mathcal{C}_{MAC}(\mathbf{P}; \mathbf{H}^*) \triangleq \bigcup_{\{\mathbf{Q}_i \geq 0, \mathrm{tr}(\mathbf{Q}_i) \leq P_i \forall i\}} \left\{ \begin{array}{l} (R_1, \ldots, R_K): \\ \sum_{i \in S} R_i \leq \log\det(\mathbf{I}_{M_T} + \sum_{i \in S} \mathbf{H}_i^* \mathbf{Q}_i \mathbf{H}_i) \forall S \subseteq \{1, \ldots, K\} \end{array} \right\}. \tag{2.14}$$

The variable S refers to a subset of $\{1, \ldots, K\}$. The ith user transmits a zero-mean Gaussian with spatial covariance matrix $\mathbf{Q}_i$. Each set of covariance matrices $(\mathbf{Q}_1, \ldots, \mathbf{Q}_K)$ corresponds to a K-dimensional polyhedron of achievable rates, i.e.

$$\{(R_1, \ldots, R_K) : \sum_{i \in S} R_i \leq \log\det(\mathbf{I}_{M_T} + \sum_{i \in S} \mathbf{H}_i^* \mathbf{Q}_i \mathbf{H}_i) \forall S \subseteq \{1, \ldots, K\}\},$$

and the capacity region is equal to the union over all covariance matrices satisfying the trace constraints of all such polyhedrons. The corner points of each polyhedron can be achieved by *successive decoding*, in which users' signals are successively decoded and subtracted out of the received signal. For the two-user case, each set of covariance matrices corresponds to a pentagon, similar in form to the capacity region of the scalar Gaussian MAC. The corner point where $R_1 = \log\det(\mathbf{I}_{M_T} + \mathbf{H}_1^*\mathbf{Q}_1\mathbf{H}_1)$ and $R_2 = \log\det(\mathbf{I}_{M_T} + \mathbf{H}_1^*\mathbf{Q}_1\mathbf{H}_1 + \mathbf{H}_2^*\mathbf{Q}_2\mathbf{H}_2) - R_1 = \log\det(\mathbf{I}_{M_T} + (\mathbf{I}_{M_T} + \mathbf{H}_1^*\mathbf{Q}_1\mathbf{H}_1)^{-1}\mathbf{H}_2^*\mathbf{Q}_2\mathbf{H}_2)$ corresponds to decoding $\mathbf{x}_2$ first while treating $\mathbf{x}_1$ as noise, then subtracting $\mathbf{x}_2$ from $\mathbf{y}_{MAC}$, and then decoding user 1. Successive decoding can reduce a complex multi-user detection problem into a series of single-user detection steps [44]. Note that capacity-achieving successive decoding is, in fact, identical to some forms of BLAST, which is a well-studied technique for single-user MIMO systems. Thus, we say that the MIMO MAC capacity can be achieved with a complexity similar to that of a single-user MIMO system.

The capacity region of a MIMO MAC for the single transmit antenna case ($M_R = 1$) is shown in Figure 2.11. When $M_R = 1$, the covariance matrix of each transmitter is a scalar equal to the transmitted power. Clearly, each user should transmit at full power. Thus, the capacity region for a K-user MAC for $M_R = 1$ is the set of all rate vectors $(R_1, \ldots, R_K)$ satisfying

$$\sum_{i\in S} R_i \le \log\det\left(\mathbf{I}_{M_T} + \sum_{i\in S} \mathbf{H}_i^* P_i \mathbf{H}_i\right) \qquad \forall S \subseteq \{1, \ldots, K\}. \tag{2.15}$$

For the two-user case, this reduces to the simple pentagon seen in Figure 2.11.

When $M_R > 1$, a union must be taken over all covariance matrices. Intuitively, the set of covariance matrices that maximize R_1 is different from the set of covariance matrices that maximize the sum rate. Furthermore, the capacity-achieving covariances cannot be simply characterized by water-filling. In Figure 2.12, a MAC capacity region for $M_R > 1$ is shown. Notice that the region is equal to the union of pentagons (with each pentagon corresponding to a different set of transmit covariance matrices), a few of which are shown with dashed lines in the figure. The boundary of the capacity region is in general curved, except at the sum rate point, where the boundary is a straight-line, and along the planes where the single-user capacities are achieved [151]. Each point on the curved

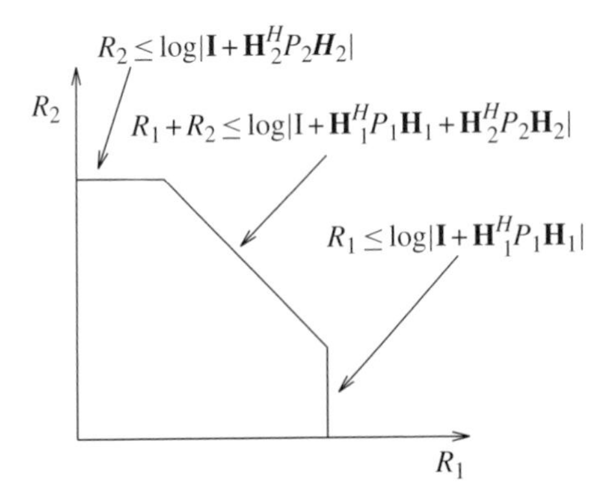

Fig. 2.11. Capacity region of the MIMO MAC for $M_R = 1$.

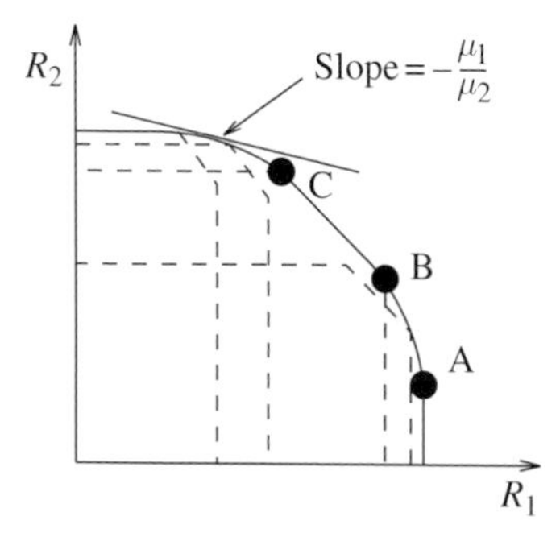

Fig. 2.12. Capacity region of the MIMO MAC for $M_R > 1$.

portion of the boundary is achieved by a *different* set of covariance matrices. At point A, user 1 is decoded last and achieves its single-user capacity by choosing $\mathbf{Q}_1$ as a water-fill of the channel $\mathbf{H}_1$ (i.e. the input that achieves the capacity of the constant MIMO channel from X_1 to Y, as described in Section 2.3.2). User 2 is decoded first, in the presence of interference from user 1, so $\mathbf{Q}_2$ is chosen as a water-fill of the channel $\mathbf{H}_2$ and the interference from user 1. The sum-rate corner points B and C are the two corner points of the pentagon corresponding to the sum-rate optimal covariance matrices $\mathbf{Q}_1^{sum}$ and $\mathbf{Q}_2^{sum}$. Note that these covariances are not individually optimal for either user, but instead are optimal in terms of the total amount of rate that the receiver can decode. At point B user 1 is decoded last whereas at point C user 2 is decoded last. Successive decoding can be used to achieve the corner points of the sum capacity plane, and time division between different decoding orders or multi-user detection can be used to achieve interior points.

Next, we focus on characterizing the optimal covariance matrices $(\mathbf{Q}_1, \ldots, \mathbf{Q}_K)$ that achieve different points on the boundary of the MIMO MAC capacity region. Since the MAC capacity region is convex, it is well known from convex theory that the boundary of the capacity region can be fully characterized by maximizing the function $\mu_1 R_1 + \cdots + \mu_K R_K$ over all rate vectors in the capacity region and for all non-negative priorities $(\mu_1, \ldots, \mu_K)$ such that $\sum_{i=1}^{K} \mu_i = 1$. For a fixed set of priorities $(\mu_1, \ldots, \mu_K)$, this is equivalent to finding the point on the capacity region boundary that is tangential to a line whose slope is defined by the priorities. See the tangent line in Figure 2.12 for an example. The structure of the MAC capacity region implies that all boundary points of the capacity region (except for the plane defining the sum capacity) are corner points of polyhedrons corresponding to different sets of covariance matrices. Furthermore, the corner point should correspond to successive decoding in order of *increasing* priority, i.e. the user with the highest priority should be decoded last and therefore sees no interference [122, 132]. Thus, the problem of finding the boundary point on the capacity region associated with priorities $\mu_1, \ldots, \mu_K$ assumed to be in descending order (users can be arbitrarily renumbered to satisfy this condition) can be written as

$$\max_{\mathbf{Q}_1, \ldots, \mathbf{Q}_K} \mu_K \log\det\left(\mathbf{I}_{M_T} + \sum_{l=1}^{K} \mathbf{H}_l^* \mathbf{Q}_l \mathbf{H}_l\right) + \sum_{i=1}^{K-1} (\mu_i - \mu_{i+1}) \log\det\left(\mathbf{I}_{M_T} + \sum_{l=1}^{i} \mathbf{H}_l^* \mathbf{Q}_l \mathbf{H}_l\right) \tag{2.16}$$

subject to power constraints on the trace of each of the covariance matrices. Note that the covariances that maximize the function above are the *optimal* covariances. The optimization problem is convex, and thus efficient numerical tools can be employed to solve it [11]. Iterative water-filling is an extremely efficient algorithm that finds the sum-rate maximizing (i.e. $\mu_1 = \cdots = \mu_K$) covariance matrices [151]. This algorithm is a coordinate-descent method, where each user greedily optimizes his own transmit covariance while treating interference from all the other users as additional noise.

A key point regarding the sum-rate capacity of the MIMO MAC is its first-order growth term. It is, in fact, easy to see that a MIMO MAC with K transmitters with M_R antennas each, and a single M_T antenna receiver, is closely related to an $M_R K$ transmit antenna, M_T receive antenna point-to-point MIMO channel. Furthermore, these two channels have the same multiplexing gain, i.e. the sum capacity of the MIMO MAC can be approximated as $\min(M_T, M_R K)\log(\mathrm{SNR})$ [68]. Thus, for systems with a large number of users, the capacity can be increased almost linearly by increasing the number of base-station antennas. This is a key benefit of MIMO in multi-user systems.

Fading channels

As in the single-user case, the capacity of the fading MIMO MAC depends on the definition of capacity and the availability of CSI and CDI at the transmitters and the receiver. The capacity with perfect CSIT and CSIR has been very well studied, as has the capacity with perfect CSIR and CDIT under the ZMSW model. However, little is known about the capacity of the MIMO MAC with CDIT under the CMI or CCI distribution models. Some results on the optimum distribution for the single-antenna case with CDIT and CDIR under the ZMSW distribution can be found in [108].

With perfect CSIR and CSIT the system can be viewed as a set of *parallel* non-interfering MIMO MACs (one for each fading state) sharing a common power constraint. Thus, the ergodic capacity region can be obtained as an average of these parallel MIMO MAC capacity regions [152], where the averaging is done with respect to the channel statistics. The iterative water-filling algorithm of [151] extends to this case, with joint space and time water-filling.

The capacity region of a single-antenna MAC with perfect CSIR and CDIT was found in [34, 109]. These results can easily be extended to MIMO channels. In this scenario, Gaussian inputs are optimal, and the ergodic capacity region is equal to the time average of the capacity obtained at each fading instant with a constant transmit policy (i.e. the input covariance matrix for each transmitter is fixed for all time). Thus, the ergodic capacity region is given by

$$\bigcup_{\{\mathbf{Q}_i \geq 0, \mathrm{tr}(\mathbf{Q}_i) \leq P_i \forall i\}} \left\{ (R_1, \ldots, R_K) : \sum_{i \in S} R_i \leq E_{\mathbf{H}} \left[\log\det\left(\mathbf{I}_{M_T} + \sum_{i \in S} \mathbf{H}_i^* \mathbf{Q}_i \mathbf{H}_i \right) \right] \right.$$
$$\left. \forall S \subseteq \{1, \ldots, K\} \right\}.$$

Note that this is identical to the expression for the capacity of the constant MIMO MAC with the addition of the expected value over the distribution of the channels. The boundary of the capacity region can be characterized by the maximization in (2.16), but taking the expectation over the fading distribution can make this problem computationally difficult.

If the channel matrices $\mathbf{H}_i$ are ZMSW and each user has the same power constraint (and there is perfect CSIR and CDIT), then the optimal covariances are scaled versions of the identity matrix, i.e. $\mathbf{Q}_i = \frac{P_i}{M_R}\mathbf{I}$ [121]. In this scenario a single choice of covariance matrices achieves the entire capacity region. This is not true, in general, for other fading distributions, i.e. different covariances may be required to achieve different points on the capacity region boundary, as is the case for the MIMO MAC with no fading. The sum-rate capacity of the MAC (achieved using scaled identity covariance matrices) is equal to

$$\mathcal{C}_{MAC}^{sum}(\mathbf{P};\mathbf{H}^*) = E_{\mathbf{H}}\left[\log\det\left(\mathbf{I}_{M_T} + \sum_{i=1}^{K}\mathbf{H}_i^*\left(\frac{P_i}{M_R}\mathbf{I}_{M_R}\right)\mathbf{H}_i\right)\right] \tag{2.17}$$

$$= E_{\mathbf{H}}\left[\log\det\left(\mathbf{I}_{M_T} + \frac{P_i}{M_R}\mathbf{H}^*\mathbf{H}\right)\right], \tag{2.18}$$

where P_i is the ith transmitter's power constraint and we have assumed $P_i = P$ for all i. Note that this expression is exactly the ergodic capacity of the single-user MIMO channel with $M_R K$ transmit antennas, M_T receive antennas, the ZMSW distribution model, and perfect CSIR and CDIT, as given in (2.10). Therefore, the lack of cooperation between the K transmitters does not reduce the capacity under this fading model, and the MAC achieves the same capacity as the fully cooperative model. This implies that the MAC sum capacity scales as $\min(M_T, M_R K)$ with perfect CSIR and CDIT. The MIMO MAC with perfect CSIR and CDIT under the CCI and CMI models has also been investigated, but only limited results are known [51, 62].

For MIMO multiple-access channels, it is generally sufficient to have perfect CSI at the receiver and CSIT is not crucial to obtain capacity benefits from MIMO. This is because for fading multiple-access channels, independent Gaussian codewords can be transmitted from each antenna (for each user). The subsequent sum rate is given by the right-hand side of (2.18). This technique is optimal when all users have the same power constraint and each channel is ZMSW, but is generally sub-optimal in other scenarios. However, this simple transmission scheme clearly achieves a sum capacity similar to the capacity of the point-to-point MIMO channel from the $M_R K$ aggregate transmit antennas to a M_T antenna receiver, i.e. the sum capacity grows as $\min(M_T, M_R K)\log(\text{SNR})$. Thus for systems with a large number of users, increasing the number of receive antennas at the base-station (M_T) while keeping the number of mobile antennas (M_R) constant can lead to linear growth.

Though CSIR and CDIT are sufficient to achieve the linear capacity growth described above, the availability of CSIT increases capacity further, though not in the first-order growth of $\min(M_T, M_R K)$. As seen for point-to-point MIMO channels, CSIT gives the largest performance increases at low SNRs, and the value of CSIT disappears at high SNR. Furthermore, if the number of base-station antennas (M_T) and the number of users (K) are taken to infinity at a fixed ratio, the gain due to CSIT relative to CDIT vanishes [137].

2.4.3 *MIMO broadcast channel*

In this section we summarize capacity results on the multiple-antenna BC. We first discuss the general broadcast channel and the single-antenna broadcast channel, followed by an explanation of dirty paper coding for the MIMO BC. We then establish a duality relationship between the MAC and the BC, which leads into sections on the capacity of the MIMO BC.

Broadcast channel capacity

Unlike the multiple-access channel, a general expression for the capacity region of the broadcast channel is unknown. In fact, this is one of the most fundamental unanswered questions in multi-user information theory. However, the capacity region for certain classes of broadcast channels is known. Amongst those is the class of degraded broadcast channels, which are channels where receivers can be absolutely ranked in terms of their channel strength [25]. The single-antenna AWGN BC falls into this class (users are ranked in terms of the absolute value of their channel strength), and the capacity region of a two-user channel (assuming, without loss of generality, $|h_1| > |h_2|$) is given by all rate pairs satisfying

$$R_1 \leq \log(1+\alpha|h_1|^2 P)$$
$$R_2 \leq \log\left(1+\frac{|h_2|^2(1-\alpha)P}{\alpha|h_2|^2 P+1}\right)$$

for some $\alpha \in [0, 1]$. In order to achieve the capacity region, the codewords for the different receivers are superimposed on one another, and successive interference cancellation is used at one of the receivers. For the AWGN channel, the transmitter generates independent Gaussian codewords for each receiver (with power αP for receiver 1 and power $(1-\alpha)P$ for receiver 2) and transmits the sum of these codewords. Receiver 1, which has the larger channel gain, first decodes the codeword intended for receiver 2, subtracts this from the received signal, and then decodes its intended codewords. Receiver 2 is not able to first decode the codeword for receiver 1 because $|h_1| > |h_2|$ by assumption. Thus receiver 2 treats the codeword for receiver 1 as additional noise (with power αP) while decoding its intended codeword.

However, when the transmitter has more than one antenna, the Gaussian broadcast channel is generally non-degraded. This is because matrix channels can only be partially ordered, i.e. there is no absolute ordering of channels. As a result, the successive decoding technique which is capacity-achieving for the single-antenna broadcast channel is not effective in the MIMO setting and an alternative technique must be used, as we describe next.

Dirty paper coding achievable rate region

As noted earlier, successive cancellation cannot effectively be performed by receivers in a MIMO BC to reduce multi-user interference. However, *dirty paper coding* (DPC) can be used at the transmitter to essentially pre-subtract multi-user interference [13, 21, 148]. The resulting rates mimic what one would expect from successive cancellation. The basic premise of dirty paper coding can be illustrated by considering a point-to-point AWGN channel with additive interference: $y = x + s + n$, where x is the transmitted signal (subject to power constraint P), y is the received signal, n is the Gaussian noise, and s is additive interference. If the transmitter but not the receiver has perfect, non-causal knowledge of the interference s, then the capacity of the channel is the same as if there was no additive interference (i.e. $\log(1+P)$). Dirty paper coding is a technique that allows non-causally known interference to be "pre-subtracted" at the transmitter, but in such a way that the transmit power is not increased. A more practical and general technique to perform interference pre-subtraction based on nested lattice codes is provided in [30].

In the MIMO BC, dirty paper coding can be applied at the transmitter when choosing codewords for different receivers. The transmitter first picks a codeword (i.e. $\mathbf{x}_1$) for receiver 1. The transmitter then chooses a codeword for receiver 2 (i.e. $\mathbf{x}_2$) with full (non-causal) knowledge of the codeword intended for receiver 1. Therefore the codeword of user 1 can be pre-subtracted such that receiver 2 does not see the codeword intended for receiver 1 as interference. Similarly, the codeword for receiver 3 is chosen such that receiver 3 does not see the signals intended for receivers 1 and 2 (i.e. $\mathbf{x}_1 + \mathbf{x}_2$) as interference. This process continues for all K receivers. If $\mathbf{x}_i$ is chosen according to $N(0, \mathbf{\Sigma}_i)$ and user $\pi(1)$ is encoded first, followed by user $\pi(2)$, etc. the following is an achievable rate vector:

$$R_{\pi(i)} = \log \frac{\det\left(\mathbf{I}_{M_R} + \mathbf{H}_{\pi(i)}(\sum_{j\geq i} \mathbf{\Sigma}_{\pi(j)})\mathbf{H}^*_{\pi(i)}\right)}{\det\left(\mathbf{I}_{M_R} + \mathbf{H}_{\pi(i)}(\sum_{j> i} \mathbf{\Sigma}_{\pi(j)})\mathbf{H}^*_{\pi(i)}\right)} \qquad i = 1, \ldots, K. \tag{2.19}$$

Different rates can be achieved by varying the input covariance matrices and the encoding order. Notice that the same rates would be achieved if receiver $\pi(i)$ knew the transmitted signals $\mathbf{x}_{\pi(1)}, \ldots, \mathbf{x}_{\pi(i-1)}$. The dirty paper region $\mathcal{C}_{DPC}(P, \mathbf{H})$ is defined as the convex hull of the union of all such rate vectors over all positive-semi-definite covariance matrices $\mathbf{\Sigma}_1, \ldots, \mathbf{\Sigma}_K$ such that $\mathrm{tr}(\mathbf{\Sigma}_1 + \cdots + \mathbf{\Sigma}_K) = \mathrm{tr}(\mathbf{\Sigma}_x) \leq P$ and over all permutations $(\pi(1), \ldots, \pi(K))$:

$$\mathcal{C}_{DPC}(P, \mathbf{H}) \triangleq Co\left(\bigcup_{\pi, \mathbf{\Sigma}_i} \mathbf{R}(\pi, \mathbf{\Sigma}_i)\right) \tag{2.20}$$

where $\mathbf{R}(\pi, \mathbf{\Sigma}_i)$ are the set of rates given by (2.19). The transmitted signal is $\mathbf{x} = \mathbf{x}_1 + \cdots + \mathbf{x}_K$ and the input covariance matrices are of the form $\mathbf{\Sigma}_i = E[\mathbf{x}_i \mathbf{x}_i^*]$. From the dirty paper result we find that $\mathbf{x}_1, \ldots, \mathbf{x}_K$ are uncorrelated, which implies $\mathbf{\Sigma}_x = \mathbf{\Sigma}_1 + \cdots + \mathbf{\Sigma}_K$.

One important feature to notice about the dirty paper rate equations in (2.19) is that the rate equations are neither a concave nor a convex function of the covariance matrices, which makes finding the dirty paper region a difficult numerical problem.

MAC–BC duality

A key component in establishing capacity results for the MIMO BC is the duality relationship between the MIMO BC and the MIMO MAC. The duality relationship refers to the fact that the dirty paper rate region of the multi-antenna BC with power constraint P is equal to the union of the capacity regions of the dual MAC, where the union is taken over all individual power constraints that sum to P [133]:

$$\mathcal{C}_{DPC}(P, \mathbf{H}) = \bigcup_{\mathbf{P}:\sum_{i=1}^{K} P_i = P} \mathcal{C}_{MAC}(P_1, \ldots, P_K, \mathbf{H}^*). \tag{2.21}$$

This is the multiple-antenna extension of the previously established duality between the scalar Gaussian broadcast and multiple-access channels [71]. In addition to the relationship between the two rate regions, for any set of covariance matrices in the MAC/BC, [133] provides an explicit set of transformations to find covariance matrices in the BC/MAC that achieve the same rates. The union of MAC capacity regions in (2.21) is easily seen to be the same expression as in (2.14) but with the constraint $\sum_{i=1}^{K} \mathrm{tr}(\mathbf{Q}_i) \leq P$ instead of $\mathrm{tr}(\mathbf{Q}_i) \leq P_i \forall i$ (i.e. a sum constraint instead of individual constraints). In establishing this duality, it is shown that every rate vector in the dual MAC capacity region can be achieved in the MIMO BC. Each set of MAC covariance matrices corresponds to the polyhedron of rates described in (2.15). Each corner point of the polyhedron is achievable in the MAC using successive decoding with a specific order, and is also achievable in the MIMO BC using DPC with the *opposite* encoding order.

The MAC–BC duality is very useful from a numerical standpoint because the dirty paper region cannot be characterized in terms of a convex optimization problem, whereas the boundary of the dual MAC capacity region can be cast as a convex optimization problem. As a result, the optimal MAC covariances can be found using standard convex optimization techniques and then transformed to the corresponding optimal BC covariances using the MAC–BC transformations given in [133].

The dirty paper rate region is shown in Figure 2.13 for a channel with two users, $M_T = 2$ and $M_R = 1$. Notice that the dirty paper rate region shown in Figure 2.13 is actually a union of MAC regions, where each MAC region corresponds to a different set of individual power constraints. Since $M_R = 1$, each of the MAC regions is a pentagon, as discussed in Section 2.4.2. Similar to the MAC capacity region, the boundary of the dirty paper coding region is curved, except at the sum-rate maximizing portion of the boundary.

Duality also allows the MIMO MAC capacity region to be expressed as an intersection of the dual dirty paper BC rate regions [133, Corollary 1]:

$$\mathcal{C}_{MAC}(P_1, \ldots, P_K, \mathbf{H}^*) = \bigcap_{\boldsymbol{\alpha}>0} \mathcal{C}_{DPC}\left(\sum_{i=1}^{K} \frac{P_i}{\alpha_i}; [\sqrt{\alpha_1}\mathbf{H}_1^T \cdots \sqrt{\alpha_K}\mathbf{H}_K^T]^T\right). \tag{2.22}$$

Constant channel capacity

In the previous section we described how the technique of dirty paper coding can be applied to the MIMO BC to pre-subtract interference at the transmitter. This strategy has,

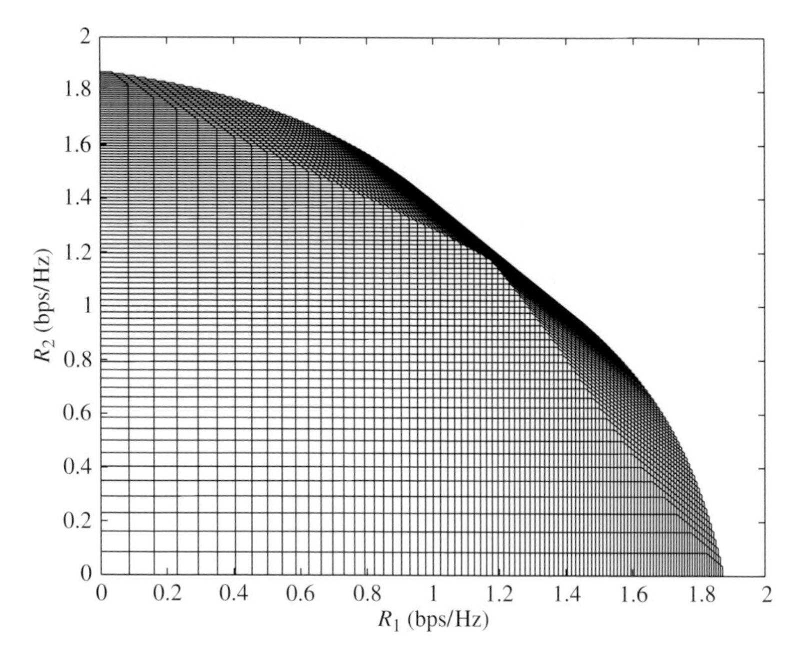

Fig. 2.13. Dirty paper rate region, $\mathbf{H}_1 = [1\ 0.5]$, $\mathbf{H}_2 = [0.5\ 1]$, $P = 10$.

in fact, been shown to achieve the capacity region of the MIMO BC [141]. The optimality of dirty paper coding was first shown for sum-rate capacity in [13, 133, 136, 149], and later extended to the full capacity region [141]. Since DPC achieves the capacity region (denoted as $\mathcal{C}_{BC}(P, \mathbf{H})$), the capacity region can be given in terms of (2.20), i.e. $\mathcal{C}_{BC}(P, \mathbf{H}) = \mathcal{C}_{DPC}(P, \mathbf{H})$. However, this form of the capacity region is difficult to find numerically because the rate equations are not concave functions of the input covariance matrices. By the MAC–BC duality explained in the previous subsection, a simpler expression of the capacity region of the MIMO BC can be given in terms of the sum power MIMO MAC:

$$\mathcal{C}_{\mathrm{BC}}(P; \mathbf{H}) = \mathcal{C}_{\mathrm{MAC}}(P; \mathbf{H}^*)$$
$$= \bigcup_{\{\mathbf{Q}_i \geq 0, \sum_{i=1}^{K} \mathrm{tr}(\mathbf{Q}_i) \leq P\}} \left\{ \begin{array}{l} (R_1, \ldots, R_K) : \\ \sum_{i \in S} R_i \leq \log\det\left(\mathbf{I}_{M_T} + \sum_{i \in S} \mathbf{H}_i^* \mathbf{Q}_i \mathbf{H}_i\right) \forall S \subseteq \{1, \ldots, K\} \end{array} \right\}, \tag{2.23}$$

where the $\mathbf{Q}_i$ are the dual MAC covariance matrices. As stated earlier, a key property of this characterization is that the rate equations are concave functions of the input covariance matrices. Thus, power convex optimization algorithms can be used to compute the MIMO BC capacity region numerically.

The capacity region of the MIMO BC is a convex region, and thus its boundary can be found by maximizing the function $\mu_1 R_1 + \cdots + \mu_K R_K$ over all rate vectors in the capacity region and for all non-negative priorities $(\mu_1, \ldots, \mu_K)$ such that $\sum_{i=1}^{K} \mu_i = 1$. Since the MIMO BC capacity region is best described in terms of the MIMO MAC (with sum power constraint) capacity region, the same intuition holds regarding finding the boundary of the capacity region. In fact, the boundary of the MIMO BC capacity region can be found by solving the maximization in (2.16) subject to a sum power constraint on the covariance matrices. Standard convex optimization techniques can be used to solve this optimization, and a specific instance of such an algorithm is provided in [139]. As mentioned earlier, the encoding order with DPC is the opposite of the decoding order in the MAC. Thus, on the boundary of the capacity region, encoding should be done in order of *decreasing* priority, i.e. the user with the lowest priority should be encoded last and benefit from the most interference cancellation. For sum-rate capacity (i.e. $\mu_1 = \cdots = \mu_K$), more efficient algorithms based on the Karush–Kuhn–Tucker (KKT) conditions exist [70, 77, 146]. Of particular interest is the sum power iterative water-filling algorithm in [70] which is related to the iterative water-filling technique for the MAC, with the difference that all the users simultaneously water-fill with respect to one power constraint.

Similar to the MIMO MAC, the sum-rate capacity of the MIMO BC grows approximately as $\min(M_T, M_R K) \log(\text{SNR})$ as SNR is taken to infinity. In other words, a MIMO BC with perfect CSIR and CSIT has a multiplexing gain of $\min(M_T, M_R K)$, i.e. the same as a point-to-point $M_T \times M_R K$ MIMO system. Additionally, if SNR is fixed but the number of transmit antennas and the number of receivers (with $M_R = 1$) are taken to infinity with $M_T \geq K$, then the sum capacity under the ZMSW model has the same growth constant as a $K \times M_T$ point-to-point MIMO channel (notice that there are K transmit antennas in the equivalent point-to-point channel, whereas K represents the receivers in the MIMO BC) with perfect CSIR and CDIT [50].

Fading channels

For fading MIMO BCs, the capacity region depends crucially on the CSI available at the transmitter and receivers. As we discuss below, the degree of channel knowledge at the transmitter is particularly important.

With perfect CSIR and CSIT, the MIMO BC can be split into parallel channels with an overall power constraint [147] (see [82] for a treatment of the single-antenna channel). Clearly, the full multiplexing gain of $\min(M_T, M_R K)$ is achievable in this scenario.

The scenario where there is perfect CSIR but only CDIT is perhaps the most practical as well as interesting situation. However, the capacity and even the multiplexing gain in this scenario is, in general, unknown. Furthermore, the technique of dirty paper coding requires perfect CSIT in order for multi-user interference to be cancelled perfectly. Thus, it is still not clear what transmission strategy should be used for such fading channels.

One special case of CSIR and CDIT for which capacity is known is when all receivers have the same channel distribution (e.g. all ZMSW) and the same number of antennas. In this situation the K channels are *statistically* identical, i.e. $p(Y_1 = y|x) = \cdots = p(Y_K = y|x)$ for all x, y.[9] This implies that if any of the K receivers can decode a codeword, then

[9] Notice that the statistical equivalence of the K receivers depends crucially on the CDIT, and the lack of CSIT. If there is CSIT, one must consider $p(y_i|x, H)$ instead of $p(y_i|x)$, and the statistical equivalence is lost, even if each of the channels follows the same model. Also note that with CDIT the statistical equivalence does not depend on the ZMSW model; the distributions of $\mathbf{H}_1, \ldots, \mathbf{H}_K$ need only be identical.

receiver 1 can also decode the same codeword. This, in turn, implies that receiver 1 can decode every message sent by the transmitter. Therefore, the sum-rate capacity of the MIMO BC is bounded by the capacity of the channel from the transmitter to any single receiver. The capacity region is therefore given by $C_{BC}(P) = \{(R_1, \ldots, R_K) : \sum_{i=1}^{K} R_i \leq C_1(P)\}$, where $C_1(P)$ is the point-to-point capacity from the transmitter to any receiver. If each of the channels follows the ZMSW model, the capacity is given by (2.11): $C_1(P) = E_{\mathbf{H}}\left[\log\det\left(\mathbf{I}_{M_R} + \frac{P}{M_T}\mathbf{H}_1\mathbf{H}_1^*\right)\right]$. The rates achievable in the MIMO BC are therefore limited by the point-to-point capacity from the transmitter to any of the receivers. In other words, there is no multi-user MIMO benefit in this scenario. Since $C_1(P)$ has a multiplexing gain of only $\min(M_T, M_R)$, the sum rate capacity also has a multiplexing gain of only $\min(M_T, M_R)$, as opposed to the $\min(M_T, M_R K)$ possible with perfect CSIT. Note that this is in contrast to the MIMO MAC, where CSIR and CDIT are sufficient in order to achieve the full multiplexing gain of $\min(M_T, M_R K)$. This idea is further investigated in [58], where it is shown that the multiplexing gain in the MIMO BC is lost whenever the channel fading distributions are spatially isotropic, i.e. the transmitter has no information regarding the spatial *direction* of each of the K channels.

If there is perfect CSIR and CDIT and the users do not have statistically identical channels, then very little is known about the capacity region. For example, consider a system under the CMI model with $K = 2$, $M_T = 2$, and $M_R = 1$. If $E[\mathbf{H}_1] \neq E[\mathbf{H}_2]$ (i.e. non-identical means), then the channels of the two users are not statistically identical. Therefore, the argument given above does not apply and it is not clear what the capacity region or sum-rate capacity of this channel is. However, the multiplexing gain of this channel has been shown to be strictly smaller than that with perfect CSIR and CSIT [81].

The addition of limited feedback from receivers to transmitters can significantly increase the MIMO BC capacity. In this setting, perfect CSIR and partial CSIT (achieved via a feedback channel) are considered. In [114], M_T random and orthogonal beamforming vectors are transmitted. The receivers measure the signal-to-interference-plus-noise ratio (SINR) on each of the beams and feed back the measurement values to the transmitter. The transmitter then transmits to the users with the highest SINR values. It is shown that this scheme achieves the full multiplexing gain when there are a large number of receivers experiencing fading according to the ZMSW model. Thus, this limited feedback is sufficient to overcome the barriers of the CSIR/CDIT model. The MIMO BC has also been considered in the finite-rate feedback model, whereby each receiver feeds back a quantized version of its channel instantiation to the transmitter (see the quantized channel information model in Section 2.3.1 for a discussion of this model in the point-to-point setting). In this setting, if the number of feedback bits per user is increased in proportion to the system SNR, then the full multiplexing gain is achievable for any number of receivers [66].

Sub-optimal methods

Though DPC is capacity-achieving for the constant MIMO BC, its complexity is rather high because implementing DPC requires the use of near-optimal vector quantizers at both the transmitter and receivers [31]. Therefore, it is also of great interest to study the behavior of less complex, perhaps sub-optimal transmission schemes. In this section we examine the performance of two classes of schemes: orthogonal transmission (i.e. TDMA), and multi-user beamforming without interference cancellation.

TDMA (time-division multiple-access) has the inherent advantage of low complexity, because the transmitter transmits to only a single-user at a time and the MIMO BC is essentially reduced to a point-to-point MIMO channel. The TDMA rate region is given by:

$$\mathcal{R}_{TDMA}(\mathbf{H}, P) \triangleq \left\{ (R_1, \ldots, R_K) : \sum_{i=1}^{K} \frac{R_i}{C(\mathbf{H}_i, P)} \leq 1 \right\}, \tag{2.24}$$

where $C(\mathbf{H}_i, P)$ denotes the single-user capacity of the ith user subject to power constraint P.

Note that other orthogonal allocations of resources such as frequency-division multiple access (FDMA) are equivalent to TDMA from a capacity standpoint. The main drawback of TDMA is that the multiplexing gain is only $\min(M_T, M_R)$, as opposed to the $\min(M_T, M_R K)$ possible using DPC. As a result of this, TDMA can yield rates up to a factor of $\min(M_T, K)$ times smaller than DPC [68]. In fact, this gap is particularly pronounced in large systems with a small number of receive antennas operating at high SNRs.

One advantage of TDMA is that it can be easily used in fading channels with CSIR and CDIT. Point-to-point MIMO capacity is well known in this scenario, and as stated earlier, TDMA is in fact optimal in this scenario if the fading distributions of each of the users are identical. In general, TDMA is expected to perform well (i.e. close to capacity) even when channels are not statistically identical but when the CDIT is not very informative (e.g. CMI with very weak line-of-sight components).

Multi-user (or linear) beamforming is another sub-optimal transmission technique for the MIMO BC. With this scheme, a different beamforming direction(s) is chosen for each receiver and the sum of independent codewords is transmitted using the different beamforming directions. Since no interference cancellation is performed, multi-user interference is treated as noise. Note that this scheme differs from capacity-achieving DPC only in that no interference pre-subtraction is performed at the transmitter; the beamforming structure with DPC is in fact optimal when $M_R = 1$. Since beamforming is typically implemented through linear transmit and receive filters, beamforming is also referred to as linear processing.

For the single-receive antenna channel ($M_R = 1$), a duality between the MAC and BC also exists for beamforming [7, 136], and the rates achievable using beamforming are most easily expressed in terms of the dual MAC as

$$C_{BF}(\mathbf{H}_1, \ldots, \mathbf{H}_K, P) = \max_{\{P_i : \sum_{i=1}^{K} P_i \leq P\}} \sum_{j=1}^{K} \log \frac{\det\left(\mathbf{I}_{M_T} + \sum_{i=1}^{K} \mathbf{H}_i^* P_i \mathbf{H}_i\right)}{\det\left(\mathbf{I}_{M_T} + \sum_{i \neq j} \mathbf{H}_i^* P_i \mathbf{H}_i\right)}. \tag{2.25}$$

This maximization, however, cannot be rephrased as a convex optimization problem, and as a result is extremely difficult to compute numerically. Though it is difficult to find the optimal beamforming strategy, it is possible to compute sub-optimal strategies, and it is easy to see that beamforming does provide the full multiplexing gain of $\min(M_T, M_R K)$. Zero-forcing beamforming, in which the transmitter multiplies the vector of data symbols by the inverse of the channel to eliminate all multi-user interference, is one example of sub-optimal multi-user beamforming. Some asymptotic results regarding beamforming are given in [68, 138, 145]. A number of works have also studied the selection of beamforming vectors when receivers have multiple antennas (cf. [119]). A novel technique combining beamforming and turbo coding has also been proposed [99, 100].

Beamforming can be used in fading channels with CSIR/CDIT, but it is difficult to choose good beamforming directions for each user unless the transmitter has a very good estimate of the channels. Thus, beamforming is not expected to perform well unless the CDIT contains a high degree of information regarding the current channel (e.g. CMI with strong line-of-sight components).

In Figure 2.14 the sum rate is plotted versus the SNR for a 10-transmit antenna, 10-receiver (each with single antennas) system with perfect CSIR and CSIT. The sum-rate capacity (achieved using DPC), zero-forcing beamforming, and TDMA rates are plotted. Notice that both the sum-rate capacity and zero-forcing curves achieve the full multiplexing gain, i.e. have a slope of $\min(M_T, K) = 10$ bits/3 dB. Though zero-forcing achieves the correct slope, there is a substantial power penalty (approximately 8.3 dB) to using this sub-optimal method [65]. TDMA, on the other hand, only achieves a multiplexing gain of unity, corresponding to a slope of only 1 bit/3 dB. Both DPC and zero-forcing provide a substantial increase over TDMA, even at moderate SNR values.

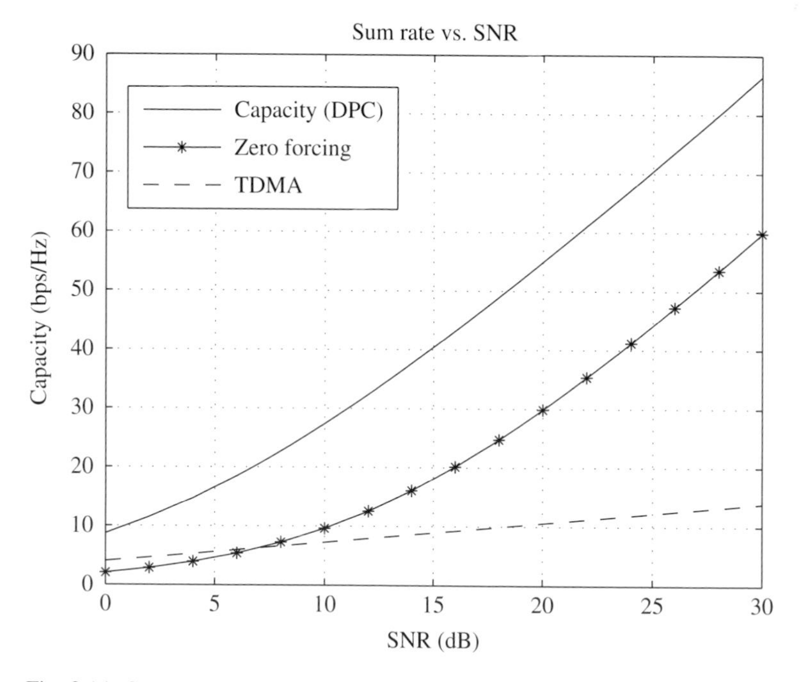

Fig. 2.14. Sum rate versus SNR for an $M_T = 10$, $M_R = 1$, and $K = 10$ system.

2.4.4 *Open problems in multi-user MIMO*

Multi-user MIMO has been a primary focus of research in recent years, mainly due to the large number of open problems in this area. Some of these are as follows.

1. MIMO BC with perfect CSIR and CDIT: capacity is only known when the channels of all users have the same distribution. When this condition is not met, however, little is known regarding the capacity.
2. CDIT and CDIR: since perfect CSI is rarely possible, a study of capacity with CDI at both the transmitter(s) and receiver(s) for both MACs and BCs is of great practical relevance.
3. Non-DPC techniques for BC: dirty paper coding is a very powerful capacity-achieving scheme, but it appears quite difficult to implement in practice. Thus, non-DPC multi-user transmission schemes for the downlink (such as downlink beamforming [103]) are also of practical relevance. In addition, performing DPC (or some variant) with imperfect CSIT or CDIT is still challenging.

2.5 Multi-cell MIMO

The MAC and the BC are information-theoretic abstractions of the uplink and the downlink of a single cell in a cellular system. However a cellular system consists of many cells with channels (timeslots, bandwidth, or codes) reused at spatially separated locations. Due to the fundamental nature of wireless propagation, transmissions in a cell are not limited to within that cell, and thus there is intercell interference between users and base-stations that use the same channels. The majority of current systems are interference limited rather than noise limited. As a result, it is not sufficient to exclusively study single-cell models and thus multi-cell environments must be explicitly considered in order to accurately assess the benefit of MIMO technology.

In this section we provide an overview of information-theoretic results for multi-cell environments. Analysis of the capacity of the cellular network explicitly taking into account the presence of multiple cells, multiple users, and multiple antennas, and the possibilities of cooperation between base-stations is inevitably a hard problem and runs into several long-standing unsolved problems in network information theory. However, such an analysis is also of utmost importance because it defines a common benchmark that can be used to gauge the efficiency of any practical scheme, in the same way that the capacity of a single-user link serves as a measure of the performance of practical schemes. Work on multi-cell environments can be grouped into two broad categories, with one group assuming that base-stations cannot cooperate (as is done in current systems), and with other work considering multi-cellular environments where base-stations are allowed some level of cooperation (i.e. cooperative transmission and/or reception). We also discuss some general system-level issues.

2.5.1 *Multi-cell MIMO without base-station cooperation*

The traditional analysis of capacity of cellular systems has assumed that neither base-stations nor users can cooperate. Note that current cellular systems do cooperate in some sense (e.g. handoff), but more so at the networking layer than at the physical layer. In this chapter we use "cooperation" to imply cooperative transmission and reception, i.e. cooperation at the physical layer. If no cooperation is allowed, the channel becomes an interference channel, in which there are multiple transmitters and multiple receivers communicating over a common medium, but each transmitter only wishes to communicate with a single receiver. Unfortunately, the Shannon capacity of channels with interference is a long-standing open problem in information theory [25, 127]; in fact, even the capacity region of a two-transmitter, two-receiver interference channel with no fading and single-antenna elements is not fully known [22]. Some results for the multiple-antenna interference channel are given in Section 2.6 on MIMO ad hoc networks.

A more promising and well-studied approach is to treat all out-of-cell interference as an additional source of Gaussian noise. The Gaussian assumption can be viewed as a worst-case assumption about the interference, since exploiting known structure of the interference can presumably help in decoding the desired signals and therefore increase capacity. By treating the interference as Gaussian noise, the capacity of both the uplink and the downlink can be determined using the single-cell analysis of Section 2.4. The capacity of a single-antenna cellular system uplink with fading based on treating interference as Gaussian noise was obtained in [109] for both one- and two-dimensional cellular grids. These capacity results show that with or without fading, when intercell interference is non-negligible, an orthogonal multiple-access method (e.g. TDMA) within a cell is optimal. This is also the case when channel-inversion power control is used within a cell. Moreover, in some cases partial or full orthogonalization of channels assigned to different cells can increase capacity. There has also been a body of work that has studied multiple transmit and receive antenna arrays in cellular systems [16, 17, 86, 143]. In this work, the spatial structure of out-of-cell interference is considered. Note that AWGN is normally considered to be spatially white. However, out-of-cell interference generally has some statistical structure, i.e. a spatial covariance, which significantly affects capacity results. One interesting conclusion that can be drawn from this work is that it is sometimes beneficial to limit the number of transmit antennas, because the receiver must simultaneously use its antennas to cancel out-of-cell interference and decode the desired signal. When performing these operations, it is beneficial to have the number of receive antennas be as large as the aggregate number of transmit antennas, i.e. including the transmit antennas of strong interfering users.

2.5.2 *Multi-cell MIMO with base-station cooperation*

A relatively new area of research has studied cellular systems where cooperation between base-stations is allowed. Since base-stations are wired and fixed in location, it may be practical to allow base-stations to cooperatively transmit and receive.

If perfect cooperation is assumed, this allows the entire network to be viewed as a single cell with a distributed antenna array at the base-station. As a result, capacity results about the single-cell uplink and downlink channel, discussed in Section 2.4, can be applied. Note, however, that the fact that the base-stations and terminals are geographically separated directly impacts the channel gains in the composite network.

The uplink capacity of cellular systems (where mobiles and base-stations are assumed to each have single antennas) under the assumption of full base-station cooperation, where signals received by all base-stations are jointly decoded, was first investigated in [47] followed by a more comprehensive treatment in [144]. In both cases propagation between the mobiles and the base-stations is characterized using an AWGN channel model (i.e. no fading is considered) with a channel gain of unity within a cell, and a gain of α, $0 \leq \alpha \leq 1$, between cells. The Wyner model of [144] considers both one- and two-dimensional arrays of cells, and derives the per-user capacity in both cases. It is also shown in both [144] and [47] that uplink capacity is achieved by using orthogonal multiple access techniques (e.g. TDMA) within each cell, and reusing these orthogonal channels in other cells, although this is not necessarily uniquely optimal.

The downlink channel can be modeled as a MIMO broadcast channel if perfect base-station cooperation is assumed. On the downlink, since the base-stations can cooperate perfectly, dirty paper coding can be used over the entire transmitted signal (i.e. across base-stations) in a straightforward manner. The application of dirty paper coding to a multiple-cell environment with cooperation between base-stations was pioneered in work by Shamai and Zaidel [110]. For one antenna at each user and each base-station, they show that a relatively simple application of dirty paper coding can enhance the capacity of the cellular downlink. While capacity computations are not the focus of [110], they do show that their scheme is asymptotically optimal at high SNRs. A number of other works have also studied the capacity of multi-cell downlink channels, in both the finite and asymptotic regimes [1, 55, 60]. The duality of the MAC and BC (discussed in Section 2.4.3) can also be applied to the multi-cell composite channels to relate uplink and downlink results.

One weakness of treating the multi-cell downlink channel as a standard MIMO broadcast channel is that the average transmit power across all antennas (and thus across all base-stations) is bounded. In practice, the power from each base-station may in fact be bounded. If these stricter constraints are enforced, it has been shown that dirty paper coding still achieves the sum capacity [77, 149]. A modified duality between the MAC and BC has also been established when such power constraints are imposed [77]. For large networks, the effect of imposing per-base power constraints is still unclear [60].

In general, results show that base-station cooperation can yield significant capacity increases relative to systems without cooperation. From a research perspective, the size of large networks makes generating numerical results difficult and as a result asymptotic results must be relied upon [1, 55]. From a practical perspective, actually achieving perfect cooperation between physically remote base-stations is still very challenging. In addition, it seems that very large-scale cooperation would require tremendous complexity. Because of these issues, it is not clear what methods of base-station cooperation are practically feasible.

2.5.3 *System level issues*

The capacity results described in this section address just a few out of many interesting questions in the design of a cellular system with multiple antennas. Multiple antennas can be used not only to enhance the capacity of the system but also to drive down the probability of error through diversity combining. As discussed in Chapter 1.3.4, work by Zheng and Tse [154] unravels a fundamental diversity versus multiplexing trade-off in MIMO systems. Also, instead of using isotropic transmit antennas on the downlink and transmitting to many users, it may be simpler to use directional antennas to divide the cell into sectors and transmit to one user within each sector. The relative impact of CDIT and/or CDIR on each of these schemes is not fully understood. Although in this paper we focus on the physical layer, smart schemes to handle CDIT can also be found at higher layers. An interesting example is the idea of opportunistic beamforming [135]. In the absence of CSIT, the transmitter randomly chooses the beamforming weights. With enough users in the system, it becomes very likely that these weights will be nearly optimal for one of the users. In other words, a random beam selected by the transmitter is very likely to be pointed towards a user if there are enough users in the system. Instead of feeding back the channel coefficients to the transmitter the users simply feed back the SNRs they see with the current choice of beamforming weights. This significantly reduces the amount of feedback required. By randomly changing the weights frequently, the scheme also treats all users fairly.

2.6 MIMO for ad hoc networks

An ad hoc wireless network is a collection of wireless mobile nodes that self-configure to form a network without the aid of any established infrastructure, as shown in Figure 2.15. Without an inherent infrastructure, the mobiles handle the necessary control and networking tasks by themselves, generally through the use of distributed control algorithms. Multi-hop routing, whereby intermediate nodes relay packets towards their final destination, is often used since it can improve the throughput and power efficiency of the network. Note that with sufficient transmit power any node in the network can transmit a signal directly to any other node. However, such transmissions over long distances will result in a low received power, and will also cause interference to other links. Thus, links with low signal-to-interference-plus-noise (SINR) power ratios are typically not used. The SINR on different links is illustrated by the different line widths in Figure 2.15.

The fundamental capacity limits of an ad hoc wireless network – the set of maximum data rates possible between all nodes – is a highly challenging problem in information theory even when the nodes only have a single antenna. For a network of K nodes, each node can communicate with $K-1$ other nodes, so the capacity region has dimension $K(K-1)$. While rate sums across any cut-set of the network are bounded by the corresponding mutual information expressions [25, Theorem 14.10.1], simplifying this formula into a tractable expression for the ad hoc network capacity region is an immensely complex problem. Given the lack of capacity results for ad hoc networks with just one antenna per node, it seems that considering multiple antennas per node will only make

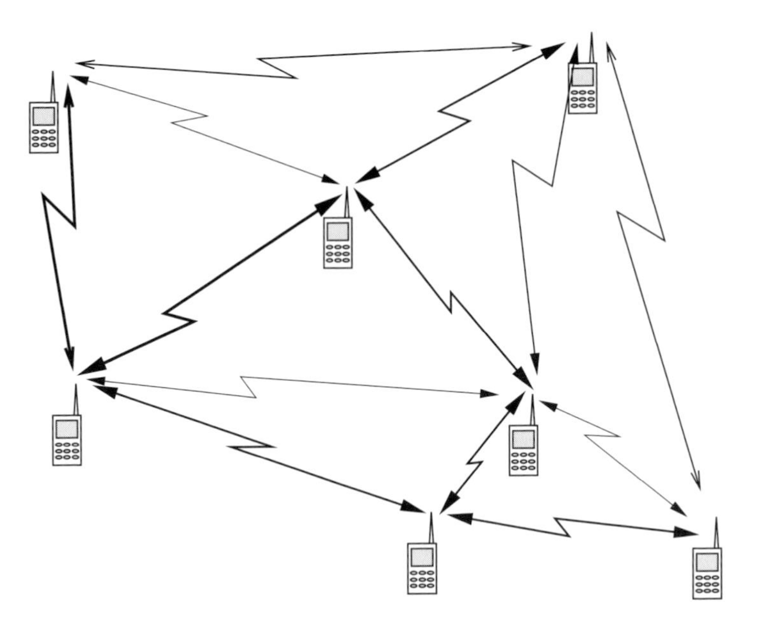

Fig. 2.15. Ad hoc network.

the problem more intractable. Moreover, as with cellular systems, multiple antennas in an ad hoc network can be used for diversity or sectorization in addition to capacity gain, and the fundamental trade-offs between these different uses are very difficult to characterize.

One approach to get around this difficulty is to study the asymptotic behavior of the net throughput as the number of nodes increases to infinity. Such an approach has been pioneered by Gupta and Kumar [45]. Various decentralized transmitter–receiver models have been considered in this context, including static, mobile, and fading channel with or without successive interference cancellation at the receiver. It has been found in most static channel cases that the capacity per node decays as $O(1/\sqrt{n})$, thus going down to zero as n grows to infinity. However, mobility and/or fading is found to enhance this performance by means of exploiting opportunism, and an $O(1)$ rate per node can now be obtained [29, 36, 43]. The advantage of this analysis technique is the intuition it provides about capacity trends, with the inter-user trade-offs being lost at this level of abstraction.

Another approach taken by researchers has been to start with networks with a few nodes that form the basic building blocks of larger ad hoc networks where the capacity analysis may be more tractable. The primary few-node components of ad hoc networks are the MAC, BC, relay, and interference channels. The MAC and BC have already been analyzed in detail earlier in this chapter, and so this section devotes itself to the discussion of the relay and interference channels.

2.6.1 *The relay channel*

The relay channel, where one node aids in the communication of another node's message, is a natural and important building block of ad hoc networks. The most elementary relay channel is modeled as a three-node system, where the transmitter communicates with the

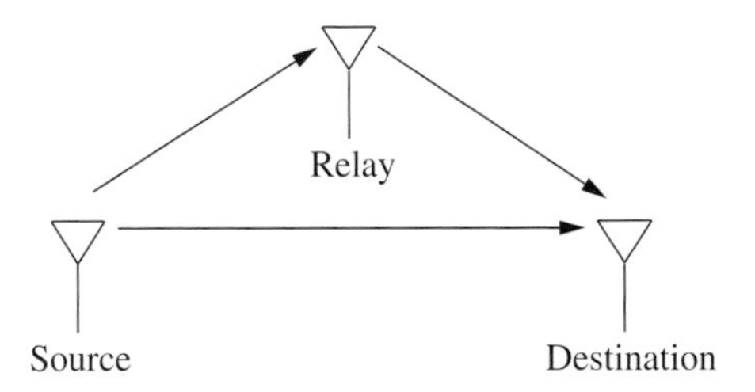

Fig. 2.16. The relay channel.

relay and the destination nodes and the relay with the destination node. The AWGN relay channel is defined to be (Figure 2.16)

$$y = h_{sd}x + h_{rd}x_1 + n_1 \tag{2.26}$$

$$y_1 = h_{sr}x + n_2, \tag{2.27}$$

where x is the signal transmitted by the source, x_1 is the signal transmitted by the relay (which is a function of the relay's previous inputs), y is the received signal at the destination, and y_1 is the received signal at the relay. The capacity of this channel is still an open problem, with the capacity known for some special cases. One such special case is the physically degraded relay channel [24], i.e. where the destination signal is a physically degraded version of the received signal at the relay. In general, only upper and lower bounds to capacity are known. The best upper bound known is the well-known cut-set upper bound, while the lower bounds are obtained using a node-cooperation strategy known as block-Markov coding [24]. In block-Markov coding, the relay uses the information received by it in the *past* to correlate its transmit signal to the transmitter's *current* signal, and the net coherent combining provides capacity enhancements over that of the point-to-point channel. This lower bound makes the assumption, however, that the relay completely decodes its received signal, and is thus called the decode-and-forward policy. In general, a relay channel's transmission schemes can be grouped into three primary categories [76].

1. *Decode-and-forward.* In the decode-and-forward approach, the relay decodes part or all of the codeword transmitted by the source during a block. Clearly, this transmission should be at a rate higher than the capacity of the direct link from source to destination. During the next block, the relay then transmits to the destination a re-encoded version of the message decoded during the previous block to assist the destination's decoding process. Since the source is aware of the relay's strategy and of the codeword it sent the previous block, the source and relay can *cooperate*, e.g. the source can transmit a signal identical to the relay's transmission (coherent combination). Note that if the relay and source cooperate, the source's transmission in each block consists of two parts: information regarding the previous block's message (to facilitate cooperation) and a new codeword. This class of strategies was first proposed in [24].
2. *Compress-and-forward.* In compress and forward, the relay transmits a compressed, or quantized version of its received signal, instead of decoding the received signal and re-encoding. This technique is related to source coding with side information.

3. *Amplify-and-forward.* In this strategy, the relay acts as a simple linear repeater, and simply amplifies its received signal instantaneously or during the next symbol or block. This clearly leads to some noise amplification, but even this simple strategy can yield significant performance improvements.

One of the best lower bounds known for relays using a mix of these strategies is given by [24, Theorem 7]. Unfortunately, none of the lower bounds proves to be tight even for the AWGN single-antenna relay channel, which is thus an open problem. However, the analysis techniques used for this simple relay channel have been extended to study more complex ad hoc networks using cooperative schemes, as explained in a later section.

2.6.2 The interference channel

In this section, we summarize what is known about the next building block: the interference channel (IFC). This is a system with two independent transmitters sending information to two receivers (Figure 2.17), where each transmitter has a message for a distinct receiver. The signals from the two transmitters interfere with one another and the resulting mixture is received at each receiver. For example, in the Gaussian IFC case, a weighted sum of the two transmit signals is assumed to arrive at each receiver in the presence of additive Gaussian noise. In the general discrete memoryless case, the interaction between the two signals is expressed in terms of the conditional probability density function $p(y_1, y_2|x_1, x_2)$. Although the capacity region of both the general case and the particular case of the Gaussian interference channel are still open problems, significant progress has been made in finding the capacity region of certain classes of these channels [23, 126].

A category of interference channels where the interference is "much stronger" than the signal, called the very-strong IFC case, was amongst the first for which capacity was determined [15]. In this case, the interference signal at each receiver is assumed to be so strong that it can be decoded out first before decoding the intended signal. The broadest class of IFCs for which the capacity is known is the strong interference channel case [23, 104], which includes the very-strong case as a sub-class. In the strong IFC case, the interference is assumed to be strong enough so that it can be decoded (not necessarily first) at each receiver. More formally, if α_i is the channel gain from transmitter i to receiver i, and β_i is the channel gain from transmitter i to the other receiver, then the strong interference condition for this single-antenna case is equivalent to $\beta_i \geq \alpha_i$ for all i.

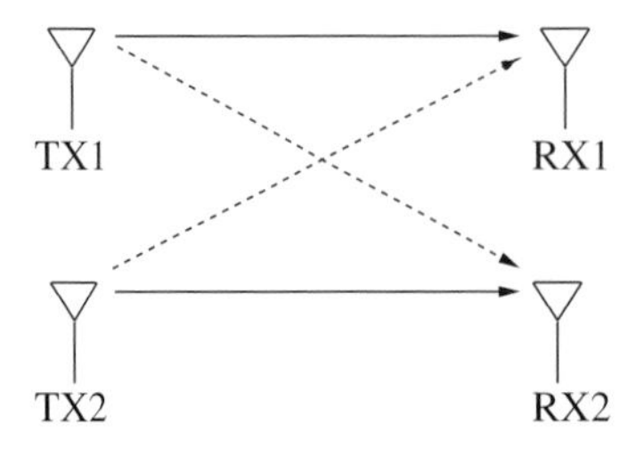

Fig. 2.17. The interference channel.

Research on the capacity limits of MIMO IFCs is still in its early stages. Interestingly, the simple strong interference result given by $\beta_i \geq \alpha_i$ for the SISO case does not generalize to the MIMO case, and a straightforward characterization of what strong interference means for MIMO IFCs is yet to be obtained. In [131] strong interference conditions for SIMO interference channels were obtained, where the transmitters possess single antennas. Finding explicit strong interference conditions for the MISO and more general MIMO IFCs is still an open problem.

A special feature of MIMO IFCs is the spatial degree of freedom that is absent in the SISO case, making the channel capacity problem much more difficult and challenging. Known results from SISO IFC analysis such as achievable regions and outer bounds require considerable reworking before they can be replicated in the MIMO domain, and in many cases do not generalize to the MIMO case at all. Summarizing, the capacity of MIMO IFCs is one of the areas in MIMO multi-user capacity analysis where very few results are known.

2.6.3 *Cooperative communication*

In this section we summarize the work on cooperative communication, i.e. using MIMO techniques in ad hoc networks. In addition to using such techniques when a node in the network possesses multiple antennas, clusters of nodes that are located close together can exchange information to create a *virtual* antenna array, leading to a distributed MIMO system [95]. In other words, nodes close together on the transmit side can exchange information to form a multiple-antenna transmitter, and nodes near each other on the receiver side can exchange information to form a multiple-antenna receiver, as shown in Figure 2.18. Since each node has a different channel to each receiver, this cooperative MIMO system has performance advantages in terms of multiplexing and diversity. In addition, the multiple antennas can be used on the transmit or receiver side to steer the beam in the direction of the intended receiver, thereby reducing interference and multi-path.

In this section we describe extensions of the basic relay channel to the case of cooperative (i.e. MIMO) communication over ad hoc networks. We consider two different settings: perfect CSIR and CSIT, and perfect CSIR and CDIT under the ZMSW model.

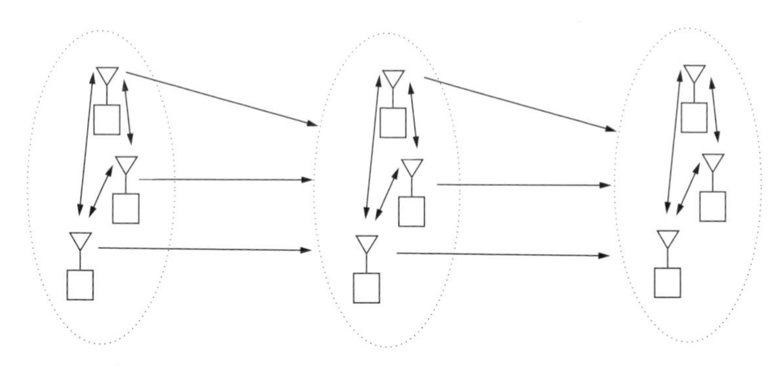

Fig. 2.18. Cooperative MIMO.

Perfect CSIR and CSIT

When the channels between the sources, relays, and destinations are fading, but are perfectly known, the different fading realizations can be treated as constant channels, thus reducing the problem to the analysis of non-fading relay channels. The capacity of MIMO relays with no fading is still an open problem. Early work in this area can be found in [140]. In this paper, the authors derive upper and lower bounds on the capacity of these channels. The upper bounds are derived based on cut-set bounds [25], while the lower bounds are based on path-diversity. The authors use convex optimization techniques to generate specialized algorithms to compute the lower and upper bounds.

Relaying has also been considered in a multiple-transmitter, multiple-receiver scenario (i.e. similar to an interference channel), where transmitting nodes help other nodes by sending their own messages as well as cooperating with other nodes. In addition, the receiving nodes have the ability to cooperate with each other as well through techniques such as amplify-and-forward. Such scenarios are beginning to be analyzed in greater detail [53, 69, 98].

Perfect CSIR and CDIT

Most work on cooperative MIMO has concentrated on the perfect CSIR and CDIT setting under the ZMSW model. This is perhaps the most practical setting for cooperation in ad hoc or cellular networks, as obtaining full CSI at each of the transmitters would require feedback from every receiver to every transmitter. Cooperation of this nature has been studied in both cellular [105–107] and ad hoc [78, 79] settings.

In a cellular uplink setting, multiple transmitting nodes (i.e. mobiles) wish to communicate to a single base-station. In traditional networks, mobiles communicate directly to the base-station over a common channel, perhaps separated by a multi-user technique such as CDMA. Notice, however, that each of the mobiles can hear the transmission of other mobiles. Therefore, a mobile can relay the messages of all other mobiles, in addition to transmitting its own message, to the base-station. The key advantage of doing so is the extra form of diversity obtained, termed *user cooperation diversity*, via the channels of other mobile devices. Even if a mobile is severely faded, with high probability one of its neighboring mobiles will not be faded and will be able to relay its message to the base-station. Thus, a cooperation scheme, such as block-Markov coding [106] increases rates relative to no cooperation. In practice, CDMA can now be combined with cooperative coding to decrease the probability of error.

Using a somewhat different problem formulation based on outage capacity, [78, 79] find cooperative communication schemes for half-duplex relays, i.e. for relays that cannot transmit and receive at the same time. These cooperative schemes achieve full-diversity, i.e. diversity order equal to the total number of transmitting nodes, but at some rate penalty. Similar to the point-to-point MIMO setting, there has been work on formalizing the connection between the diversity and multiplexing (i.e. rate) gain that cooperative MIMO can provide [3, 101].

Though most work on cooperative communication has concentrated on the outage formulation, there has also been some work on ergodic capacity of cooperative channels.

In particular, [76] has established the ergodic capacity of a class of cooperative MIMO channels. Their results prove that the decode-and-forward strategy is capacity-achieving under the ZMSW model with perfect CSIR and CDIT when the relay is sufficiently close to the source, i.e. the average SNR from the source to relay is much higher than the SNR from the source to the destination or the relay to the destination. In fact, the capacity is equal to the capacity of the channel when the source and relay nodes are assumed to fully cooperate, i.e. to act as a transmit antenna array. Note that this result is surprising because the source and relay are in fact not co-located, and can only cooperate over a noisy channel.

In this situation, the lack of CSIT, and more specifically the lack of phase information, makes it optimal for the relay to decode the entire message from the source, and then re-encode it and transmit it independently of the source transmission. In other words, the source transmits Gaussian codewords at rate R, which is equal to the rate assuming full cooperation between the source and relay. The relay decodes the codeword transmitted by the source (the condition on the relay being close to the source implies that the relay–source capacity is large enough to allow this). In the next block, the relay re-encodes the message decoded in the previous block, while the source transmits an entirely new message. The destination uses the information from the current and previous block to then decode the message sent by the source in the previous block, and so forth. This result can also be generalized to a channel where there are multiple relays.

Diversity–multiplexing trade-offs

Researchers have investigated the diversity and multiplexing gains in distributed MIMO (also known as virtual MIMO) systems in [54, 56, 153]. Specifically, these works studied whether distributed MIMO obtained by wireless relaying can mimic a multi-antenna system in terms of its diversity–multiplexing trade-off. It can be easily seen by applying the min-cut max-flow bound that the spatial multiplexing gain of a distributed MIMO system is limited by the number of antennas at the initial source and the final destination regardless of the number of intermediate relays [56, 153]. Thus, for example, if the source or the destination has only one antenna, the multiplexing gain is limited to one. This is in contrast with the fact that full-diversity gain can be obtained through relays in such systems [78]. Note that with multiple sources, relays and destinations, and half-duplex operation it has been shown that the full multiplexing gain (subject to a factor of one-half due to the half-duplex operation) can be obtained in a distributed MIMO system with single antennas at all nodes [9]. Thus, the nature of the diversity and multiplexing trade-off is quite different from that in true MIMO systems [54].

2.7 Summary

We have summarized results on the capacity of MIMO channels for both single- and multi-user systems. The great capacity gains predicted for such systems can be realized in some cases, but realistic assumptions about channel knowledge and the underlying channel model can significantly mitigate these gains. For single-user systems the capacity under perfect CSI at the transmitter and receiver is relatively straightforward and predicts that capacity grows linearly with the number of antennas. Backing off from the perfect CSI assumption makes the capacity calculation much more difficult, and the capacity gains are highly dependent on the nature of the CSI/CDI, the channel SNR, and the antenna element correlations. Specifically, assuming perfect CSIR, CSIT provides significant capacity gain at low SNRs but not much at high SNRs. The insight here is that at low SNRs it is important to put power into the appropriate eigenmodes of the system. Interestingly, with perfect CSIR and CSIT, antenna correlations are found to increase capacity at low SNRs and decrease capacity at high SNRs. Finally, under CDIT and CDIR for a zero-mean spatially white channel, at high SNRs the capacity grows relative to only the double-log of the SNR with the number of antennas as a constant additive term. This rather poor capacity gain would not typically justify adding more antennas. However, at moderate SNRs the growth relative to the number of antennas is less pessimistic.

We also examined the capacity of MIMO broadcast and multiple-access channels. The capacity region of the MIMO MAC is well known and can be characterized as a convex optimization problem. The MAC–BC duality greatly simplifies the capacity region calculations for the MIMO BC that could otherwise lead to non-convex optimization problems. These capacity and achievable regions are only known for ergodic capacity under perfect CSIT and CSIR. Relatively little is known about the MIMO MAC and BC regions under more realistic CSI assumptions. A multi-cell system with base-station cooperation can be modeled as a MIMO BC (downlink) or MIMO MAC (uplink) where the antennas associated with each base-station are pooled by the system. Exploiting this antenna structure leads to significant capacity gains over HDR transmission strategies. We also describe the rather limited results on the capacity of ad hoc networks with multiple antennas, as well as the capacity gain associated with node cooperation to form virtual MIMO channels from single-antenna nodes.

There are many open problems in this area. For single-user systems the problems are mainly associated with CDI only at either the transmitter or the receiver. Most capacity regions associated with multi-user MIMO channels remain unsolved, especially the ergodic capacity and the capacity versus outage for the MIMO BC under perfect receiver CSI only. There are very few existing results for CDI at either the transmitter or the receiver for any multi-user MIMO channel. The capacity of cellular systems with multiple antennas remains a relatively open area, in part because the single-cell problem is mostly unsolved, and in part because the Shannon capacity of a cellular system is not well defined and depends heavily on frequency assumptions and propagation models. Other fundamental trade-offs in MIMO cellular designs such as whether antennas should be used for sectorization, capacity gain, or diversity are not well understood. Similar trade-offs exist in ad hoc networks, where the capacity associated with multiple antennas is a wide open problem. In short, we have only scratched the surface in understanding the fundamental capacity limits of systems with multiple transmitter and receiver antennas as well as the implications of these limits for practical system designs. This area of research is likely to remain timely, important, and fruitful for many years to come.

This chapter has provided the fundamental capacity limits of MIMO channels as well as design insights into how these limits may be approached in practice. The remainder of the book is devoted to exploring practical techniques that exploit the benefits of MIMO to achieve large capacity gains as well as robustness through diversity. In particular, Chapter 3 describes how transmitter precoding improves both the capacity and the error rate performance, and also describes practical techniques for obtaining the transmitter CSI necessary for precoding. Chapter 4 provides an overview of practical space–time coding and decoding techniques that can approach the Shannon theoretic capacity bounds of MIMO channels. Optimal detection of MIMO transmission is typically prohibitively complex, as it requires joint maximum-likelihood detection across all receive antennas. Chapter 5 addresses this issue by outlining practical receiver techniques that approach the performance of maximum-likelihood detection with much lower complexity. Chapter 6 extends these ideas to multi-user receivers, where signals of all users in the system must be detected simultaneously.

2.8 Bibliographical notes

Research on the capacity limits of MIMO channel was sparked by the initial works of Foschini and Gans [33] and Telatar [121] in the mid 1990's. Since then, there has been a tremendous amount of research on MIMO capacity, both for single-user and multi-user channels. For additional results on MIMO capacity, readers may wish to consult MIMO tutorial articles [28, 38, 39]. The well known information theory books by Gallager [34] and Cover and Thomas [25] are excellent references for additional material on channel capacity. In addition, [4] provides an excellent survey of channel capacity results for general single antenna channels. The monograph by Tulino and V́erdu on applications of random matrix theory to wireless communication [123] provides an excellent summary of general concepts of random matrix theory and their use in the study of the capacity of MIMO systems with an asymptotically large number of antennas. Readers interested in the capacity of MIMO channels in either the wideband (i.e. low SNR) regime or at high SNR should consider papers by Tulino, Lozano, and Verdú [88, 89]. For readers interested in multi-user capacity, Ch. 14 of [25] serves as an excellent starting point, as does the survey paper of El Gamal and Cover [35].

References

[1] D. Aftas, M. Bacha, J. Evans, and S. Hanly, "On the sum capacity of multi-user MIMO channels," in *Proceedings of Intl. Symp. on Inform. Theory and its Applications*, pp. 1013–1018, Oct. 2004.

[2] R. Ahlswede, "Multi-way communication channels," in *Proceedings of Intl. Symp. Inform. Theory*, pp. 23–52, 1973.

[3] K. Azarian, H. E. Gamal, and P. Schniter, "On the achievable diversity–multiplexing tradeoff in half-duplex cooperative channels," *IEEE Trans. Inform. Theory*, vol. 51, no. 12, pp. 4152–4172, Dec. 2005.

[4] E. Biglieri, J. Proakis, and S. Shamai (Shitz), "Fading channels: information theoretic and communication aspects," *IEEE Trans. Inform. Theory*, vol. 44, no. 6, pp. 2619–2692, Oct. 1998.

[5] D. Blackwell, L. Breiman, and A. J. Thomasian, "The capacity of a class of channels," *Ann. Math. Stat.*, vol. 30, pp. 1229–1241, Dec. 1959.

[6] H. Boche and E. Jorswieck, "Outage probability of multiple antenna systems: optimal transmission and impact of correlation," *Proceedings of Intl. Zurich Seminar Commun.*, 2004, pp. 116–119.

[7] H. Boche and M. Schubert, "A general duality theory for uplink and downlink beamforming," in *Proceedings of IEEE Vehicular Tech. Conf.*, vol. 1, pp. 87–91, 2002.

[8] H. Bölcskei, D. Gesbert, and A. J. Paulraj, "On the capacity of OFDM-based spatial multiplexing systems," *IEEE Trans. Commun.*, vol. 50, no. 2, pp. 225–234, Feb. 2002.

[9] H. Bölcskei, R. Nabar, O. Oyman, and A. Paulraj, "Capacity scaling laws in MIMO relay networks," *IEEE Trans. Wireless Commun.*, vol. 5, no. 6, pp. 1433–1444, June 2006.

[10] S. Borst and P. Whiting, "The use of diversity antennas in high-speed wireless systems: capacity gains, fairness issues, multi-user scheduling," *Bell Labs. Tech. Mem.*, 2001, download available at http://mars.bell-labs.com.

[11] S. Boyd and L. Vandenberghe, *Convex Optimization*. Cambridge University Press, 2003.

[12] G. Caire and S. Shamai, "On the capacity of some channels with channel state information," *IEEE Trans. Inform. Theory*, vol. 45, no. 6, pp. 2007–2019, Sept. 1999.

[13] G. Caire and S. Shamai, "On the achievable throughput of a multiantenna Gaussian broadcast channel," *IEEE Trans. Inform. Theory*, vol. 49, no. 7, pp. 1691–1706, July 2003.

[14] G. Caire, G. Taricco, and E. Biglieri, "Optimum power control over fading channels," *IEEE Trans. Inform. Theory*, vol. 45, no. 5, pp. 1468–1489, July 1999.

[15] A. B. Carliel, "A case where interference does not reduce capacity," *IEEE Trans. Inform. Theory*, vol. 21, no. 5, pp. 569–570, Sept. 1975.

[16] S. Catreux, P. Driessen, and L. Greenstein, "Simulation results for an interference-limited multiple-input multiple-output cellular system," *IEEE Commun. Letters*, vol. 4, no. 11, pp. 334–336, Nov. 2000.

[17] S. Catreux, P. Driessen, and L. Greenstein, "Attainable throughput of an interference-limited multiple-input multiple-output (MIMO) cellular system," *IEEE Trans. Commun.*, vol. 49, no. 8, pp. 1307–1311, Aug. 2001.

[18] M. Chiani, M. Z. Win, and Z. Zanella, "On the capacity of spatially correlated MIMO Rayleigh-fading channels," *IEEE Trans. Inform. Theory*, vol. 49, no. 10, pp. 2363–2371, Oct. 2003.

[19] D. Chizhik, J. Ling, P. Wolniansky, R. Valenzuela, N. Costa, and K. Huber, "Multiple input multiple output measurements and modeling in Manhattan," in *Proceedings of IEEE Vehicular Tech. Conf.*, 2002, pp. 107–110.

[20] C. Chuah, D. Tse, J. Kahn, and R. Valenzuela, "Capacity scaling in MIMO wireless systems under correlated fading," *IEEE Trans. Inform. Theory*, vol. 48, no. 3, pp. 637–650, March 2002.

[21] M. Costa, "Writing on dirty paper," *IEEE Trans. Inform. Theory*, vol. 29, no. 3, pp. 439–441, May 1983.

[22] M. Costa and A. El Gamal, "The capacity region of the discrete memoryless interference channel with strong interference," *IEEE Trans. Inform. Theory*, vol. 33, no. 5, pp. 710–711, Sept. 1987.

[23] M. H. M. Costa and A. A. E. Gamal, "The capacity region of the discrete memoryless interference channel with strong interference," *IEEE Trans. Inform. Theory*, vol. 33, no. 5, pp. 710–711, Sept. 1987.

[24] T. Cover and A. E. Gamal, "Capacity theorems for the relay channel," *IEEE Trans. Inform. Theory*, vol. 25, no. 5, pp. 572–584, Sept. 1979.

[25] T. M. Cover and J. A. Thomas, *Elements of Information Theory*, 2nd edn. Wiley.

[26] I. Csiszár, "Arbitrarily varying channels with general alphabets and states," *IEEE Trans. Inform. Theory*, vol. 38, no. 6, pp. 1725–1742, Nov. 1992.

[27] I. Csiszár and J. Körner, *Information Theory: Coding Theorems for Discrete Memoryless Systems*. Academic Press, 1997.

[28] S. Diggavi, N. Al-Dahir, A. Stamoulis, and R. Calderbank, "Great expectations: the value of spatial diversity in wireless networks," *Proceedings of the IEEE*, vol. 92, no. 2, pp. 219–270, Feb. 2004.

[29] S. Diggavi, M. Grossglauser, and D. Tse, "Even one-dimensional mobility increases ad hoc wireless capacity," in *Proceedings of Intl. Symp. Inform. Theory*, June 2002, p. 352.

[30] U. Erez, S. Shamai, and R. Zamir, "Capacity and lattice strategies for cancelling known interference," in *Proceedings of Intl. Symp. Inform. Theory and its Applications*, Nov. 2000, pp. 681–684.

[31] U. Erez and S. ten Brink, "Approaching the dirty paper limit for cancelling known interference," in *Proceedings of 41st Annual Allerton Conf. on Commun., Control and Computing*, Oct. 2003, pp. 799–808.

[32] G. J. Foschini, "Layered space-time architecture for wireless communication in fading environments when using multi-element antennas," *Bell Labs Techn. J.*, pp. 41–59, Autumn 1996.

[33] G. J. Foschini and M. J. Gans, "On limits of wireless communications in a fading environment when using multiple antennas," *Wireless Personal Commun.*, vol. 6, pp. 311–335, 1998.

[34] R. G. Gallager, "An inequality on the capacity region of multiaccess fading channels," *Communication and Cryptography – Two Sides of One Tapestry*. Kluwer, 1994, pp. 129–139.

[35] A. El Gamal and T. Cover, "Multiple user information theory," *Proceedings of the IEEE*, vol. 68, no. 12, pp. 1466–1483, Dec. 1980.

[36] A. E. Gamal, J. Mammen, B. Prabhakar, and D. Shah, "Throughput-delay trade-off in energy constrained wireless networks," in *Proceedings of Intl. Symp. Inform. Theory*, July 2004, p. 439.

[37] M. J. Gans, N. Amitay, Y. S. Yeh, H. Xu, T. Damen, R. A. Valenzuela, T. Sizer, R. Storz, D. Taylor, W. M. MacDonald, C. Tran, and A. Adamiecki, "Outdoor

BLAST measurement system at 2.44 GHz: calibration and initial results," *IEEE J. Select. Areas Commun.*, vol. 20, no. 3, pp. 570–581, April 2002.

[38] D. Gesbert, M. Shafi, D. Shiu, P. J. Smith, and A. Naguib, "From theory to practice: an overview of MIMO space-time coded wireless systems," *IEEE J. Select Areas Commun.*, vol. 21, no. 3, pp. 281–302, June 2003.

[39] A. Goldsmith, S. Jafar, N. Jindal, and S. Vishwanath, "Capacity limits of MIMO channels," *IEEE J. Select Areas Commun.*, vol. 21, no. 3, pp. 684–702, June 2003.

[40] A. Goldsmith and P. Varaiya, "Capacity of fading channels with channel side information," *IEEE Trans. Inform. Theory*, vol. 43, no. 6, pp. 1986–1992, Nov. 1997.

[41] A. J. Goldsmith, *Wireless Communications*. Cambridge University Press, 2005.

[42] P. K. Gopala and H. El Gamal, "On the throughput–delay tradeoff in cellular multicast," *Proc. IEEE Intl. Conf. on Commun. (ICC)*, vol. 2, June 2005, pp. 1401–1406.

[43] M. Grossglauser and D. Tse, "Mobility increases the capacity of ad hoc wireless networks," *IEEE/ACM Trans. Networking*, vol. 10, no. 4, pp. 477–486, Aug. 2002.

[44] T. Guess and M. K. Varanasi, "Multi-user decision-feedback receivers for the general Gaussian multiple-access channel," in *Proceedings of 34th Allerton Conf. on Commun., Control, and Computing*, Oct 1996, pp. 190–199.

[45] P. Gupta and P. R. Kumar, "The capacity of wireless networks," *IEEE Trans. Inform. Theory*, vol. 46, no. 2, pp. 388–404, Mar. 2000.

[46] S. Hanly and D. Tse, "Multiaccess fading channels – part II: Delay-limited capacities," *IEEE Trans. Inform. Theory*, vol. 44, no. 7, pp. 2816–2831, Nov. 1998.

[47] S. V. Hanly and P. Whiting, "Information theory and the design of multi-receiver networks," in *Proceedings of IEEE 2nd Intl. Symp. on Spread-spectrum Technical Applications (ISSTA)*, Nov. 1992, pp. 103–106.

[48] B. Hassibi and B. Hochwald, "How much training is needed in multiple-antenna wireless links?" *IEEE Trans. Inform. Theory*, vol. 49, no. 4, pp. 951–963, April 2003.

[49] B. Hochwald, T. L. Marzetta, and V. Tarokh, "Multi-antenna channel-hardening and its implications for rate feedback and scheduling," *IEEE Trans. Inform. Theory*, vol. 50, no. 9, pp. 1893–1909, Sept. 2004.

[50] B. Hochwald and S. Vishwanath, "Space–time multiple access: linear growth in sum rate," in *Proceedings of 40th Annual Allerton Conf. on Commun., Control and Computing*, Oct. 2002, pp. 387–396.

[51] D. Hoesli, Y.-H. Kim, and A. Lapidoth, "Monotonicity results for coherent MIMO Rician channels," *IEEE Trans. Inform. Theory*, vol. 51, no. 12, pp. 4334–4339, Dec. 2005.

[52] D. Hoesli and A. Lapidoth, "The capacity of a MIMO Ricean channel is monotonic in the singular values of the mean," in *Proceedings of the 5th International ITG Conference on Source and Channel Coding (SCC), Erlangen, Nuremberg*, January 2004, pp. 287–292.

[53] A. Host-Madsen, "On the achievable rate for receiver cooperation in ad-hoc networks," in *Proceedings of IEEE Intl. Symp. Inform. Theory*, June 2004, p. 272.

[54] A. Host-Madsen and Z. Yang, "Interference and cooperation in multi-source wireless networks," in *IEEE Communication Theory Workshop*, May 2005.

[55] H. Huang and S. Venkatesan, "Asymptotic downlink capacity of coordinated cellular networks," in *Proceedings of Asilomar Conf. on Signals, Systems, & Computing*, Nov. 2004, pp. 850–855.

[56] S. Jafar, "Degrees of freedom in distributed MIMO communications," in *IEEE Communication Theory Workshop*, May 2005.

[57] S. Jafar and A. Goldsmith, "Transmitter optimization and optimality of beamforming for multiple antenna systems with imperfect feedback," *IEEE Trans. Wireless Commun.*, vol. 3, no. 4, pp. 1165–1175, July 2004.

[58] S. Jafar and A. Goldsmith, "Isotropic fading vector broadcast channels: the scalar upperbound and loss in degrees of freedom," *IEEE Trans. Inform. Theory*, vol. 51, no. 3, pp. 848–857, March 2005.

[59] S. Jafar and A. Goldsmith, "Multiple-antenna capacity in correlated Rayleigh fading with channel covariance information," *IEEE Trans. Wireless Commun.*, vol. 4, no. 3, pp. 990–997, May 2005.

[60] S. A. Jafar and A. Goldsmith, "Transmitter optimization for multiple antenna cellular systems," in *Proceedings of Intl. Symp. on Information Theory*, June 2002, p. 50.

[61] S. A. Jafar and A. J. Goldsmith, "On optimality of beamforming for multiple antenna systems with imperfect feedback," in *Proceedings of Intl. Symp. on Information Theory*, June 2001, p. 321.

[62] S. A. Jafar and A. J. Goldsmith, "Vector MAC capacity region with covariance feedback," in *Proceedings of Intl. Symp. on Information Theory*, June 2001, p. 321.

[63] S. A. Jafar, S. Vishwanath, and A. J. Goldsmith, "Channel capacity and beamforming for multiple transmit and receive antennas with covariance feedback," in *Proceedings of Intl. Conf. Commun.*, vol. 7, pp. 2266–2270, 2001.

[64] S. Jayaweera and H. V. Poor, "Capacity of multiple-antenna systems with both receiver and transmitter channel state information," *IEEE Trans. Inform. Theory*, vol. 49, no. 10, pp. 2697–2709, Oct. 2003.

[65] N. Jindal, "High SNR analysis of MIMO broadcast channels," in *Proceedings of Intl. Symp. on Information Theory*, Sept. 2005, pp. 2310–2314.

[66] N. Jindal, "MIMO broadcast channels with finite rate feedback," in *Proceedings of IEEE Globecom*, Nov. 2005, pp. 1520–1524.

[67] N. Jindal and A. Goldsmith, "Capacity and dirty paper coding for Gaussian broadcast channels with common information," in *Proceedings of Intl. Symp. on Inform. Theory*, July 2004, p. 215.

[68] N. Jindal and A. Goldsmith, "Dirty paper coding vs. TDMA for MIMO broadcast channels," *IEEE Trans. Inform. Theory*, vol. 51, no. 5, pp. 1783–1794, May 2005.

[69] N. Jindal, U. Mitra, and A. Goldsmith, "Capacity of ad-hoc networks with node cooperation," in *Proceedings of Intl. Symp. on Inform. Theory*, July 2004, p. 271.

[70] N. Jindal, W. Rhee, S. Vishwanath, S. Jafar, and A. Goldsmith, "Sum power iterative water-filling for multi-antenna Gaussian broadcast channels," *IEEE Trans. Inform. Theory*, vol. 51, no. 4, pp. 1570–1580, April 2005.

[71] N. Jindal, S. Vishwanath, and A. Goldsmith, "On the duality of Gaussian multiple-access and broadcast channels," *IEEE Trans. Inform. Theory*, vol. 50, no. 5, pp. 768–783, May 2004.

[72] E. Jorswieck and H. Boche, "Optimal transmission with imperfect channel state information at the transmit antenna array," *Wireless Personal Commun.*, vol. 27, pp. 33–56, 2003.

[73] E. Jorswieck and H. Boche, "Channel capacity and capacity-range of beamforming in MIMO wireless systems under correlated fading with covariance feedback," *IEEE Trans. Wireless Commun.*, vol. 3, no. 5, pp. 1543–1553, Sept. 2004.

[74] E. Jorswieck and H. Boche, "Optimal transmission strategies and impact of correlation in multi-antenna systems with different types of channel state information," *IEEE Trans. Signal Process.*, vol. 52, no. 12, pp. 3440–3453, December 2004.

[75] A. Khisti, U. Erez, and G. Wornell, "A capacity theorem for co-operative multicasting in large wireless networks," in *Proceedings of 42nd Allerton Conf. on Commun., Control, and Computing*, Oct 2004, pp. 522–531.

[76] G. Kramer, M. Gastpar, and P. Gupta, "Cooperative strategies and capacity theorems for relay networks," *IEEE Trans. Inform. Theory*, vol. 51, no. 9, pp. 3037–3063, Sept. 2005.

[77] T. Lan and W. Yu, "Input optimization for multi-antenna broadcast channels and per-antenna power constraints," in *Proceedings of IEEE Globecom*, 2004, pp. 420–424.

[78] N. Laneman, D. N. C. Tse, and G. W. Wornell, "Cooperative diversity in wireless networks: efficient protocols and outage behavior," *IEEE Trans. Inform. Theory*, vol. 50, no. 12, pp. 3062–3080, Dec. 2004.

[79] N. Laneman and G. W. Wornell, "Distributed space–time-coded protocols for exploiting cooperative diversity in wireless networks," *IEEE Trans. Inform. Theory*, vol. 49, no. 10, pp. 2415–2425, Oct. 2003.

[80] A. Lapidoth and S. M. Moser, "Capacity bounds via duality with applications to multi-antenna systems on flat-fading channels," *IEEE Trans. Inform. Theory*, vol. 49, no. 10, pp. 2426–2467, Oct. 2003.

[81] A. Lapidoth, S. Shamai, and M. Wigger, "On the capacity of fading MIMO broadcast channels with imperfect transmitter side-information," in *Proceedings of 43rd Annual Allerton Conf. on Commun., Control and Computing*, Sept. 2005.

[82] L. Li and A. Goldsmith, "Capacity and optimal resource allocation for fading broadcast channels – part I: Ergodic capacity," *IEEE Trans. Inform. Theory*, vol. 47, no. 3, pp. 1083–1102, March 2001.

[83] L. Li, N. Jindal, and A. Goldsmith, "Outage capacities and optimal power allocation for fading multiple-access channels," *IEEE Trans. Inform. Theory*, vol. 51, no. 4, pp. 1326–1347, April 2005.

[84] H. Liao, "Multiple access channels," Ph.D. dissertation, Dept. of Electrical Engineering, University of Hawaii, 1972.

[85] D. J. Love, R. W. Heath Jr., and T. Strohmer, "Grassmannian beamforming for multiple-input multiple-output wireless systems," *IEEE Trans. Inform. Theory*, vol. 49, no. 10, pp. 2735–2747, Oct 2003.

[86] A. Lozano and A. Tulino, "Capacity of multiple-transmit multiple-receive antenna architectures," *IEEE Trans. Inform. Theory*, vol. 48, no. 12, pp. 3117–3128, Dec. 2002.

[87] A. Lozano, A. Tulino, and S. Verdú, "Multiantenna capacity: myths and realities," in *Space–Time Wireless Systems: From Array Processing to MIMO Communications*, ed. H. Bölcskei, D. Gesbert, C. Papadias, and A. J. van der Veen, 2005.

[88] A. Lozano, A. Tulino, and S. Verdú, "Multiple-antenna capacity in the low-power regime," *IEEE Trans. Inform. Theory*, vol. 49, no. 10, pp. 2527–2544, Oct. 2003.

[89] A. Lozano, A. Tulino, and S. Verdú, "High-SNR power offset in multi-antenna communication," *IEEE Trans. Inform. Theory*, vol. 51, no. 12, pp. 4134–4151, Dec. 2005.

[90] T. Marzetta and B. Hochwald, "Capacity of a mobile multiple-antenna communication link in Rayleigh flat-fading," *IEEE Trans. Inform. Theory*, vol. 45, no. 1, pp. 139–157, Jan 1999.

[91] T. Marzetta and B. Hochwald, "Unitary space–time modulation for multiple-antenna communications in Rayleigh flat-fading," *IEEE Trans. Inform. Theory*, vol. 46, no. 2, pp. 543–564, March 2000.

[92] A. Molisch, M. Stienbauer, M. Toeltsch, E. Bonek, and R. S. Thoma, "Capacity of MIMO systems based on measured wireless channels," *IEEE J. Select. Areas Commun.*, vol. 20, no. 3, pp. 561–569, April 2002.

[93] A. Moustakas and S. Simon, "Optimizing multiple-input single-output (MISO) communication systems with general Gaussian channels: nontrivial covariance and nonzero-mean," *IEEE Trans. Inform. Theory*, vol. 49, no. 10, pp. 2770–2780, Oct. 2003.

[94] K. Mukkavilli, A. Sabharwal, E. Erkip, and B. Aazhang, "On beamforming with finite rate feedback in multiple antenna systems," *IEEE Trans. Inform. Theory*, vol. 49, no. 10, pp. 2562–2579, Oct. 2003.

[95] R. U. Nabar, H. Bölcskei, and F. W. Kneubühler, "Fading relay channels: performance limits and space–time signal design," *IEEE J. Select. Areas Commun.*, vol. 22, no. 6, pp. 1099–1109, Aug. 2004.

[96] A. Narula, M. Trott, and G. Wornel, "Performance limits of coded diversity methods for transmitter antenna arrays," *IEEE Trans. Inform. Theory*, vol. 45, no. 7, pp. 2418–2433, Nov. 1999.

[97] A. Narula, M. J. Lopez, M. D. Trott, and G. W. Wornell, "Efficient use of side information in multiple antenna data transmission over fading channels," *IEEE J. Select. Areas Commun.*, vol. 16, no. 8, pp.1423–1436, Oct. 1998.

[98] C. Ng and A. Goldsmith, "Transmitter cooperation in ad-hoc wireless networks: does dirty-paper coding beat relaying?" in *Proceedings of IEEE Inform. Theory Workshop*, Oct. 2004, pp. 277–282.

[99] C. Peel, B. Hochwald, and L. Swindlehurst, "A vector-perturbation technique for near-capacity multiantenna multi-user communication – Part I: Channel inversion and regularization," *IEEE Trans. Commun.*, vol. 53, no. 1, pp. 195–202, Jan. 2005.

[100] C. Peel, B. Hochwald, and L. Swindlehurst, "A vector-perturbation technique for near-capacity multiantenna multi-user communication – Part II: Perturbation," *IEEE Trans. Commun.*, vol. 53, no. 3, pp. 537–544, March 2005.

[101] N. Prasad and M. K. Varanasi, "Diversity and multiplexing tradeoff bounds for cooperative diversity schemes," in *Proceedings of IEEE Intl. Symp. Inform. Theory*, June 2004, p. 268.

[102] G. Raleigh and J. M. Cioffi, "Spatio-temporal coding for wireless communication," *IEEE Trans. Commun.*, vol. 46, no. 3, pp. 357–366, March 1998.

[103] F. Rashid-Farrokhi, K. R. Liu, and L. Tassiulas, "Transit beamforming and power control for cellular wireless systems," *IEEE J. Select. Areas Commun.*, vol. 16, no. 8, pp. 1437–1450, Oct. 1998.

[104] H. Sato, "The capacity of Gaussian interference channel under strong interference (corresp.)," *IEEE Trans. Inform. Theory*, pp. 786–788, Nov. 1981.

[105] A. Sendonaris, E. Erkip, and B. Aazhang, "Increasing uplink capacity via user cooperation diversity," in *Proceedings of Intl. Symp. Inform. Theory*, Aug. 1994, p. 156.

[106] A. Sendonaris, E. Erkip, and B. Aazhang, "User cooperation diversity – part I: System description," *IEEE Trans. Commun.*, vol. 51, no. 11, pp. 1927–1938, Nov. 2003.

[107] A. Sendonaris, E. Erkip, and B. Aazhang, "User cooperation diversity – part II: Implementation aspects and performance analysis," *IEEE Trans. Commun.*, vol. 51, no. 11, pp. 1939–1948, Nov. 2003.

[108] S. Shamai and T. L. Marzetta, "Multi-user capacity in block fading with no channel state information," *IEEE Trans. Inform. Theory*, vol. 48, no. 4, pp. 938–942, April 2002.

[109] S. Shamai and A. D. Wyner, "Information-theoretic considerations for symmetric, cellular, multiple-access fading channels: Part I," *IEEE Trans. Inform. Theory*, vol. 43, pp. 1877–1894, Nov. 1997.

[110] S. Shamai and B. M. Zaidel, "Enhancing the cellular downlink capacity via co-processing at the transmitting end," in *Proceedings of IEEE Vehicular Tech. Conf.*, May 2001, pp. 1745–1749.

[111] C. Shannon, "A mathematical theory of communication," *Bell Sys. Tech. J.*, pp. 379–423, 623–656, 1948.

[112] C. Shannon, "Communications in the presence of noise," in *Proceedings of IRE*, pp. 10–21, 1949.

[113] C. Shannon and W. Weaver, *The Mathematical Theory of Communication*. Univ. Illinois Press, 1949.

[114] M. Sharif and B. Hassibi, "On the capacity of MIMO broadcast channels with partial side information," *IEEE Trans. Inform. Theory*, vol. 51, no. 2, pp. 506–522, Feb. 2005.

[115] H. Shin and J. Lee, "Capacity of multiple-antenna fading channels: spatial fading correlation, double scattering and keyhole," *IEEE Trans. Inform. Theory*, vol. 49, no. 10, pp. 2636–2647, Oct. 2003.

[116] N. Sidiropoulos, T. Davidson, and Z. Q. Luo, "Transmit beamforming for physical layer multicasting," *IEEE Trans. Sig. Proc.*, vol. 54, no. 6, pp. 2239–2251, June 2006.

[117] S. Simon and A. Moustakas, "Optimizing MIMO antenna systems with channel covariance feedback," *IEEE J. Select. Areas Commun.*, vol. 21, no. 3, pp. 406–417, April 2003.

[118] P. J. Smith and M. Shafi, "On a Gaussian approximation to the capacity of wireless MIMO systems," in *Proceedings of IEEE Intl. Conf. Commun.*, April 2002, pp. 406–410.

[119] Q. Spencer, L. Swindlehurst, and M. Haardt, "Zero-forcing methods for downlink spatial multiplexing in multi-user MIMO channels," *IEEE Trans. Sig. Proc.*, vol. 52, no. 2, pp. 461–471, Feb. 2004.

[120] S. Srinivasa and S. Jafar, "Vector channel capacity with quantized feedback," in *Proceedings of IEEE Intl. Conf. on Commun.*, May 2005.

[121] E. Telatar, "Capacity of multi-antenna Gaussian channels," *European Trans. on Telecomm. ETT*, vol. 10, no. 6, pp. 585–596, Nov. 1999.

[122] D. Tse and S. Hanly, "Multiaccess fading channels – part I: polymatroid structure, optimal resource allocation and throughput capacities," *IEEE Trans. Inform. Theory*, vol. 44, no. 7, pp. 2796–2815, Nov. 1998.

[123] A. Tulino and S. Verdú, "Random matrix theory and wireless communications," *Foundations Trends in Commun. Inform. Theory*, vol. 1, no. 1, 2004.

[124] A. Tulino, A. Lozano, and S. Verdú, "Capacity-achieving input covariance for single-user multi-antenna channels," *IEEE Trans. Wireless Commun.*, vol. 5, no. 3, pp. 662–671, March 2006.

[125] A. Tulino, A. Lozano, and S. Verdú, "Impact of antenna correlation on the capacity of multiantenna channels," *IEEE Trans. Inform. Theory*, vol. 51, no. 7, pp. 2491–2509, July 2005.

[126] E. C. van der Meulen, "A survey of multi-way channels in information theory: 1961–1976," *IEEE Trans. Inform. Theory*, vol. 23, no.1, pp. 1–37, Jan. 1977.

[127] E. C. van der Meulen, "Some reflections on the interference channel," in *Communications and Cryptography: Two Sides of One Tapestry*, ed. R. E. Blahut, D. J. Costello, and T. Mittelholzer. Kluwer, 1994, pp. 409–421.

[128] S. Venkatesan, S. Simon, and R. Valenzuela, "Capacity of a Gaussian MIMO channel with nonzero mean," in *Proceedings of IEEE Vehicular Tech. Conf.*, Oct. 2003, pp. 1767–1771.

[129] S. Verdú, *Multi-user Detection*. Cambridge University Press, 1998.

[130] S. Verdú, "Spectral efficiency in the wideband regime," *IEEE Trans. Inform. Theory*, vol. 48, no. 6, pp. 1319–1343, June 2002.

[131] S. Vishwanath and S. Jafar, "On the capacity of vector interference channels," in *Proceedings of IEEE Inform. Theory Workshop*, Oct. 2004.

[132] S. Vishwanath, S. Jafar, and A. Goldsmith, "Optimum power and rate allocation strategies for multiple access fading channels," in *Proceedings of Vehicular Tech. Conf.*, May 2000, pp. 2888–2892.

[133] S. Vishwanath, N. Jindal, and A. Goldsmith, "Duality, achievable rates, and sum-rate capacity of MIMO broadcast channels," *IEEE Trans. Inform. Theory*, vol. 49,

no. 10, pp. 2658–2668, Oct. 2003.

[134] E. Visotsky and U. Madhow, "Space–time transmit precoding with imperfect feedback," *IEEE Trans. Inform. Theory*, vol. 47, no. 6, pp. 2632–2639, Sept. 2001.

[135] P. Viswanath, D. Tse, and R. Laroia, "Opportunistic beamforming using dumb antennas," *IEEE Trans. Inform. Theory*, vol. 48, no. 6, pp. 1277–1294, June 2002.

[136] P. Viswanath and D. N. Tse, "Sum capacity of the vector Gaussian broadcast channel and uplink-downlink duality," *IEEE Trans. Inform. Theory*, vol. 49, no. 8, pp. 1912–1921, Aug. 2003.

[137] P. Viswanath, D. N. Tse, and V. Anantharam, "Asymptotically optimal water-filling in vector multiple-access channels," *IEEE Trans. Inform. Theory*, vol. 47, no. 1, pp. 241–267, Jan. 2001.

[138] H. Viswanathan and S. Venkatesan, "Asymptotics of sum rate for dirty paper coding and beamforming in multiple-antenna broadcast channels," in *Proceedings of 41st Allerton Conf. on Commun. Control, and Computing*, Oct. 2003.

[139] H. Viswanathan, S. Venkatesan, and H. C. Huang, "Downlink capacity evaluation of cellular networks with known interference cancellation," *IEEE J. Select. Areas Commun.*, vol. 21, pp. 802–811, June 2003.

[140] B. Wang, J. Zhang, and A. Host-Madsen, "On the capacity of MIMO relay channels," *IEEE Trans. Inform. Theory*, vol. 51, no. 1, pp. 29–43, Jan. 2005.

[141] H. Weingarten, Y. Steinberg, and S. Shamai, "Capacity region of the degraded MIMO broadcast channel," *IEEE Trans. Inform. Theory*, vol. 52, no. 9, pp. 3936–3964, Sept. 2006.

[142] J. Winters, "On the capacity of radio communication systems with diversity in a Rayleigh fading environment," *IEEE J. Select. Areas Commun.*, vol. 5, no.5, pp. 871–878, June 1987.

[143] J. Winters, J. Salz, and R. Gitlin, "The impact of antenna diversity on the capacity of wireless communication systems," *IEEE Trans. Commun.*, vol. 42, no. 2, pp. 1740–1751, Feb. 1994.

[144] A. Wyner, "Shannon-theoretic approach to a Gaussian cellular network," *IEEE Trans. Inform. Theory*, vol. 40, no. 6, pp. 1713–1727, Nov. 1994.

[145] T. Yoo and A. Goldsmith, "On the optimality of multi-antenna broadcast scheduling using zero-forcing beamforming," *IEEE J. Select. Areas Commun.*, special issue on 46 wireless systems, vol. 24, no. 3, pp. 528–541, March 2006.

[146] W. Yu, "A dual decomposition approach to the sum power Gaussian vector multiple access channel sum capacity problem," in *Proceedings of Conf. on Information Sciences and Systems (CISS)*, March 2003.

[147] W. Yu, "Spatial multiplex in downlink multi-user multiple-antenna wireless environments," in *Proceedings of IEEE Global Commun. Conf.*, Nov. 2003, pp.1887–1891.

[148] W. Yu and J. Cioffi, "Trellis precoding for the broadcast channel," in *Proceedings of Global Commun. Conf.*, Oct. 2001, pp. 1344–1348.

[149] W. Yu and J. M. Cioffi, "Sum capacity of Gaussian vector broadcast channels," *IEEE Trans. Inform. Theory*, vol. 50, no. 9, pp. 1875–1892, Sept. 2004.

[150] W. Yu, G. Ginis, and J. Cioffi, "An adaptive multi-user power control algorithm for VDSL," in *Proceedings of Global Commun. Conf.*, Oct. 2001.

[151] W. Yu, W. Rhee, S. Boyd, and J. Cioffi, "Iterative water-filling for Gaussian vector multiple access channels," *IEEE Trans. Inform. Theory*, vol. 50, no. 1, pp. 145–152, Jan. 2004.

[152] W. Yu, W. Rhee, and J. Cioffi, "Optimal power control in multiple access fading channels with multiple antennas," in *Proceedings of Intl. Conf. Commun.*, 2001.

[153] M. Yuksel and E. Erkip, "Can virtual MIMO mimic a multi-antenna system: diversity–multiplexing tradeoff for wireless relay networks," in *IEEE Commun. Theory Workshop*, May 2005.

[154] L. Zheng and D. Tse, "Diversity and multiplexing: a fundamental tradeoff in multiple-antenna channels," *IEEE Trans. Inform. Theory*, vol. 49, no. 5, pp. 1073–1096, May 2005.

[155] L. Zheng and D. N. Tse, "Packing spheres in the Grassmann manifold: a geometric approach to the non-coherent multi-antenna channel," *IEEE Trans. Inform. Theory*, vol. 48, no. 2, pp. 359–383, Feb. 2002.

3 Precoding design

Armed with the theoretical limits of MIMO wireless performance from Chapter 2, we now embark on the design of specific system blocks. At the transmitter, two major MIMO processing components at the symbol level are precoding and space–time coding. Precoding, the last digital processing block at the transmitter (see Figure 1.2), is a technique that exploits the channel information available at the transmitter. Such information is generally referred to as transmit channel side information, or CSIT (this definition is more general than that in Chapter 2). In MIMO wireless, spatial CSIT is particularly useful in enhancing system performance [39]. Space–time coding, on the other hand, assumes no CSIT and focuses on enhancing reliability through diversity [49]. In addition to these two components, regular channel coding is required for bit-level protection. This chapter focuses on precoding design, and space–time coding is discussed in Chapter 4.

CSIT helps to increase the transmission rate, to enhance coverage, and to reduce receiver complexity in MIMO wireless systems. Many forms of CSIT exist. Exact channel knowledge at each time instance, or perfect CSIT, is ideal; but it is often difficult to acquire in a time-selective fading channel. CSIT is more likely to be available as a channel estimate with an associated error covariance, which reduce in the limit to the channel statistics, such as the channel mean and covariance [60]. Such CSIT encompasses several models discussed in Chapter 2, including perfect CSIT and CDIT. Other partial CSIT forms can involve only parametric channel information, such as the channel condition number or the Ricean K factor. In this chapter, we focus on the CSIT given by a channel estimate with a known error covariance. Most analyses assume that the channel is known perfectly at the receiver; however, the impact of receive channel side information (CSIR) will also be briefly discussed.

Precoding design for MIMO wireless has been an active research area in recent years [18, 21, 25, 28–30, 38, 41, 53, 54, 57, 61, 67, 68] and is now finding applications in emerging wireless standards [24]. This chapter provides an overview of precoding design, combining both theoretical foundations and practical issues. We focus on linear precoding, known to be capacity-optimal when causal CSIT is available in the form of channel estimates [8, 44, 46]. Functionally, a linear precoder is a transmit operation matching to the input signal covariance on one side and to the channel on the other. The structure is essentially a beamformer with single or multiple beams, each with a defined direction and power loading.

The chapter starts with a discussion of the principles for acquiring CSIT in a wireless channel and the derivation of a CSIT model as a channel estimate and its error covariance. Information-theoretic results for optimal signaling given the CSIT are then analyzed, leading to the linear precoding solution. Next, a transmitter structure is established, comprising a precoder and different encoding architectures, such as space–time coding and spatial multiplexing. The chapter then examines several linear precoder design criteria. Corresponding precoder designs follow for various CSIT scenarios: perfect CSI, correlation CSI, mean CSI, and general CSI – both mean (or channel estimate) and covariance. Extensive performance simulation results are provided with discussion on the precoding gain. The chapter continues with an overview of precoding applications in practical wireless systems, covering open- and closed-loop channel acquisition methods in TDD and FDD systems, the codebook design in closed-loop systems, impacts of CSIR, and a survey of the current status of precoding in emerging wireless standards. Finally, the conclusion includes discussions of other types of CSIT, open problems, and alternate methods for exploiting CSIT.

3.1 Transmit channel side information

This section discusses the MIMO wireless channel model and the principles for obtaining channel information at the transmitter. A CSIT model is then established for use throughout this chapter.

3.1.1 *The MIMO channel*

A wireless channel exhibits time, frequency, and space selective variations, known as fading. This fading arises due to Doppler, delay, and angle spreads in the scattering environment [27, 39, 40]. We focus on a time-varying frequency flat channel in this chapter. A frequency-flat solution can be applied per sub-carrier in a frequency-selective channel with orthogonal frequency-division multiplexing (OFDM).

The wireless channel in a multi-path environment can be modeled as a complex Gaussian random variable. In the presence of a direct line-of-sight, the channel may exhibit a non-zero mean. The MIMO channel between M_T transmit and M_R receive antennas is an $M_R \times M_T$ complex Gaussian random matrix $\mathbf{H}$, which can be decomposed as

$$\mathbf{H} = \mathbf{H}_m + \tilde{\mathbf{H}}, \tag{3.1}$$

where $\mathbf{H}_m$ is the complex channel mean and $\tilde{\mathbf{H}}$ is a zero-mean complex Gaussian random matrix. The elements of $\tilde{\mathbf{H}}$ are complex random variables, of which the real and the imaginary parts are independent and identically distributed as a zero-mean Gaussian with the same variance. The channel covariance is an $M_R M_T \times M_R M_T$ matrix defined as

$$\mathbf{R}_h = E\big[\tilde{\mathbf{h}}\tilde{\mathbf{h}}^*\big], \tag{3.2}$$

where $\tilde{\mathbf{h}} = \mathrm{vec}(\tilde{\mathbf{H}})$, and $(\cdot)^*$ denotes a conjugate transpose. $\mathbf{R}_h$ is the covariance of the $M_R M_T$ scalar channels between the M_T transmit and M_R receive antennas, and is a complex Hermitian positive semi-definite matrix. The ratio of the power in the channel mean and the average power in the channel variable component is the channel K factor, or the Ricean factor, defined as

$$K = \frac{||\mathbf{H}_m||_F^2}{\mathrm{tr}(\mathbf{R}_h)}, \tag{3.3}$$

where $||.||_F$ is the matrix Frobenius norm and $\mathrm{tr}(\cdot)$ is the trace of a matrix.

The channel covariance $\mathbf{R}_h$ is often assumed to have a simpler separable Kronecker structure [45]. The Kronecker model assumes that the covariance of the scalar channels seen from all M_T transmit antennas to a single receive antenna (corresponding to a row of $\mathbf{H}$) is the same for any receive antenna (any row) and is equal to $\mathbf{R}_t$ $(M_T \times M_T)$. Similarly, the covariance of the scalar channels seen from a single transmit antenna to all M_R receive antennas (corresponding to a column of $\mathbf{H}$) is the same for any transmit antenna (any column) and is equal to $\mathbf{R}_r$ $(M_R \times M_R)$. The channel covariance can then be decomposed as

$$\mathbf{R}_h = \mathbf{R}_t^T \otimes \mathbf{R}_r, \tag{3.4}$$

where $\otimes$ denotes the Kronecker product [17]. Both covariance matrices $\mathbf{R}_t$ and $\mathbf{R}_r$ are complex Hermitian positive semi-definite. The channel can then be written as

$$\mathbf{H} = \mathbf{H}_m + \mathbf{R}_r^{1/2}\mathbf{H}_w\mathbf{R}_t^{1/2}, \tag{3.5}$$

where $\mathbf{H}_w$ is an $M_R \times M_T$ matrix with zero-mean unit-variance i.i.d. complex Gaussian entries. Here $\mathbf{R}_t^{1/2}$ is the unique square-root of $\mathbf{R}_t$, such that $\mathbf{R}_t^{1/2}\mathbf{R}_t^{1/2} = \mathbf{R}_t$, and similarly for $\mathbf{R}_r^{1/2}$.

Other more general channel covariance structures have been proposed in the literature [42, 62], where the transmit covariances $\mathbf{R}_t$ corresponding to different reference receive antennas are assumed to have the same eigenvectors, but not necessarily the same eigenvalues; similarly for $\mathbf{R}_r$. In this chapter, we will use only the simpler Kronecker correlation structure. Furthermore, since precoding is primarily affected by transmit correlation, we assume that $\mathbf{R}_r = \mathbf{I}$ in most cases, unless otherwise specified. The Kronecker correlation model has been experimentally verified in indoor environments for up to 3×3 antenna configurations [32, 64] and in outdoor environments for up to 8×8 configurations [5].

3.1.2 Methods of obtaining CSIT

We use the term CSIT here loosely to mean any channel information available to the transmitter. In the next section, we will define a specific CSIT model. The transmitter can only acquire CSIT indirectly, since the signal enters the channel only after leaving the transmitter. The receiver, however, can estimate the channel directly from the channel-modified received signal. Pilots are usually inserted in the transmitted signal to facilitate channel estimation by the receiver. The transmitter can then indirectly acquire CSIT by either invoking the reciprocity principle or using feedback from the receiver.

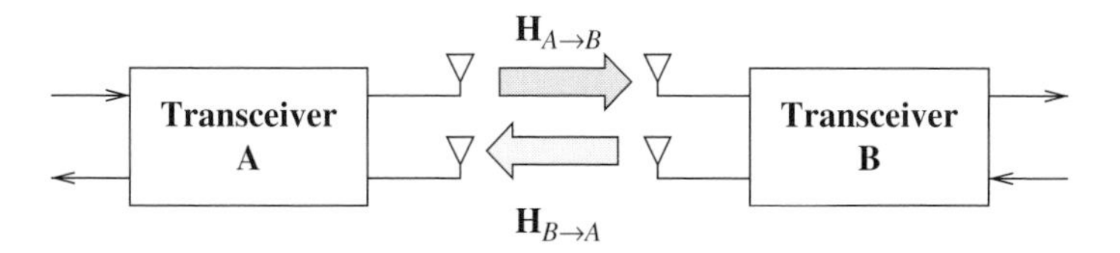

Figure 3.1. Obtaining CSIT using reciprocity.

The reciprocity principle in wireless communication states that the channel from an antenna A to another antenna B is identical to the transpose of the channel from B to A. Reciprocity holds, provided that both forward and reverse links occur at the same frequency, the same time, and the same antenna locations. Since communication systems are often full-duplex, the reciprocity principle suggests that the transmitter can obtain the forward (A to B) channel from the reverse (B to A) channel measurements, which the receiver (at A) can measure, as illustrated in Figure 3.1.

In practical full-duplex communications, however, the forward and reverse links cannot use all identical frequency, time, and spatial instances. The reciprocity principle may still hold approximately if the difference in any of these dimensions is relatively small, compared to the channel variation across the referenced dimension. In the temporal dimension, this condition implies that any time lag Δ_t between the forward and reverse transmissions must be much smaller than the channel coherence time T_c:

$$\Delta_t \ll T_c. \tag{3.6}$$

Similarly, any frequency offset Δ_f must be much smaller than the channel coherence bandwidth B_c:

$$\Delta_f \ll B_c, \tag{3.7}$$

and the antenna location differences on the two links must be much smaller than the channel coherence distance D_c [39].

Practical channel acquisition based on reciprocity is referred to as the open-loop method and may be applicable in time-division-duplex (TDD) systems. While TDD systems often have identical forward and reverse frequency bands and antennas, there is a time lag between the forward and reverse links (e.g. the ping-pong period in voice systems). In asynchronous data systems, the time lag is between the reception of a signal from a reference user and the next transmission to that user. Such time lags must be negligible compared to the channel coherence time. In a frequency-division-duplex (FDD) system, the temporal and spatial dimensions may be identical, but the frequency offset between the forward and reverse links is usually much larger than the channel coherence bandwidth. Therefore, reciprocity is usually not applicable in FDD systems.

One complication in using reciprocity methods is that the principle only applies to the radio channel between the antennas, while in practice, the "channel" is measured and used at the baseband processor. This fact means that different transmit and receive radio-frequency (RF) hardware chains become part of the forward and reverse channels.

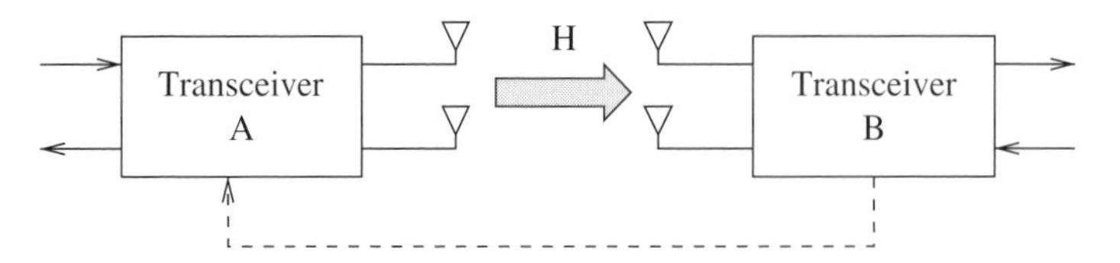

Figure 3.2. Obtaining CSIT using feedback.

Since these chains have different frequency transfer characteristics, reciprocity requires transmit–receive chain calibration, discussed in Section 3.7.

Another method of obtaining CSIT is using feedback from the receiver of the forward link, depicted in Figure 3.2. The channel is measured at the receiver at B during the forward link (A to B) transmission, and the information is sent to the transmitter at A on the reverse-link. Feedback is not limited by the reciprocity requirements. However, the time lag Δ_{lag} between the channel measurement at B and its use by the transmitter at A can be a source of error, unless it is much smaller than the channel coherence time:

$$\Delta_{\text{lag}} \ll T_c.$$

Feedback can also be used to send channel statistics that change much slower in time compared to the channel itself. In such cases, the time lag requirement for valid feedback can be relaxed significantly.

Channel acquisition using feedback is referred to as the closed-loop method and is more common in FDD systems. Although not subjected to transmit–receive calibration, feedback imposes another system overhead by using up transmission resources. Therefore, methods of reducing feedback overhead, such as quantizing feedback information, are both important and necessary. In Section 3.7, we will discuss practical issues for both open- and closed-loop methods in more detail. Further references can be found in [4].

3.1.3 A dynamic CSIT model

We assume that the channel is a stationary Gaussian stochastic process. We consider CSIT at the transmit time s in the form of a channel estimate and its error covariance. This estimate and its error covariance are derived from a channel measurement at time 0 and relevant channel statistics. This model is applicable to both open- and closed-loop methods.

The main source of irreducible error in channel estimation is the random channel time-variation, or equivalently, the Doppler spread. We assume that the channel measurement at time 0 is error-free; the error in channel estimates depends only on the time lag s between this initial measurement and its use by the transmitter. A CSIT model can then be written as

$$\begin{aligned}\mathbf{H}(s) &= \hat{\mathbf{H}}(s) + \mathbf{E}(s),\\ \mathbf{R}_e(s) &= E\left[\mathbf{e}(s)\mathbf{e}(s)^*\right],\end{aligned} \tag{3.8}$$

where the notation (s) denotes the time dependence. Here, $\mathbf{H}(s)$ is the channel, $\hat{\mathbf{H}}(s)$ is a channel estimate, and $\mathbf{E}(s)$ is the estimation error at time s. $\mathbf{R}_e(s)$ is the error correlation matrix, where $\mathbf{e}(s) = \text{vec}(\mathbf{E}(s))$. Assuming unbiased estimates, we can model $\mathbf{E}(s)$ as a zero-mean complex Gaussian random matrix, and $\mathbf{R}_e(s)$ now becomes an error covariance matrix, which depends on s and the Doppler spread. CSIT consists of the estimate $\hat{\mathbf{H}}(s)$ and its error covariance $\mathbf{R}_e(s)$. By our assumption, at time zero, $\mathbf{E}(0) = \mathbf{0}$ and $\mathbf{R}_e(0) = \mathbf{0}$, i.e., we have perfect CSIT.

The correlation between the channel at time 0 and the channel at time s is characterized by the channel auto-covariance, defined as

$$\mathbf{R}_h(s) = E\left[\mathbf{h}(s)\mathbf{h}(0)^*\right] - \mathbf{h}_m\mathbf{h}_m^*, \tag{3.9}$$

where $\mathbf{h}_m = \text{vec}(\mathbf{H}_m)$ and $\mathbf{h} = \text{vec}(\mathbf{H})$. This auto-covariance measures how rapidly $\mathbf{H}(s)$ decorrelates with time; it eventually decays to zero as s becomes large. At a zero time lag ($s = 0$), $\mathbf{R}_h(0) = \mathbf{R}_h$ as in (3.2).

When the time lag s is large compared to the channel coherence time, channel estimates based on $\mathbf{H}(0)$ are no longer meaningful; therefore, short-term statistics, such as the channel mean $\mathbf{H}_m$ (3.1) and the covariance $\mathbf{R}_h$ (3.2) (or $\mathbf{R}_t$ and $\mathbf{R}_r$ in (3.4)), are relevant. Physical models of wireless channels indicate that short-term channel statistics are stable over periods much longer than the channel coherence time. These statistics, including $\mathbf{R}_h(s)$, obtained by time averaging over a short window (about 10 times the coherence time) are valid for 10s to 100s of coherence time periods. Thus, the receiver can make $\mathbf{H}_m$, $\mathbf{R}_h$, and $\mathbf{R}_h(s)$ available to the transmitter with negligible lag-induced errors.

A framework to dynamically estimate the CSIT – $\hat{\mathbf{H}}(s)$ and $\mathbf{R}_e(s)$ – can be developed. Assume that the initial channel measurement $\mathbf{H}(0)$ with the channel statistics $\mathbf{H}_m$, $\mathbf{R}_h$, and $\mathbf{R}_h(s)$ are available to the transmitter. Then the CSIT at time s follows directly from the standard minimum mean-square error (MMSE) estimation theory [31]

$$\begin{aligned} \hat{\mathbf{h}}(s) &= E\left[\mathbf{h}(s)|\mathbf{h}(0)\right] = \mathbf{h}_m + \mathbf{R}_h\mathbf{R}_h(s)^{-1}\left[\mathbf{h}(0) - \mathbf{h}_m\right] \\ \mathbf{R}_e(s) &= \text{cov}\left[\mathbf{h}(s)|\mathbf{h}(0)\right] = \mathbf{R}_h - \mathbf{R}_h(s)\mathbf{R}_h^{-1}\mathbf{R}_h(s), \end{aligned} \tag{3.10}$$

where $\hat{\mathbf{h}}(s) = \text{vec}(\hat{\mathbf{H}}(s))$. A similar model was proposed in [15] for estimating a scalar time-varying channel from a vector of outdated estimates. CSIT formulations conditioned on noisy channel estimates were also studied in [28, 38].

The channel covariance $\mathbf{R}_h$ captures the spatial correlation between all the transmit and receive antennas, while the auto-covariance at a non-zero delay $\mathbf{R}_h(s)$ captures both the spatial and temporal correlations of the channel. If we assume that the temporal correlation is homogeneous and identical for any channel element, then we can separate these two correlation effects and write

$$\mathbf{R}_h(s) = \rho(s)\mathbf{R}_h, \tag{3.11}$$

where $\rho(s)$ is the temporal correlation of a scalar channel. In other words, all the $M_R M_T$ scalar channels between the M_T transmit and M_R receive antennas have the same temporal correlation function. This assumption is based on the premise that the channel temporal statistics can be expected to be the same for all antenna pairs. Similar assumptions for MIMO Doppler decorrelation have also been used in [32, 62]. Using Jakes' model [27] for a time-varying scalar channel for example, $\rho(s) = J_0(2\pi f_d s)$, where f_d is the channel Doppler spread and $J_0(\cdot)$ is the zeroth-order Bessel function of the first kind.

The estimated channel and its error covariance at time s (3.10) can now be simplified to

$$\begin{aligned} \hat{\mathbf{H}}(s) &= \rho(s)\mathbf{H}(0) + (1-\rho(s))\,\mathbf{H}_m, \\ \mathbf{R}_e(s) &= \left(1-\rho(s)^2\right)\mathbf{R}_h. \end{aligned} \tag{3.12}$$

For the Kronecker covariance model (3.4), the estimated channel has the effective antenna covariance of

$$\begin{aligned} \mathbf{R}_t(s) &= \left(1-\rho(s)^2\right)^{1/2}\mathbf{R}_t, \\ \mathbf{R}_r(s) &= \left(1-\rho(s)^2\right)^{1/2}\mathbf{R}_r. \end{aligned} \tag{3.13}$$

In these expressions, ρ functions as an estimate quality. The estimated channel $\hat{\mathbf{H}}(s)$ ranges from perfect channel measurement, when $\rho = 1$, to pure statistics, when $\rho = 0$. As ρ approaches zero, the influence of the initial channel measurement $\mathbf{H}(0)$ diminishes, and the estimate moves toward the channel mean $\mathbf{H}_m$. In parallel, the error covariance is zero when $\rho = 1$, and grows to $\mathbf{R}_h$ as ρ approaches zero. Figure 3.3 illustrates this estimate evolution as a function of the time lag s.

In the subsequent development, we divide CSIT into two categories: *perfect CSIT* when $\rho = 1$, and *channel estimate CSIT* when $0 \le \rho < 1$. The latter consists of a channel estimate $\hat{\mathbf{H}}(s)$, or an effective mean, and a non-zero error covariance $\mathbf{R}_e(s)$, or an effective covariance. Since the precoder design algorithms for channel estimate CSIT are the same for all values of ρ, we derive all precoding results using a representative $\rho = 0$, where $\hat{\mathbf{H}} = \mathbf{H}_m$ and $\mathbf{R}_e = \mathbf{R}_h$. This case is referred to as *statistical CSIT*. Special cases of statistical CSIT include *correlation CSIT* when $\mathbf{R}_h$ is arbitrary but $\mathbf{H}_m = 0$, and *mean CSIT* when $\mathbf{H}_m$ is arbitrary but $\mathbf{R}_h = \mathbf{I}$. The results for $0 < \rho < 1$ can be established by using the estimate $\hat{\mathbf{H}}(s)$ and the covariance $\mathbf{R}_e(s)$ as the channel statistics. Finally, we

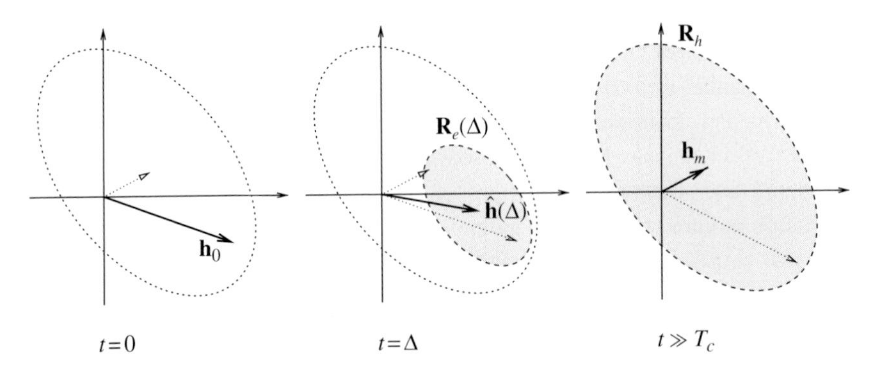

Figure 3.3. Time-lag-dependent channel estimate (bold vector) and its error covariance (shaded ellipse).

define *no CSIT* as the condition when $\rho = 0$, $\mathbf{H}_m = 0$, and $\mathbf{R}_h = \mathbf{I}$, corresponding to an i.i.d. Rayleigh fading channel with no channel information at the transmitter.

3.2 Information-theoretic foundation for exploiting CSIT

This section highlights the capacity gain due to CSIT and examines the information-theoretic foundation for optimal signaling with CSIT.

3.2.1 *Value of CSIT in MIMO systems*

In a frequency-flat MIMO channel, CSIT can be exploited in the temporal and spatial dimensions; but in a frequency-flat SISO channel, only the temporal CSIT is relevant. It is well-known that the temporal CSIT – channel information across multiple time instances – provides negligible channel capacity gain at medium-to-high SNRs, and this gain disappears at SNRs approximately above 15 dB [16]. On the other hand, spatial CSIT in a MIMO channel can offer a potentially significant improvement in channel capacity at all SNRs. In this chapter, we focus on exploiting spatial CSIT in the form of a channel estimate and its error covariance at a time instance s (3.12).

Consider the perfect and no CSIT cases. The optimal signal for achieving the channel capacity is zero-mean complex Gaussian distributed with the covariance determined by the CSIT, as established in Chapter 2. With perfect CSIT, the optimal signal covariance has its eigenvectors matched to the channel eigen-directions and its eigenvalues determined by a water-filling solution [9]. With no CSIT, the optimal signal covariance is a scaled identity matrix, equivalent to transmitting with equi-power in all directions [50]. At low SNRs, the capacity with perfect CSIT is always higher than that without CSIT for all antenna configurations, since the water-filling power allocation drops weak modes. CSIT increases the capacity multiplicatively at low SNRs. As the SNR decreases, eventually only the strongest channel eigen-mode is used, resulting in a capacity ratio gain as [55]

$$r_1 = \frac{C_{\text{perfect CSIT}}}{C_{\text{no CSIT}}} = \frac{M_T E[\lambda_{\max}(\mathbf{H}^*\mathbf{H})]}{E[\text{tr}(\mathbf{H}^*\mathbf{H})]}, \tag{3.14}$$

where tr$(\cdot)$ is the trace of a matrix. For an i.i.d. Rayleigh fading channel, the low-SNR asymptotic capacity ratio gain from perfect CSIT, in the limit of a large number of antennas, approaches

$$r_1 \overset{M_T, M_R \to \infty}{\longrightarrow} \left(1 + \sqrt{\frac{M_T}{M_R}}\right)^2. \tag{3.15}$$

This quantity is always larger than 1 and can be significant in systems with more transmit than receive antennas ($M_T > M_R$). Examples of the capacity ratio gain versus the SNR are given in Figure 3.4. Note that this ratio increases with a lower SNR and with a larger number of antennas.

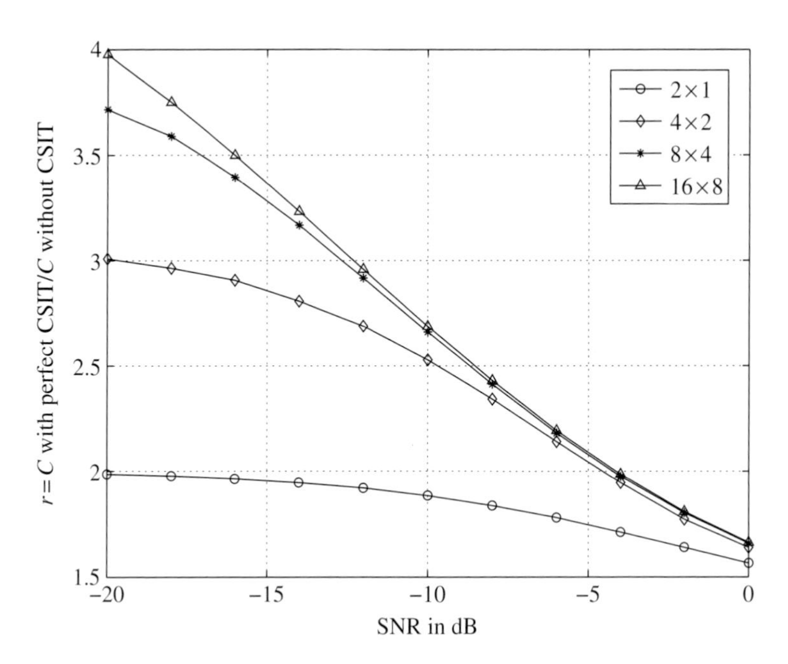

Figure 3.4. Capacity ratio gain from perfect CSIT at low SNRs.

At high SNRs, systems with equal or fewer transmit than receive antennas have a diminishing capacity gain due to CSIT, as the water-filling solution approaches equi-power. In systems with more transmit than receive antennas, however, CSIT increases the capacity even at high SNRs, since the channel rank is smaller than the number of transmit antennas. The incremental capacity gain with perfect CSIT at high SNRs is [56]

$$\Delta\mathcal{C}_H = \max\left\{M_R \log\left(\frac{M_T}{M_R}\right), 0\right\}. \tag{3.16}$$

The gain scales linearly with the number of receive antennas if $M_T > M_R$. This expression is accurate at high SNRs, but is somewhat optimistic at low SNRs. Figure 3.5 shows examples of the capacity with and without CSIT for several antenna configurations, with the incremental gain (3.16) superimposed for comparison.

Statistical CSIT can also increase the capacity. Consider the correlation CSIT with a known transmit antenna correlation $\mathbf{R}_t$, assuming uncorrelated receive antennas ($\mathbf{R}_r = \mathbf{I}$). This CSIT suggests transmission along the eigenvectors of $\mathbf{R}_t$, with a water-filling type power allocation [25, 54]. At low SNRs, such an allocation may lead to null-power in certain directions; whereas the optimal strategy without knowing $\mathbf{R}_t$ is to always transmit with equal power in every direction. Thus, the correlation CSIT always helps to increase the capacity at low SNRs. As the SNR decreases, the ratio between the capacity with correlation CSIT and that without CSIT becomes

$$r_2 = \frac{C_{\text{corr. CSIT}}}{C_{\text{no CSIT}}} = \frac{M_T \lambda_{\max}(\mathbf{R}_t)}{\text{tr}(\mathbf{R}_t)}. \tag{3.17}$$

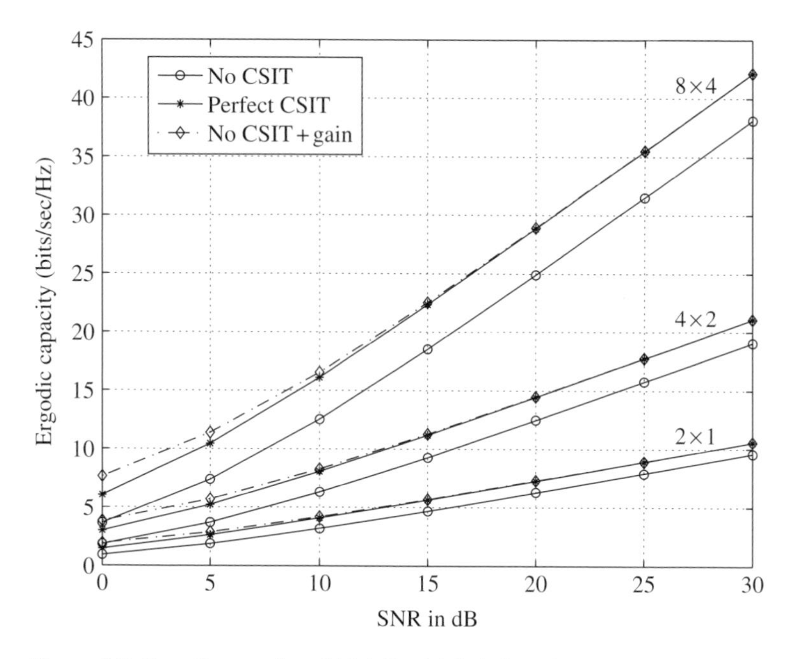

Figure 3.5. Ergodic capacity of i.i.d. Rayleigh fading channels with perfect and no CSIT.

For a rank-one correlated Rayleigh channel for example, this ratio equals the number of transmit antennas M_T.

At high SNRs, the capacity advantage depends on the rank of $\mathbf{R}_t$ and the relative number of transmit and receive antennas. In systems with equal or fewer transmit than receive antennas, if $\mathbf{R}_t$ is full-rank, the capacity gain by the correlation CSIT diminishes at high SNRs, similar to the symmetric antenna configuration in the perfect CSIT case. In systems with more transmit than receive antennas, however, full-rank correlation CSIT can still increase the capacity. On the other hand, if $\mathbf{R}_t$ is rank-deficient, then the correlation CSIT helps to increase the capacity at all SNRs for all antenna configurations. The incremental capacity gain by the correlation CSIT at high SNRs in systems with equal or fewer transmit than receive antennas is [56]

$$\Delta \mathcal{C}_{\mathbf{R}_t} = \max \left\{ K_t \log \frac{M_T}{K_t}, 0 \right\}, \tag{3.18}$$

where K_t is the rank of $\mathbf{R}_t$. In Figure 3.6, we plot the capacity with and without correlation CSIT for rank-one correlated channels for various antenna configurations at 10 dB SNR. Note that for rank-one correlation, having more transmit antennas helps to increase the capacity with correlation CSIT, but does not without CSIT.

3.2.2 *Optimal signaling with CSIT*

We now review the information theory background for a fading channel with causal side information. The theory can be established by first examining a scalar channel.

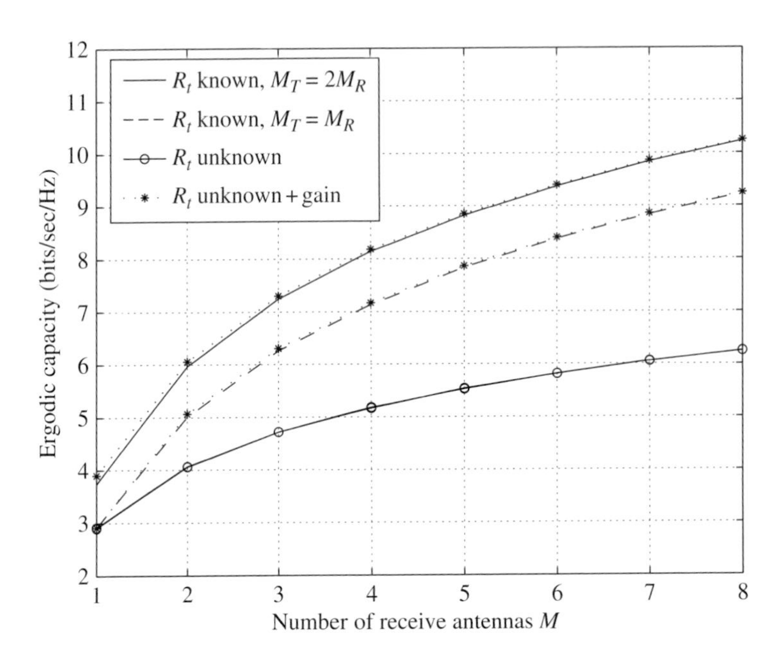

Figure 3.6. Capacity of channels with a rank-one transmit correlation at SNR = 10 dB.

The scalar fading channel is studied in [8] as a particular case of a time-varying channel with channel-state information at both the transmitter and the receiver. Let $h(s)$ be the state of a fading channel, and U_q and V_q be the side information available to the transmitter and the receiver at time s, respectively, as shown in Figure 3.7(a). The channel state and the CSI at time s are assumed to be independent of past channel inputs; therefore, the model applies to a frequency-flat channel without inter-symbol interference. Given the current CSIT U_s, the channel state $h(s)$ is assumed to be independent of the past CSIT $U_1^{s-1} = \{U_1, U_2, \ldots, U_{s-1}\}$:

$$\Pr\left(h(s)|U_1^s\right) = \Pr\left(h(s)|U_s\right). \tag{3.19}$$

This condition enables the capacity to be a stationary function of the CSIT, not depending on the entire history of CSIT; it covers the perfect CSIT, noisy or delayed channel estimates, channel prediction, or the statistical CSIT. The receiver is assumed to know how the CSIT is used. This assumption is reasonable, since the receiver can obtain channel information more readily than the transmitter, and they can both agree on a precoding algorithm. Furthermore, in practical systems, it is the receiver that usually decides the transmit precoding function. The receiver CSI is in the form of a perfectly known received SNR, which encompasses perfect receive channel knowledge ($V_s = h(s)$) as a special case. Under these conditions, the channel capacity with an average input power constraint $E[|X_s|^2] \leq P$ is

$$C = \max_f E\left[\frac{1}{2}\log\left(1 + hf(U)\right)\right], \tag{3.20}$$

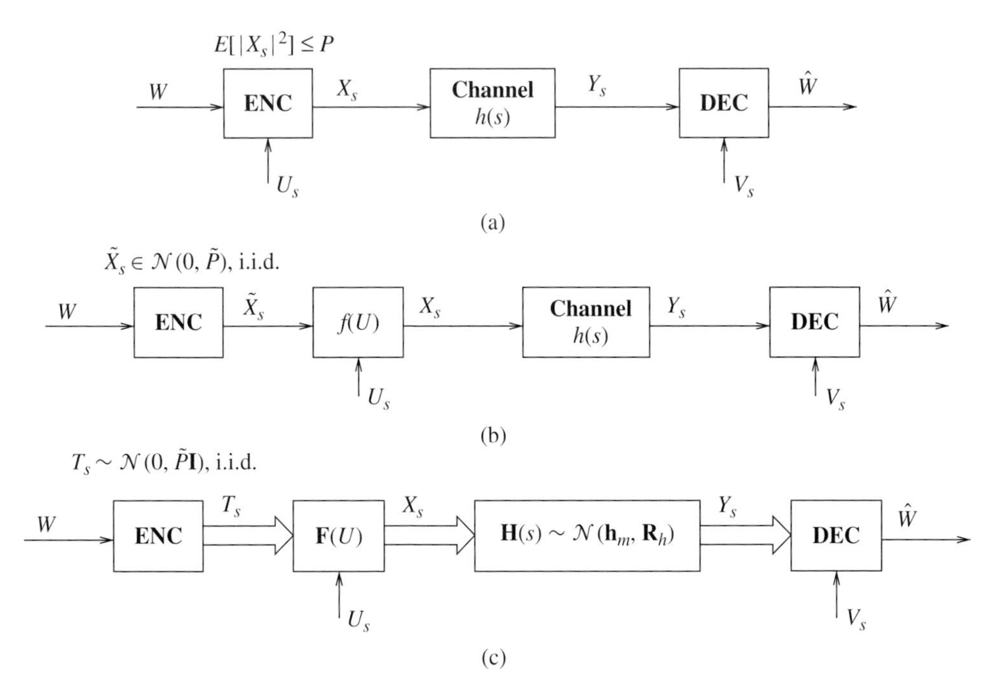

Figure 3.7. (a) A system with a time-varying channel state, CSIT, and CSIR. The optimal configuration for exploiting CSIT in (b) a scalar fading channel and (c) a MIMO fading channel.

where the expectation is over the joint distribution of h and U; and $f(U)$ is a power allocation function satisfying the constraint $E[f(U)] \leq P$.

This is a significant result. It implies that the capacity of a channel with CSIT can be achieved by a single Gaussian codebook designed for a channel without CSIT, provided that the code symbols are dynamically scaled by an appropriate power-allocation function $f(U)$ determined by the CSIT. Therefore, in exploiting the CSIT, it is optimal to separate channel coding and the CSIT-dependent function $f(U)$, as shown in Figure 3.7(b). This CSIT-dependent function, combined with the channel, creates an effective channel, outside of which coding can be applied as if the transmitter had no channel side information. This insight, in fact, can be traced back to Shannon in [44]. For a scalar channel, $f(U)$ is simply dynamic power-allocation.

Subsequently, this result has been extended to the MIMO fading channel [46]. The channel state $\mathbf{H}(s)$ is now a matrix. A similar condition on CSIT as (3.19) applies; i.e. previous side information at the transmitter is independent of the current channel state, given the current transmitter knowledge. Therefore, for a memoryless channel, the CSIT function depends only on the current CSIT U_s, instead of the complete previous information U_1^s. The receiver again is assumed to know the channel perfectly: $V_s = \mathbf{H}(s)$. Under these conditions, the capacity-optimal input signal can be decomposed as

$$X(U_1^s, W) = \mathbf{F}(U_s)T(W). \tag{3.21}$$

Here, $T(W)$ is a codeword optimal for an i.i.d. Rayleigh fading MIMO channel without CSIT, generated from a complex Gaussian distribution with zero-mean and an appropriate covariance $\tilde{P}\mathbf{I}$. The CSIT dependent function $\mathbf{F}(U_s)$ is now a weighting matrix – a linear precoder. In other words, the capacity-achieving signal is zero-mean Gaussian-distributed with covariance $\mathbf{F}\mathbf{F}^*$. This optimal configuration is shown in Figure 3.7(c).

These results establish important properties of capacity-optimal signaling for a fading channel with CSIT. First, it is optimal to separate the CSIT dependent function $\mathbf{F}(U)$ and channel coding, in which the coding is designed for a channel without CSIT. Second, a linear $\mathbf{F}(U)$ is optimal. The separation and linearity properties are the guiding principles for precoder designs exploiting CSIT throughout this chapter.

3.3 A transmitter structure

We focus on a communication system with precoding as depicted in Figure 3.8. The transmitter encompasses an encoder and a linear precoder $\mathbf{F}$. The encoder can include channel coding or space–time coding, or both. These structures are discussed below.

3.3.1 *Encoding structure*

An encoder contains a channel coding and a symbol-mapping block, which delivers vector symbols to the precoder. The encoder influences the precoder design. We classify two broad structures for the encoder block.

The first is a spatial multiplexing structure, in which independent bit streams are generated by demultiplexing the output of the channel encoder and the bit interleaver. These streams are then mapped into vector symbols, which are fed into the precoder, as shown in Figure 3.9(a). Since each stream has its own SNR, per-stream rate adaptation can be used.

Another structure is space–time (ST) coding, in which the bit stream, after channel coding and interleaving, is mapped into symbols. These symbols are then processed by an ST coder, which outputs vector symbols to the precoder, as shown in Figure 3.9 (b). From Section 3.2, if the space–time code is capacity lossless for a channel with no CSIT, then this structure is capacity-optimal for the channel with a CSIT. This structure has a single data stream; hence, only a single rate-adaptation is necessary. The rate is controlled by the outer-code rate and the constellation design.

In the precoding analysis followed, however, we will refer to both of these structures by a single code-block $\mathbf{C}$. The multiplexing structure is represented by a special single

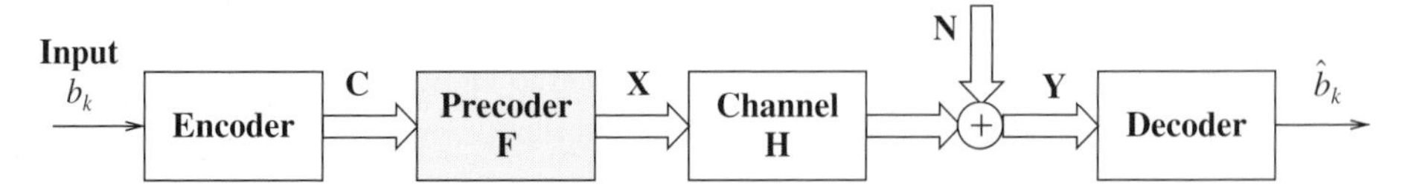

Figure 3.8. A general precoding system.

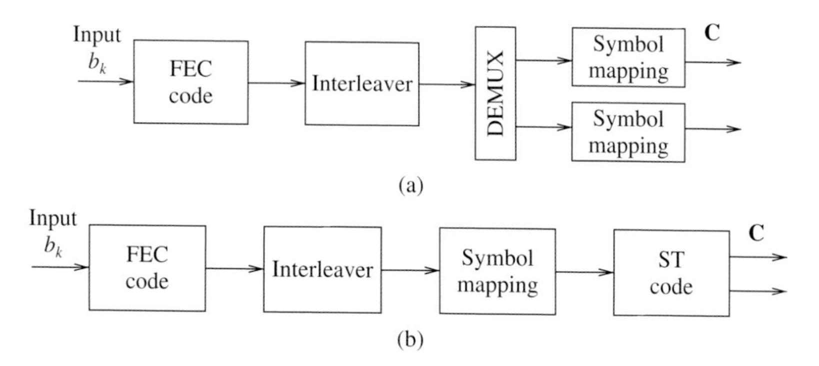

Figure 3.9. Encoder structures: (a) spatial multiplexing; (b) space–time coding.

vector-symbol block. Assuming a Gaussian-distributed codeword $\mathbf{C}$ of size $M_T \times T$ with a zero-mean, we define the codeword covariance matrix as

$$\mathbf{Q} = \frac{1}{T} E\left[\mathbf{C}\mathbf{C}^*\right], \tag{3.22}$$

where the expectation is over the codeword distribution. When $\mathbf{C}$ is spatial multiplexing, $\mathbf{Q} = \mathbf{I}$. $\mathbf{C}$ can also be an ST block code (STBC), which we discuss below.

Space–time block codes

STBCs are usually designed to capture the diversity in the spatial channel, assuming no CSIT (Chapter 4). Diversity determines the slope of the error probability versus the SNR and is related to the number of spatial links that are not fully correlated [27, 52]. A full-diversity code achieves the maximum diversity order of $M_T M_R$ available in the channel. Not all STBCs offer full-diversity, however. High diversity is useful in a fading link since it reduces the so-called fade margin, which is needed to meet a required link reliability.

An STBC can also be characterized by its spatial rate, which is the average number of distinct symbols sent per symbol time-period. Rate-one STBCs average one symbol per symbol-period, independent of the number of transmit antennas. Orthogonal STBCs [48] have a rate of 1 or less. An STBC with a rate greater than 1 is called a high-rate code; the highest rate can be $\min(M_T, M_R)$. Spatial multiplexing can be viewed as a special STBC with a full spatial rate but no transmit diversity. A higher STBC rate does not imply reduced diversity; many new codes have high rates and full-diversity [11, 43].

There is, however, a fundamental trade-off between diversity and multiplexing in ST coding [66]. The multiplexing order is defined by the scale at which the transmission rate increases asymptotically with the SNR. A fixed-rate system has a zero multiplexing-order. The STBC design achieving the optimal diversity–multiplexing trade-off is an active research area [3, 10, 63]. Our framework, focusing on extracting the gain from the CSIT by precoding, is independent of, and complementary to, the diversity–multiplexing trade-offs for ST codes.

The combination of an STBC and a precoder makes the system robust against fading, while exploiting the CSIT. Certain STBCs (e.g. the Alamouti code [1]) can achieve the ergodic capacity of a channel with no CSIT. The combination of such optimal STBCs and a linear precoder can achieve the channel capacity with CSIT.

3.3.2 Linear precoding structure

A linear precoder functions as an input shaper and a beamformer with one or multiple beams with per-beam power allocation. Consider the singular value decomposition of the precoder matrix $\mathbf{F}$

$$\mathbf{F} = \mathbf{U}_F \mathbf{D} \mathbf{V}_F^*. \tag{3.23}$$

The orthogonal beam directions (patterns) are the left singular vectors $\mathbf{U}_F$; the beam power loadings are the squared singular values $\mathbf{D}^2$. The right singular vectors $\mathbf{V}_F$, termed the input shaping matrix, combine the input symbols from the encoder to feed into each beam, as shown in Figure 3.10. The beam directions and power loadings are influenced by the CSIT, the design criterion, and in many cases the SNR.

To ensure a constant average sum transmit power from all the antennas, the precoder must satisfy the power constraint

$$\mathrm{tr}(\mathbf{F}\mathbf{F}^*) = 1. \tag{3.24}$$

We presume that the input codeword $\mathbf{C}$ is orthogonal and power-normalized. The covariance of the precoder output signal becomes

$$\Phi = \mathbf{F}\mathbf{Q}\mathbf{F}^*. \tag{3.25}$$

This is the covariance of the transmitted signal.

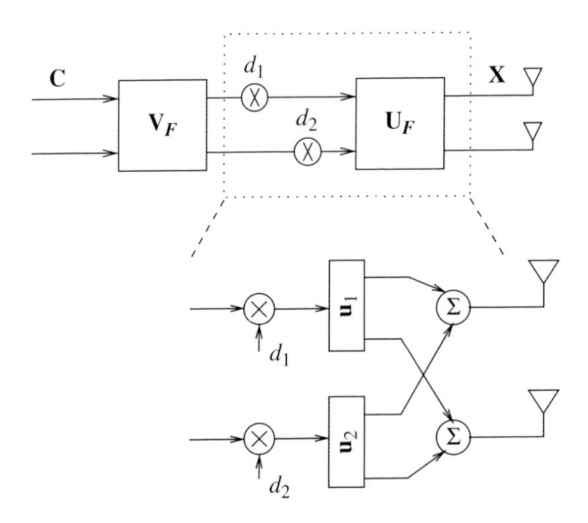

Figure 3.10. A linear precoder as a beamformer.

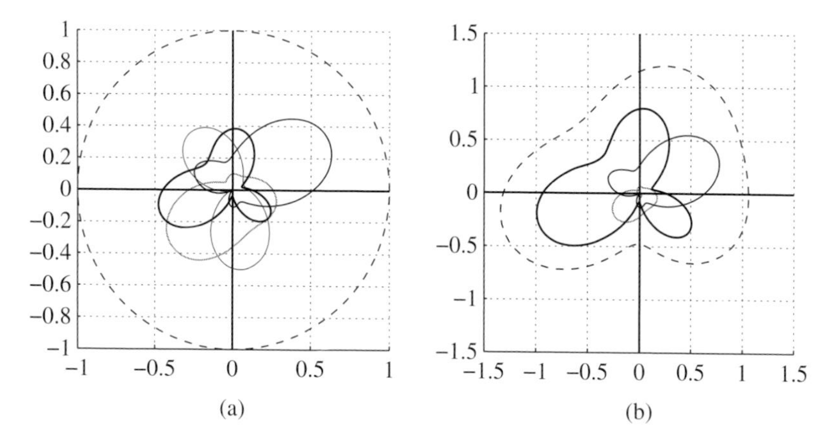

Figure 3.11. Precoded orthogonal eigen-beam patterns with equal (left) and unequal (right) beam power. The dotted line is the total radiation pattern from the antenna array.

A precoder therefore has two effects: decoupling the input signal into orthogonal spatial modes in the form of eigen-beams, and allocating power over these beams, based on the CSIT. If the precoded orthogonal spatial modes match the channel eigen-directions, there will be no interference among signals sent on these modes, creating parallel channels and allowing transmission of independent signal streams. This effect, however, requires perfect CSIT. With partial CSIT, the precoder performs its best to approximately match its eigen-beams to the channel eigen-directions, reducing the interference among signals sent on these beams. This is the decoupling effect. Moreover, the precoder allocates power on these beams. For orthogonal eigen-beams, if the beams all have equal power, the radiation pattern of the transmit antenna array is isotropic, as shown on the left in Figure 3.11 for example. If the beam powers are different, however, the overall transmit radiation pattern will have a specific shape, as on the right in Figure 3.11. By allocating power, the precoder effectively creates a radiation shape matching to the channel based on the CSIT, so that more power is sent in the directions where the channel is strong and less or no power where it is weak. More transmit antennas will increase the ability to finely shape the radiation pattern and therefore are likely to deliver more precoding gain.

The received signal of the system in Figure 3.8 is

$$\mathbf{Y} = \sqrt{\mathcal{E}}\mathbf{HFC} + \mathbf{N}, \tag{3.26}$$

where $\mathcal{E}$ is the transmit signal power and $\mathbf{N}$ is the additive white Gaussian noise. This expression is the system model used throughout this chapter, which encompasses both precoding with ST block coding and with spatial multiplexing.

3.4 Precoding design criteria

We now examine alternate precoding design criteria, including both fundamental performance measures, such as the ergodic capacity and the error exponent, and practical measures, such as the pairwise error probability (PEP) and detection mean-square error (MSE). The capacity formulation generally assumes ideal channel coding. Ergodic capacity implies that the channel evolves through all possible realizations over arbitrarily long codewords. The error exponent, on the other hand, is applicable for finite codeword lengths. For the practical criteria, we perform uncoded analysis without channel coding, assuming a quasi-static block fading channel. The choice of the precoder design criterion depends on the system setup, operating parameters, and the channel (fast or slow fading). For examples, systems with strong channel coding, like turbo or low-density parity check codes with long codeword lengths, may operate at close to the capacity limit and, therefore, are qualified to use a capacity-based performance criterion; while systems with weaker channel codes, such as convolutional codes with small free distances, are more suitable for uncoded analysis. The operating SNR is also important in deciding the criterion. A lower SNR usually favors uncoded analysis, while at high SNRs, coded criteria can yield better performance.

3.4.1 *Information and system capacity*

The ergodic capacity criterion aims at maximizing the transmission rate with a vanishing error, assuming asymptotically long codewords. From Section 3.2, we know that for both perfect and statistical CSIT, the system-capacity-optimal signal is zero-mean Gaussian-distributed with the covariance $\mathbf{FQF}^*$. With Gaussian signaling and Gaussian additive noise, the mutual information between the channel input and output can be obtained explicitly as [9]

$$\mathcal{I}(X, Y) = \log\det(\mathbf{I} + \gamma\mathbf{HFQF}^*\mathbf{H}^*),$$

where γ is the ratio of the transmit signal power to the receiver noise power. The ergodic capacity and optimal signaling are established by maximizing this mutual information subject to the transmit power constraint, using the following formulation:

$$\max E_{\mathbf{H}}\left[\log\det(\mathbf{I} + \gamma\mathbf{HFQF}^*\mathbf{H}^*)\right] \tag{3.27}$$
$$\text{subject to} \operatorname{tr}(\mathbf{FF}^*) = 1.$$

When the encoder in Figure 3.8 is spatial multiplexing (hence, $\mathbf{Q} = \mathbf{I}$), the above formulation coincides with the channel information capacity used in Chapter 2. When $\mathbf{C}$ is an STBC, it provides the system capacity. Subsequently, we will refer to this formulation as the capacity criterion.

Achieving the capacity, however, requires asymptotically long codewords and an ideal receiver, both of which are difficult to meet in practice. Therefore, a precoder design needs to be evaluated for more practical system configurations.

3.4.2 *Error exponent*

The error exponent characterizes the error probability with a finite code-block length for a given transmission rate and is derived from a random code upper-bound error analysis [13]. At an information rate of R (nats/sec/Hz) using block codes of length L, there exists a code, such that the error performance with maximum-likelihood (ML) decoding is upper-bounded by

$$\bar{P}_e \leq \exp(-LE_r(R)), \tag{3.28}$$

where $E_r(R)$ is the random coding exponent defined as

$$E_r(R) = \max_{0\leq\alpha\leq 1} \max_{Q} \left[E_0(\alpha, Q) - \alpha R\right]. \tag{3.29}$$

Here, Q is the channel input distribution; $E_0(\alpha, Q)$ is a function of Q, $\alpha \in [0,\ 1]$, and the channel transition probabilities.

For a MIMO fading channel, using a Gaussian input signal with the covariance $\mathbf{FQF}^*$ leads to a relatively simple expression for $E_0(\alpha, \mathbf{FQF}^*)$ [50],

$$E_0(\alpha, \mathbf{FQF}^*) = -\ln E_{\mathbf{H}}\left[\det\left(I + \gamma(1+\alpha)^{-1}\mathbf{HFQF}^*\mathbf{H}^*\right)^{-\alpha}\right].$$

The objective then is to find the covariance matrix $\mathbf{FQF}^*$ that maximizes the coding exponent, leading to the following optimization problem:

$$\begin{aligned} &\min_{\alpha}\min_{\mathbf{F}}\ \ln E_{\mathbf{H}}\left[\det\left(I + \frac{\gamma}{1+\alpha}\mathbf{HFQF}^*\mathbf{H}^*\right)^{-\alpha}\right] + \alpha R \\ &\text{subject to } \operatorname{tr}(\mathbf{FF}^*) = 1 \\ &\qquad\qquad 0 \leq \alpha \leq 1, \end{aligned} \tag{3.30}$$

where the two constraints correspond to power conservation of the precoder and the defined range of α. The error exponent criterion includes the transmission rate and code length and, hence, is closer to practice than the capacity criterion. However, the bound is derived from a random coding analysis; thus, the actual code used in a system may not satisfy this performance bound.

3.4.3 *Pairwise error probability*

The error probability, averaged over channel fading, is a common system performance measure. For the system in Figure 3.8, the average system error probability is

$$\bar{P}_e = E_{\mathbf{H}}\left[\sum_i p_i \Pr\left(\bigcup_{j\neq i}(\mathbf{C}_i \to \mathbf{C}_j)\right)\right],$$

where p_i is the probability of the codeword $\mathbf{C}_i$, and $(\mathbf{C}_i \to \mathbf{C}_j)$ is the event that $\mathbf{C}_i$ is misdetected as $\mathbf{C}_j$. Since this error expression involves the union of all misdetection events, the probability of which can only be obtained empirically in most cases, especially for a large code space, it offers little useful insight for constructing a precoder analytically. A simpler approach is to use the pairwise codeword error probability (PEP), an error measure that is related loosely to system error performance via the union bound. The PEP is the probability of a codeword $\mathbf{C}$ having a better metric than the codeword $\hat{\mathbf{C}}$, leading to a potential decoding error. The PEP analysis can be applied to systems with or without channel coding.

Assuming a quasi-static block fading channel and ML detection at the receiver:

$$\hat{\mathbf{C}} = \arg\min_{\mathbf{C}\in\mathcal{C}} ||\mathbf{Y} - \sqrt{\mathcal{E}_s}\mathbf{HFC}||_F^2,$$

the PEP can then be upper-bounded by the well-known Chernoff bound [49]

$$\mathrm{P}(\mathbf{C} \to \hat{\mathbf{C}}) \leq \exp\left(-\frac{\gamma}{4}||\mathbf{HF}(\mathbf{C} - \hat{\mathbf{C}})||_F^2\right), \tag{3.31}$$

where $||\cdot||_F$ denotes the Frobenius norm. The Chernoff bound is fairly tight and simplifies the derivation of an optimal precoder. Note that the bound (3.31) depends on the distance between the specific codeword pair $(\mathbf{C}, \hat{\mathbf{C}})$. We define the following expression as the codeword distance product matrix:

$$\mathbf{A} = (\mathbf{C} - \hat{\mathbf{C}})(\mathbf{C} - \hat{\mathbf{C}})^*. \tag{3.32}$$

We can choose to minimize the PEP, either for a chosen codeword distance, or averaged over the codeword distribution, referred to as the PEP per-distance and the average PEP, respectively. In both cases we are interested in the performance averaged over channel fading.

PEP per-distance criterion

The PEP per-distance relates to the system performance through the choice of the codeword distance and the SNR. A common choice is the minimum codeword distance. At high SNRs, minimum-distance codeword pairs dominate the errors, and the PEP for such pairs has the same slope versus the SNR as that of the system error probability. At lower SNRs, however, the minimum-distance effect may not be dominant, and other choices of a codeword distance for the PEP can lead to a comparable system performance, while simplifying the precoder design process. For example, another choice is the average codeword distance, defined as $\bar{\mathbf{A}} = E[\mathbf{A}]/T$, where the expectation is over the codeword distribution. From (3.32), assuming that $\mathbf{C}$ and $\hat{\mathbf{C}}$ are independent, we have

$$\bar{\mathbf{A}} = \frac{2}{T}E[\mathbf{CC}^*] = 2\mathbf{Q}, \tag{3.33}$$

where $\mathbf{Q}$ is the codeword covariance defined in (3.22).

With a chosen distance matrix $\mathbf{A}$ and assuming that only transmit antenna correlation exists, averaging the Chernoff bound over the channel fading statistics (3.5), we obtain an upper bound on the averaged PEP [28],

$$E_{\mathbf{H}}[\text{PEP}(\mathbf{A})] \leq \frac{\exp\left[\text{tr}\left(\mathbf{H}_m(\mathbf{R}_t\Phi\mathbf{R}_t)^{-1}\mathbf{H}_m^*\right)\right]}{\det(\Phi)^{M_R}\det(\mathbf{R}_t)^{M_R}}\exp\left[-\text{tr}(\mathbf{H}_m\mathbf{R}_t^{-1}\mathbf{H}_m^*)\right], \tag{3.34}$$

with

$$\Phi = \frac{\gamma}{4}\mathbf{FAF}^* + \mathbf{R}_t^{-1}.$$

Ignoring the constant terms, the PEP per-distance precoder design problem can be formulated as

$$\begin{aligned} \min_{\mathbf{F}} J &= \text{tr}\left(\mathbf{H}_m(\mathbf{R}_t\Phi\mathbf{R}_t)^{-1}\mathbf{H}_m^*\right) - M_R \log\det(\Phi) \\ \text{subject to } \Phi &= \frac{\gamma}{4}\mathbf{FAF}^* + \mathbf{R}_t^{-1} \\ \text{tr}(\mathbf{FF}^*) &= 1. \end{aligned} \tag{3.35}$$

This formulation aims to find the precoder $\mathbf{F}$ to minimize the averaged PEP at the codeword distance $\mathbf{A}$, using the Chernoff bound approximation, subject to the transmit power constraint and involves only deterministic functions.

Average PEP criterion

Another formulation is obtained by averaging the PEP over both the codeword distribution and the fading statistics. The average PEP calculated in this way is independent of the codeword distance matrix $\mathbf{A}$. Assuming a Gaussian codeword distribution and noting (3.33), the Chernoff bound on the average PEP depends only on the codeword covariance matrix $\mathbf{Q}$ [65],

$$E_{\text{C}}[\text{PEP}] \leq \det\left(\frac{\gamma}{2}\mathbf{HFQF}^*\mathbf{H}^* + \mathbf{I}\right)^{-M_R}.$$

The precoder optimization problem in this case becomes

$$\begin{aligned} &\min_{\mathbf{F}} E_{\mathbf{H}}\left[\det\left(\frac{\gamma}{2}\mathbf{HFQF}^*\mathbf{H}^* + \mathbf{I}\right)^{-M_R}\right] \\ &\text{subject to } \text{tr}(\mathbf{FF}^*) = 1. \end{aligned} \tag{3.36}$$

This formulation aims to find the precoder $\mathbf{F}$ to minimize the PEP, averaged over all codeword distances and the channel statistics, subject to transmit power constraint. Note the similarity between this formulation and the capacity and error exponent criteria, where all the objective functions are expectations of a convex (or concave) function of the Hermitian positive semi-definite (PSD) matrix variable $\mathbf{HFQF}^*\mathbf{H}^*$.

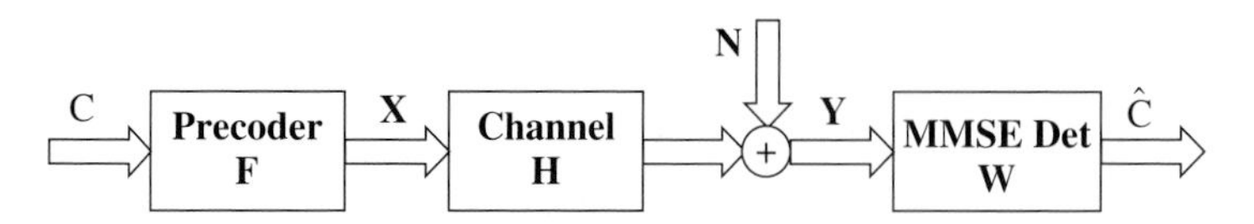

Figure 3.12. Precoding with MMSE receiver.

3.4.4 *Detection mean-squared error*

In many systems, ML detection can be too computationally demanding; therefore, a linear minimum mean-squared error receiver is preferred, as shown in Figure 3.12. The input signal **C** can be an ST code block, or a vector-symbol in the spatial multiplexing case. We examine an uncoded criterion with the MMSE receiver, which treats the combination of the precoder **F** and the channel **H** as an effective channel. Assuming a quasi-static block fading channel, the receiver detects $\hat{\mathbf{C}}$ by minimizing the mean-squared error (MSE)

$$\min_{\mathbf{W}} E||\hat{\mathbf{C}} - \mathbf{C}||^2 = E||(\mathbf{WHF} - \mathbf{I})\mathbf{C} + \mathbf{WN}||_F^2,$$

where the expectation is over the input signal and noise distributions. Assume that the signal has a zero mean with the covariance given in (3.22). The optimum MMSE receiver is then

$$W = \gamma \mathbf{Q}\mathbf{F}^*\mathbf{H}^* \left(\gamma \mathbf{HFQF}^*\mathbf{H}^* + \mathbf{I}\right)^{-1}.$$

Note that this MMSE receiver depends only on the first and second moments of the input codeword distribution. The covariance matrix of the detection error is

$$\mathbf{R}_\epsilon = \mathbf{Q} - \gamma \mathbf{Q}\mathbf{F}^*\mathbf{H}^* \left(\gamma \mathbf{HFQF}^*\mathbf{H}^* + \mathbf{I}\right)^{-1} \mathbf{HFQ}.$$

For a MIMO system, the normalized MSE provides a norm-measure of the MSE. This quantity is defined as the trace of the error covariance normalized by the input signal covariance, $\mathbf{Q}^{-1/2}\mathbf{R}_\epsilon\mathbf{Q}^{-1/2}$ [30]. Averaging the normalized MSE over the channel statistics, we obtain

$$\text{MSE} = M_T - M_R + E_{\mathbf{H}}\left[\text{tr}\left(\left[\gamma \mathbf{HFQF}^*\mathbf{H}^* + \mathbf{I}\right]^{-1}\right)\right]. \tag{3.37}$$

The precoder that minimizes the average MSE subject to the transmit power constraint is given by the optimization problem

$$\begin{aligned} &\min_{\mathbf{F}}\ E_{\mathbf{H}}\left[\text{tr}\left(\left[\gamma \mathbf{HFQF}^*\mathbf{H}^* + \mathbf{I}\right]^{-1}\right)\right] \\ &\text{subject to } \text{tr}(\mathbf{FF}^*) = 1. \end{aligned} \tag{3.38}$$

Unlike the PEP criterion, which can be applied to both coded and uncoded formulations, the MSE criterion is used only in uncoded performance analysis. MSE minimization is a common criterion, although linking the MSE to the actual error performance is not straightforward. Interested readers are referred to [2], where an error bound using the MSE is established.

3.4.5 *Criteria grouping*

We divide the above criteria into two groups, based on the objective function being stochastic or deterministic. The first group, which has stochastic objective functions, consists of the capacity, the error exponent, the average PEP, and the MSE criteria. The optimization problems in this group can be written in a general form as

$$\min\ E\big[f(\mathbf{I}+a\gamma\mathbf{HFQF}^{*}\mathbf{H}^{*})\big] \quad \text{subject to } \operatorname{tr}(\mathbf{FF}^{*})=1, \tag{3.39}$$

where $f(\cdot)$ is a convex function of PSD matrix variables, and a is a criterion dependent constant. For example, f can be $\log\det(\cdot)^{-1}$, $\det(\cdot)^{-\alpha}$, $\det(\cdot)^{-M_R}$, or $\operatorname{tr}(\cdot)^{-1}$, corresponding to the capacity, error exponent, average PEP, or MMSE criterion, respectively.

The second group has a deterministic objective function and contains the PEP per-distance criterion (3.35). These two groups have different solvability properties, as shown in the next section.

There are other precoding design criteria studied in the literature; for example, maximizing the received SNR [38], or minimizing the uncoded symbol error probability [67]. We will, however, focus on the above criteria in this chapter.

3.5 Linear precoder designs

In this section, we discuss precoding solutions for all the outlined criteria. We first examine the impact of the encoder, the STBC if used, on the precoder optimal input-shaping matrix. We then discuss the designs of the optimal beam directions and power allocation for different CSIT scenarios: perfect CSI, correlation CSI, mean CSI, and general statistical CSI – both mean and correlation.

3.5.1 *Optimal precoder input-shaping matrix*

The encoder shapes the covariance of the signal input to the precoder; in response, the precoder chooses its right singular vectors to match this covariance. It can be shown that, for all criteria in group one (3.39), the optimal input-shaping matrix – the precoder right singular vectors – is given by the eigenvectors of the codeword covariance matrix $\mathbf{Q}$ (3.22) [55]

$$\mathbf{V}_{\mathrm{F}}=\mathbf{U}_{\mathrm{Q}}, \tag{3.40}$$

where $\mathbf{Q}=\mathbf{U}_{\mathrm{Q}}\Lambda_{\mathrm{Q}}\mathbf{U}_{\mathrm{Q}}^{*}$ is the eigenvalue decomposition. For the PEP per-distance criterion (3.35), the relevant eigenvectors are those of the codeword difference product matrix $\mathbf{A}$ (3.32); that is, $\mathbf{V}_{\mathrm{F}}=\mathbf{U}_{\mathrm{A}}$, where $\mathbf{A}=\mathbf{U}_{\mathrm{A}}\Lambda_{\mathrm{A}}\mathbf{U}_{\mathrm{A}}^{*}$ is the eigenvalue decomposition of $\mathbf{A}$ [61]. By matching the input signal covariance, the precoder optimally collects the input signal energy. When the STBC produces an identity signal covariance ($\mathbf{Q}=\mathbf{I}$), which also applies to spatial multiplexing, $\mathbf{V}_{\mathrm{F}}$ is an arbitrary unitary matrix and can usually be omitted.

Unlike the optimal input-shaping matrix (the precoder right singular vectors), which is independent of the CSIT, the optimal beam directions (the left singular vectors) and the power allocation (the squared singular values) depend on the CSIT, and are discussed for each CSIT scenario next.

3.5.2 *Precoding on perfect CSIT*

Given perfect CSIT, the MIMO channel can be decomposed into r independent parallel additive white noise channels [9], where $r = \text{rank}(\mathbf{H}) \leq \min(M_R, M_T)$. This decomposition was also discussed in Chapter 2 in terms of capacity; here, we analyze this from a signal processing viewpoint. To demonstrate this, let the singular value decomposition (SVD) of the channel be

$$\mathbf{H} = \mathbf{U_H} \Sigma \mathbf{V}^*_{\mathbf{H}}. \tag{3.41}$$

Then, multiplying the signal at the transmitter with $\mathbf{V_H}$ and at the receiver with $\mathbf{U}^*_{\mathbf{H}}$ results in the parallel channels, corresponding to the original channel eigen-modes, as shown in Figure 3.13. The r parallel channels can be processed independently, each with independent modulation and coding; thus, allowing per-mode rate control. The parallel channel decomposition helps to significantly simplify the receiver signal processing, as the receiver now needs to perform only scalar rather than complex joint detection and decoding.

Optimal beam directions

The parallel channel decomposition implies that the precoder left singular vectors (3.23), or the beam directions, are matched to the channel right singular vectors:

$$\mathbf{U_F} = \mathbf{V_H}. \tag{3.42}$$

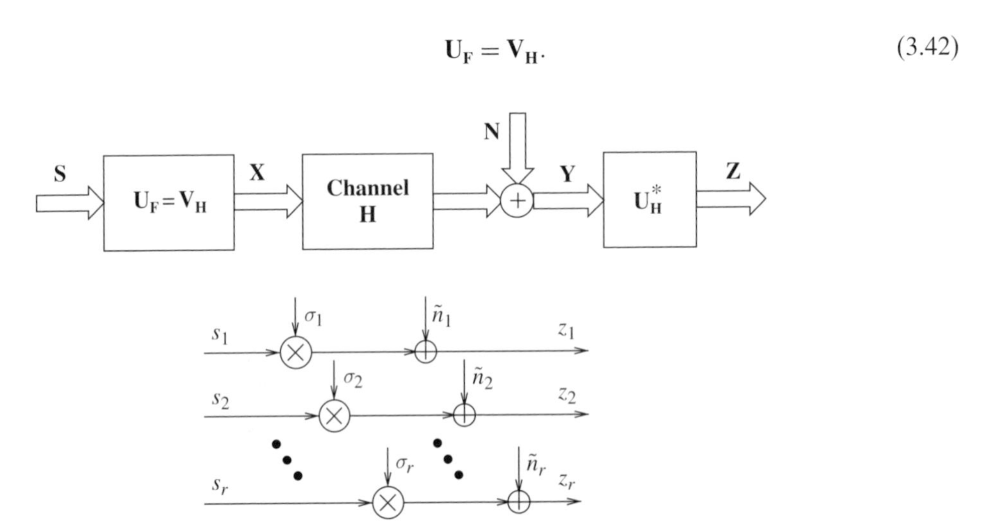

Figure 3.13. Singular value channel decomposition with perfect CSIT.

These beam directions are optimal for all criteria, including the capacity, the error exponent, the average PEP, the PEP per-distance, and the MSE. The optimality can be established using matrix inequalities that show function extrema obtained when the matrix variables have the same eigenvectors [36]. Therefore, the optimal beam directions are given by the eigenvectors of $\mathbf{H}^*\mathbf{H}$, or the *channel eigen-directions*. For multiple-input single-output (MISO) systems, the solution reduces to the well-known scheme: channel matched single-mode beamforming [27]. In all systems, the optimal beam directions with perfect CSIT are independent of the SNR.

Note that, independent of the criteria and the SNR, the left and right singular vectors of the precoder matrix are determined separately by the eigenvectors of the channel gain $\mathbf{H}^*\mathbf{H}$ and the input codeword covariance $\mathbf{Q}$, respectively. Therefore, the precoder matches both sides. It effectively re-maps the spatial signal directions from the input code into those optimally matched to the channel given the CSIT, as shown in Figure 3.14.

Optimal power allocation

In contrast to the beam directions, the optimal power allocation across the beams varies for each design criterion and is a function of the SNR. The power p_i on the beams are the eigenvalues of $\mathbf{FF}^*$, normalized for unit sum. The precoder singular values (3.23) can be established from the beam power as $d_i = \sqrt{p_i}$.

For the first criterion group (3.39), the optimization problem for the power allocation becomes

$$\begin{aligned} \min \quad & f(\mathbf{I} + a\gamma\Lambda_{\mathbf{H}}\Lambda_{\mathbf{F}}\Lambda_{\mathbf{Q}}) \\ \text{subject to} \quad & \text{tr}(\Lambda_{\mathbf{F}}) = 1\,, \quad \Lambda_{\mathbf{F}} \geq 0, \end{aligned} \tag{3.43}$$

where $\Lambda_{\mathbf{H}}$, $\Lambda_{\mathbf{F}}$, and $\Lambda_{\mathbf{Q}}$ are diagonal matrices of the eigenvalues of $\mathbf{H}^*\mathbf{H}$, $\mathbf{FF}^*$, and $\mathbf{Q}$, respectively. The expectation operator in (3.39) is removed in this formulation due to the perfect CSIT. Thus, the power allocation is a function of the channel and codeword-covariance eigenvalues. For convenience, we define

$$\lambda_i = \lambda_i(\mathbf{H}^*\mathbf{H})\lambda_i(\mathbf{Q}), \tag{3.44}$$

where $\lambda_i(\mathbf{H}^*\mathbf{H}) = \sigma_i^2$ in Figure 3.13, and $\lambda_i(\mathbf{Q})$ is the ith eigenvalue of $\mathbf{Q}$, sorted in the same order.

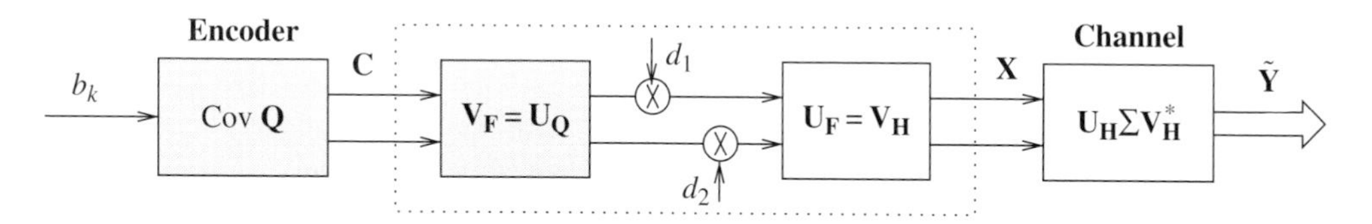

Figure 3.14. The precoder matches both the channel and the input code structure.

For the capacity criterion (3.27), power is allocated among the M_T beams via the standard water-filling solution [9]. The power allocated to channel i is

$$p_i = \left(\kappa - \frac{N_0}{\lambda_i}\right)_+, \tag{3.45}$$

where κ is chosen such that $\sum_i p_i = P$, the total transmit power, and N_0 is the noise power per spatial dimension. The notation $(\cdot)_+$ means that the expression takes the value inside the parentheses if this value is positive; otherwise, it is zero.

For the error exponent criterion (3.30), the optimal power allocation becomes

$$p_i = \left(\kappa - \frac{N_0(1+\alpha)}{\lambda_i}\right)_+, \tag{3.46}$$

where κ satisfies $\sum_i p_i = P$. The error exponent solution is similar to water-filling for capacity, but with the noise scaled by $1+\alpha$. For a particular rate R, the optimal α can be found numerically. Since $0 \le \alpha \le 1$, the effective noise is larger; hence, power will be allocated more selectively using this criterion. At low SNRs, more modes are dropped; as the SNR increases, this solution approaches equi-power more slowly than does the solution for the capacity criterion.

Similarly, for the average PEP criterion (3.36), the optimal power allocation is equivalent to capacity-based water-filling (3.45) with the noise scaled-up by a factor of 2. Again, this solution exhibits a more selective power allocation scheme. At low SNRs, weak modes tend to have a high error rate; therefore, dropping these modes and allocating power to stronger modes leads to better overall system error performance. As the SNR increases, power is allocated across more modes, but again, at a slower rate than is the case for the capacity solution.

The solution for the MMSE criterion (3.38) also resembles water-filling, but with a slight modification. The power allocated to mode i is [18]

$$p_i = \left(\frac{\kappa}{\sqrt{\lambda_i}} - \frac{N_0}{\lambda_i}\right)_+, \tag{3.47}$$

where κ satisfies $\sum_i p_i = P$. Water-filling effectively applies to $p_i\sqrt{\lambda_i}$, which has the water-level κ.

For the PEP per-distance criterion (3.35) in the second group, the solution is to allocate all the power to the strongest eigen-mode of the channel, effectively reducing to single-mode beamforming. This scheme is an extreme case of selective power allocation. The precoder becomes a rank-one matrix with the left singular vector matched to the dominant eigen-direction of the channel, and the right singular vector matched to the dominant eigenvector of the codeword distance product matrix (3.32). Although single-mode beamforming is optimal here for any number of receive antennas, the power gain of this mode (the largest eigenvalue of the channel gain $\mathbf{H}^*\mathbf{H}$) increases with more receive antennas. This solution also maximizes the received SNR.

3.5.3 *Precoding on correlation CSIT*

We now consider correlation CSIT in the form of transmit antenna correlation. The channel is assumed to be Rayleigh fading with zero-mean ($\mathbf{H}_m = 0$) and has the Kronecker correlation structure (3.4). For most of this section, the receive antennas are assumed to be uncorrelated: $\mathbf{R}_r = \mathbf{I}$; although we will briefly mention the effect of receive correlation on the precoder design under the capacity criterion.

The channel model (3.5) can now be written as

$$\mathbf{H} = \mathbf{H}_w \mathbf{R}_t^{1/2}. \tag{3.48}$$

The transmit antenna correlation $\mathbf{R}_t$, a complex Hermitian positive semi-definite matrix, is known at the transmitter.

Optimal beam directions

For all criteria in both groups, we can show that the optimal beam directions of the precoder (3.23) are now the eigenvectors of $\mathbf{R}_t$

$$\mathbf{U}_\mathrm{F} = \mathbf{U}_t, \tag{3.49}$$

where $\mathbf{R}_t = \mathbf{U}_t \Lambda_t \mathbf{U}_t^*$ is the eigenvalue decomposition. This result was established for the channel capacity criterion (with $\mathbf{Q} = \mathbf{I}$) in [54] for MISO systems, and in [25] for MIMO systems. It can be shown, however, that the result also applies to all other criteria under study and holds for arbitrary input codeword covariance $\mathbf{Q}$ [55]. Comparing to the solution for perfect CSIT (3.42), the eigenvectors of $\mathbf{R}_t$ have replaced the channel eigen-directions. The precoder now relies on the statistically preferred directions (the eigenvectors of $\mathbf{R}_t$) to direct its power into the channel.

Several remarks follow. First, when $\mathbf{R}_t = \mathbf{I}$ (uncorrelated transmit antennas), the eigenvectors of $\mathbf{R}_t$ can be any set of M_T orthogonal vectors, and no precoding is necessary. Second, in the presence of receive correlation ($\mathbf{R}_r \neq \mathbf{I}$), assuming the separable Kronecker antenna correlation structure (3.4), the precoder optimal beam directions are still given by the eigenvectors of $\mathbf{R}_t$ (3.49). This result was established for the channel capacity criterion in [29]. The receive correlation, however, does affect the beam power allocation. Finally, we note that both receive and transmit antenna correlations generally reduce the channel ergodic capacity at high SNR, compared to an i.i.d. channel. At low SNR however, transmit correlation can help to increase the capacity [58].

Optimal power allocation

Unlike the beam directions that remain invariant with the criterion, the optimal beam power allocation is different for each criterion. We will examine group one and group two criteria separately.

Group one

For group one criteria (3.39), the optimal power allocation is the solution of the following optimization problem:

$$\begin{aligned} &\min E_{\mathbf{H}_w}\left[f(\mathbf{I}+a\gamma\Lambda_{\mathbf{Q}}\Lambda_{\mathbf{F}}\Lambda_t\mathbf{H}_w^*\mathbf{H}_w)\right] \\ &\operatorname{tr}(\Lambda_{\mathbf{F}})=1,\quad \Lambda_{\mathbf{F}}\geq 0, \end{aligned} \tag{3.50}$$

formulated based on the unitary invariant property of $\mathbf{H}_w$. Here, $\Lambda_{\mathbf{Q}}$, $\Lambda_{\mathbf{F}}$, and Λ_t are diagonal matrices of the eigenvalues of $\mathbf{Q}$, $\mathbf{FF}^*$, and $\mathbf{R}_t$, respectively.

The optimal power solutions are more complicated than those for perfect CSIT, due to the expectation operator. The power allocation, however, follows the water-filling principle: stronger eigenmodes of $\mathbf{R}_t$ are given more power, and weak modes are dropped, depending on the SNR. Numerical techniques are usually required to compute the optimal power allocation. Fortunately, the problem formulation is convex in all group one criteria, allowing efficient numerical algorithms to be implemented (see [7], chapter 11).

When the transmit correlation is strong, the solution can reduce to single-mode beamforming, in which all the transmit power is allocated to the strongest eigenmode of $\mathbf{R}_t$. The condition for the optimality of single-mode beamforming at a given SNR, under the channel capacity criterion ($\mathbf{Q}=\mathbf{I}$), was studied in [25] and [29]. They showed that single-mode beamforming is optimal when the two largest eigenvalues of $\mathbf{R}_t$ satisfy an inequality related to the dominance of the largest eigenvalue. If this largest mode is sufficiently dominant, then water-filling will drop all other modes. At higher SNRs, the required eigenvalue dominance must increase, implying a more correlated channel. A similar trend is observed for an increasing number of receive antennas. However, note that if $\mathbf{R}_t$ is full-rank, then for systems with equal or fewer transmit than receive antennas, the capacity-optimal power allocation asymptotically approaches equi-power as the SNR increases.

A sub-optimal power solution for group one criteria is to perform the allocation on the averaged channel gain. This scheme is a special case of the general group one solution (3.60) discussed in Section 3.5.5.

Group two

For the PEP per-distance criterion (3.35), the power allocation has a closed-form analytical solution. Using the optimal left and right precoder singular vectors, the problem becomes

$$\begin{aligned} &\max_{\mathbf{F}}\ \log\det\left(\Lambda_t^{-1}+\frac{\gamma}{4}\Lambda_{\mathbf{A}}\Lambda_{\mathbf{F}}\right) \\ &\text{subject to } \operatorname{tr}(\Lambda_{\mathbf{F}})=1,\quad \Lambda_{\mathbf{F}}\geq 0. \end{aligned}$$

The solution follows from standard water-filling (see [41]) as

$$p_i = \left(\kappa - \frac{4}{\gamma} \lambda_i^{-1}(\mathbf{A}) \lambda_i^{-1}(\mathbf{R}_t) \right)_+ , \tag{3.51}$$

where κ satisfies $\sum_i p_i = 1$, and p_i are the diagonal values of $\Lambda_\mathbf{F}$.

In general, the stronger the channel correlation (measured by, for example, the condition number of the correlation matrix), the larger the precoding gain (the SNR advantage) from correlation CSIT. In contrast, note that strong correlation usually reduces the channel capacity.

3.5.4 *Precoding on mean CSIT*

We proceed to examine an uncorrelated channel ($\mathbf{R}_h = \mathbf{I}$, or $\mathbf{R}_t = \mathbf{I}$ and $\mathbf{R}_r = \mathbf{I}$) with a non-zero mean $\mathbf{H}_m$ (the Ricean component). The channel model (3.5) can now be written as

$$\mathbf{H} = \mathbf{H}_m + \mathbf{H}_w . \tag{3.52}$$

The channel mean $\mathbf{H}_m$ is an arbitrary complex matrix known to the transmitter. This mean CSIT also covers the channel estimate $\hat{\mathbf{H}}(s)$ in the framework (3.12), assuming uncorrelated errors ($\mathbf{R}_e(s) = \mathbf{I}$). We will discuss precoding solutions using model (3.52) as the representative case.

Optimal beam directions

For all criteria, the optimal beam directions for mean CSIT are given by the eigenvectors of $\mathbf{H}_m^* \mathbf{H}_m$, i.e., the right singular vectors of the channel mean:

$$\mathbf{U}_\mathbf{F} = \mathbf{V}_m , \tag{3.53}$$

where $\mathbf{H}_m = \mathbf{U}_m \Sigma_m \mathbf{V}_m^*$ is the channel mean singular value decomposition. This result was established for the channel capacity criterion (with $\mathbf{Q} = \mathbf{I}$) in [54] for MISO, in [21, 53] for MIMO fading channels, and for the MSE criterion in [30]. The analysis can easily be extended to cover the average PEP criterion. The solution for the PEP per-distance criterion was established in [28].

For mean CSIT, since the channel is uncorrelated, the eigenvectors of the averaged channel gain $E[\mathbf{H}^*\mathbf{H}]$ are the same as those of $\mathbf{H}_m^* \mathbf{H}_m$. Thus, the channel mean eigen-directions are the average channel eigen-directions. Beamforming along these average directions is optimal for the mean CSIT.

Optimal power allocation

The optimal power allocation varies according to the criterion. Again, we consider group one and group two criteria separately.

Group one

The optimal power allocation for group one criteria usually requires numerical search and can be formulated from (3.39) as

$$\min\ E_{\mathbf{H}_w}\left[f\left(\mathbf{I}+a\gamma(\Sigma_m+\mathbf{H}_w)\Lambda_{\mathbf{Q}}\Lambda_{\mathbf{F}}(\Sigma_m+\mathbf{H}_w)^*\right)\right] \quad \text{subject to } \operatorname{tr}(\Lambda_{\mathbf{F}})=1\,,\quad \Lambda_{\mathbf{F}}\geq 0. \tag{3.54}$$

Although no closed-form power allocation solution exists for these criteria, general properties can be established. The power allocation follows the water-filling principle by allocating more power to the stronger modes of the channel mean. Only the singular values, but not the singular vectors, of the channel mean affect the power allocation. The influence of the channel mean on the power allocation can be characterized by the channel K factor (3.3). A larger K factor causes the power allocation to depend strongly on the channel mean. For example, an ill-conditioned or rank-one mean then is likely to result in single-mode beamforming. As K goes to infinity, the mean CSIT is equivalent to a perfect CSIT. The role of the channel mean, however, diminishes as the K factor decreases. If K goes to zero, the precoder becomes an arbitrary unitary matrix, implying equal power allocation, and can be omitted.

The effect of the channel mean on the channel ergodic capacity is analyzed in [21]. They show that the capacity is a monotonically increasing function of the singular values of the channel mean. If all but one of the singular values of the channel means are equal for two channels, then the channel with the larger non-dual singular value has higher capacity.

Similar to correlation CSIT, a sub-optimal power allocation with mean CSIT can be developed, based on the averaged channel gain. This group one criteria solution (3.60) is discussed in Section 3.5.5.

Group two

On the other hand, the PEP per-distance criterion has an analytical power allocation solution. The problem formulation can be deduced from (3.35) as

$$\min_{\lambda_i}\sum_i\left(1+\frac{\gamma}{4}\lambda_i\right)^{-1}\lambda_{m,i}-M_R\sum_i\log\left(1+\frac{\gamma}{4}\lambda_i\right)$$
$$\text{subject to}\sum_i\alpha_i\lambda_i=1,\quad \lambda_i\geq 0,$$

where $\lambda_i=\lambda_i(\mathbf{FF}^*)\lambda_i(\mathbf{A})$, $\alpha_i=1/\lambda_i(\mathbf{A})$ (for non-zero eigenvalues), and $\lambda_{m,i}$ are the eigenvalues of $\mathbf{H}_m^*\mathbf{H}_m$, where all the eigenvalues are sorted in the same order. This problem is convex and can be solved using the standard Lagrange multiplier technique to arrive at

$$\lambda_i=\left[\frac{\lambda_i(\mathbf{A})}{2\nu}\left(M_R+\sqrt{M^2+16\nu\frac{\lambda_{m,i}}{\gamma\lambda_i(\mathbf{A})}}\right)-\frac{4}{\gamma}\right]_+, \tag{3.55}$$

where ν is the Lagrange multiplier satisfying the constraint $\sum_i \alpha_i \lambda_i = 1$. Depending on the number of dropped modes k (where $0 \le k \le M_T - 1$), this ν value can be found using a one-dimensional binary numerical search between the following two bounds:

$$\nu_{\text{upper}} = \frac{4\tilde{\lambda}_M}{\gamma \zeta_k} + \frac{M_R}{\zeta_k}, \qquad \nu_{\text{lower}} = \frac{4\tilde{\lambda}_1}{\gamma \zeta_k} + \frac{M_R}{\zeta_k}, \tag{3.56}$$

where $\tilde{\lambda}_M$ and $\tilde{\lambda}_1$ are the largest and smallest values, respectively, of the set $\{\lambda_{m,i}/\lambda_i(\mathbf{A}) | \lambda_i(\mathbf{A}) \neq 0, \ i = 1, \ldots, M_T\}$, and

$$\zeta_k = \frac{1}{M_T - k}\left(1 + \frac{4}{\gamma} \sum_{i=k+1}^{M_T} \frac{1}{\lambda_i(\mathbf{A})}\right).$$

The beam powers are then

$$p_i = \frac{\lambda_i}{\lambda_i(\mathbf{A})}.$$

This result was first derived for the case $\mathbf{A} = \mu \mathbf{I}$, where μ is a scalar constant, in [28], and later obtained for an arbitrary positive semi-definite $\mathbf{A}$ in [61].

3.5.5 Precoding on both mean and correlation CSIT

Finally, we study a general statistical CSIT model with both a non-zero channel mean $\mathbf{H}_m$ and a transmit antenna correlation $\mathbf{R}_t$. This CSIT also covers the channel estimate model (3.12) with an estimate $\hat{\mathbf{H}}(s)$ and an effective transmit covariance $\mathbf{R}_t(s)$ (3.13). However, we will discuss precoding solutions for the statistical CSIT as the representative case. The channel can be written as

$$\mathbf{H} = \mathbf{H}_m + \mathbf{H}_w \mathbf{R}_t^{1/2}. \tag{3.57}$$

From the statistical parameters, the channel K factor can be established as in (3.3).

Group one precoder

For group one criteria (3.39), the precoder design problem now becomes

$$\begin{aligned} &\max\, E_{\mathbf{H}_w}\left[f\left(\mathbf{I} + a\gamma\left(\mathbf{H}_m + \mathbf{H}_w \mathbf{R}_t^{1/2}\right)\mathbf{FQF}^*\left(\mathbf{H}_m + \mathbf{H}_w \mathbf{R}_t^{1/2}\right)^*\right)\right] \\ &\text{subject to } \operatorname{tr}\left(\mathbf{FF}^*\right) = 1. \end{aligned} \tag{3.58}$$

The optimal precoders for these criteria remain an open problem, in which both the optimal beam directions and the optimal power allocations are unknown. The optimal beam directions depend on both the channel mean and covariance, and are complicated functions of the channel K factor and the SNR. At high K, the channel mean tends to dominate the beam directions. As K drops, the influence of the channel covariance becomes more pronounced. On the other hand, as the SNR increases, the precoder for systems with equal or fewer transmit than receive antennas asymptotically approaches an arbitrary unitary matrix with equal power in all directions and can be omitted, provided that the transmit correlation is full-rank. The interplay between the channel mean and the covariance at different SNRs, therefore, complicates the precoder solution.

A sub-optimal precoder solution can be established by precoding on the averaged channel gain $E[\mathbf{H}^*\mathbf{H}] = \mathbf{H}_m^*\mathbf{H}_m + M_R\mathbf{R}_t$. This solution is obtained by replacing the objective function in (3.58) with its Jensen upper bound [9]. The solution is equivalent to a perfect CSIT solution with $\mathbf{H}^*\mathbf{H}$ being replaced by $\mathbf{H}_m^*\mathbf{H}_m + M_R\mathbf{R}_t$ in the formulation. For this solution, the precoder beam directions are independent of the SNR and are the eigenvectors of $\mathbf{H}_m^*\mathbf{H}_m + M_R\mathbf{R}_t$:

$$\mathbf{U_F} = \mathbf{U_R}, \tag{3.59}$$

where $\mathbf{H}_m^*\mathbf{H}_m + M_R\mathbf{R}_t = \mathbf{U_R}\Lambda_\mathbf{R}\mathbf{U}_\mathbf{R}^*$ is the eigenvalue decomposition. Note that when there is only a correlation or mean CSIT, these beam directions coincide with the optimal beam directions in each respective case.

The sub-optimal power allocation depends on the SNR and follows the analysis for perfect CSIT in Section 3.5.2, where the λ_i in (3.44) now becomes

$$\lambda_i = \lambda_i(\mathbf{H}_m^*\mathbf{H}_m + M_R\mathbf{R}_t)\lambda_i(\mathbf{Q}). \tag{3.60}$$

This solution provides a closed-form power allocation for all group one criteria; it covers both the correlation and mean CSIT cases.

This sub-optimal precoder can be evaluated in terms of capacity [59]. For a system with equal or fewer transmit than receive antennas, the proposed solution results in little or no capacity penalty, which occurs primarily at mid-range SNRs. Systems with more transmit than receive antennas and with strongly correlated transmit antennas (an ill-conditioned correlation matrix, for example) or strong mean (a high K factor), however, may suffer a capacity loss at high SNR. Figure 3.15 shows examples of the system capacity with optimal precoding and with the proposed precoder solution, together with the corresponding power allocations, for two different antenna configurations: 4×4 and 4×2. The system capacity without precoding, equivalent to equal power allocation, is also included for comparison.

Group two precoder

Consider the PEP per-distance criterion (3.35). We distinguish between two cases: when $\mathbf{A}$ is a scaled-identity matrix and when it is not.

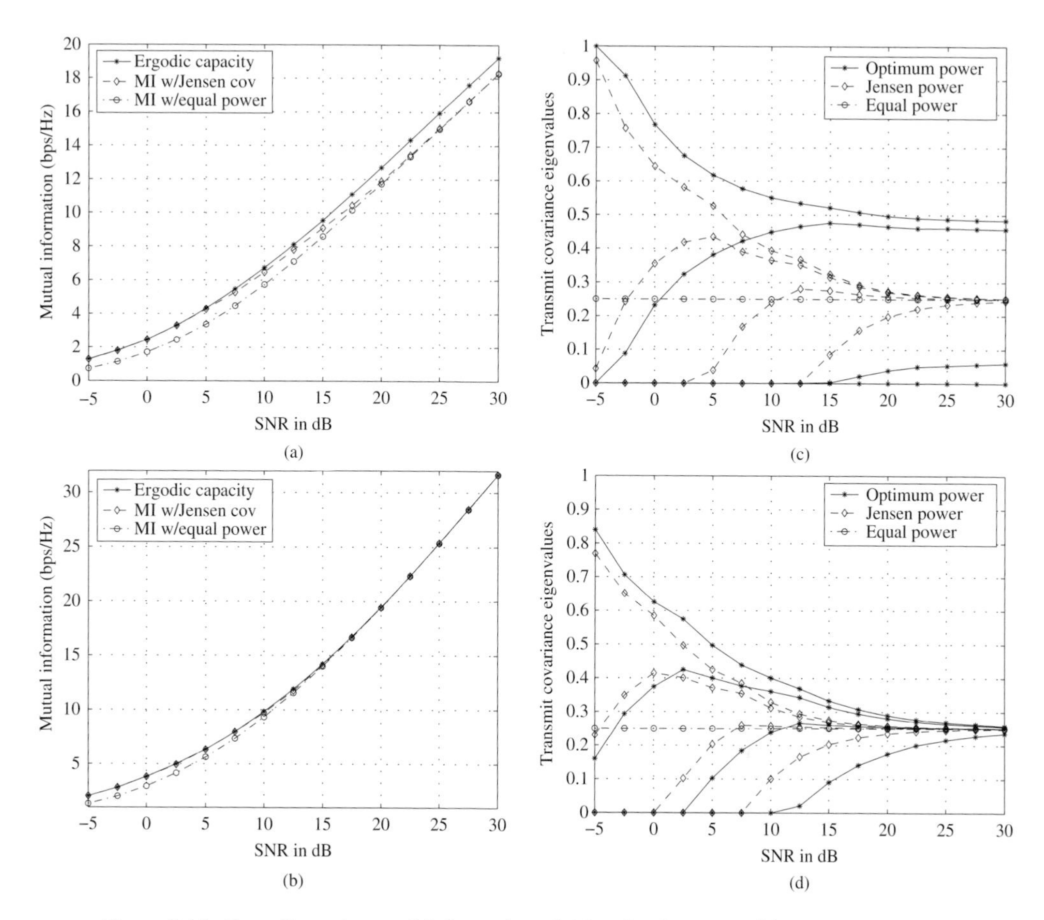

Figure 3.15. Capacity and mutual information of (a) a 4×2 system, (b) a 4×4 system; with the corresponding power allocations (c) and (d). The channel mean and transmit correlation parameters are specified in Appendix 3.1.

For a scaled-identity $\mathbf{A} = \mu_0 \mathbf{I}$, the optimal precoder solution can be established by solving the following convex problem:

$$\begin{aligned}
&\min_{\mathbf{F}} \ \operatorname{tr}\left(\mathbf{H}_m (\mathbf{R}_t \Phi \mathbf{R}_t)^{-1} \mathbf{H}_m^*\right) - M_R \log\det(\Phi) \\
&\text{subject to } \operatorname{tr}\left(\Phi - \mathbf{R}_t^{-1}\right) = \frac{\gamma\mu_0}{4} \\
&\Phi - \mathbf{R}_t^{-1} \geq 0,
\end{aligned}$$

where the precoder $\mathbf{F}$ can be deduced from the solution for Φ as

$$\mathbf{F}\mathbf{F}^* = \frac{4}{\gamma\mu_0}\left(\Phi - \mathbf{R}_t^{-1}\right). \tag{3.61}$$

The Φ solution is given by

$$\Phi = \frac{1}{2\nu}\left[M_R\mathbf{I} + \left(M_R^2\mathbf{I} + 4\nu\mathbf{R}_t^{-1}\mathbf{H}_m^*\mathbf{H}_m\mathbf{R}_t^{-1}\right)^{1/2}\right], \tag{3.62}$$

where ν is the Lagrange multiplier associated with the power equality constraint.

Solving for ν is carried out using a dynamic water-filling process. This process involves mode-dropping due to the PSD constraint $\Phi - \mathbf{R}_t^{-1} \geq 0$, and consists of two steps. In the first step, the precoder is assumed to have full-mode, and ν is found by solving the equation $\mathrm{tr}\left(\Phi - \mathbf{R}_t^{-1}\right) = \frac{1}{4}\gamma\mu_0$. If the ν solution does not produce $\Phi - \mathbf{R}_t^{-1} \geq 0$, then we proceed to the second step. In this step, we drop the weakest mode of $\Phi - \mathbf{R}_t^{-1}$, and re-solve for ν using the equation

$$\sum_{i=2}^{M_T} \lambda_i\left(\Phi - \mathbf{R}_t^{-1}\right) = \frac{\gamma\mu_0}{4}, \tag{3.63}$$

where λ_i are the eigenvalues sorted in increasing order. By dropping the weakest eigenmode, we distribute power over only the $M_T - 1$ stronger modes; hence, the eigenvalue sum in (3.63) starts at 2. We keep repeating this second step, dropping the weakest non-zero mode of $\Phi - \mathbf{R}_t^{-1}$, until the ν solution produces $\Phi - \mathbf{R}_t^{-1} \geq 0$.

Solving for ν, in the general case of the k mode being dropped $(0 \leq k \leq M_T - 1)$, can be performed using a simple one-dimensional binary search between the following bounds:

$$\nu_{\text{upper}} = \frac{\lambda_M}{\beta_k^2} + \frac{M_R}{\beta_k}, \qquad \nu_{\text{lower}} = \frac{\lambda_1}{\beta_k^2} + \frac{M_R}{\beta_k}, \tag{3.64}$$

where λ_M and λ_1 are the maximum and minimum eigenvalues of $\mathbf{R}_t^{-1}\mathbf{H}_m^*\mathbf{H}_m\mathbf{R}_t^{-1}$, respectively, and

$$\beta_k = \frac{1}{M_T - k}\left(\frac{\mu_0\gamma}{4} + \sum_{i=k+1}^{M_T}\frac{1}{\lambda_i(\mathbf{R}_t)}\right).$$

The mode-dropping process above is similar to the standard water-filling, in that at each iteration, the weakest mode may be dropped, and the total transmit power is re-allocated among the remaining modes. However, there is a significant difference in that the mode directions here also evolve at each iteration. This difference arises due to the interaction between the channel mean and transmit correlation. To illustrate, re-write the expression for $\mathbf{FF}^*$ in the following form:

$$\mathbf{FF}^* = \left[\frac{M_R}{2\nu}\mathbf{I}_N + \left(\frac{1}{2\nu}\Psi(\nu)^{1/2} - \mathbf{R}_t^{-1}\right)\right]\frac{4}{\mu_0\gamma},$$

where $\Psi(\nu) = M_R^2\mathbf{I}_N + 4\nu\mathbf{R}_t^{-1}\mathbf{H}_m^*\mathbf{H}_m\mathbf{R}_t^{-1}$ is dependent on ν. The "water-level" is $2M_R/\nu\mu_0\gamma$; the mode directions are determined by the eigenvectors of $\frac{1}{2\nu}\Psi(\nu)^{1/2} - \mathbf{R}_t^{-1}$. Thus, when ν changes at each iteration, both the water-level (hence, the power allocation) and the mode directions change. For this reason, we refer to this process as a *dynamic* water-filling process. Note that since ν depends on the SNR, both the mode (beam) directions and the power allocation are dependent on the SNR in this precoding solution.

When $\mathbf{A}$ is not a scaled-identity matrix, the problem formulation (3.35) cannot be transformed into a convex problem in terms of $\mathbf{FF}^*$; hence, finding an optimal analytical solution is challenging. Several relaxation methods are proposed in [61]. One method is to replace $\mathbf{A}$ with an identity matrix scaled by the smallest non-zero eigenvalue of $\mathbf{A}$. This solution is equivalent to optimizing the PEP for a smaller codeword distance; hence, the precoding gain is likely to be pessimistic.

The asymptotic behavior of this precoder is worth noting. Consider the asymptotic cases when the channel K factor and the SNR approach infinity. When $K \to \infty$, the precoder converges to a solution that depends on the channel mean alone; furthermore, it becomes a single-mode beamformer aligned to the dominant right singular vector of $\mathbf{H}_m$. However, at high SNRs, the precoder approaches the solution for correlation CSIT (3.51), and the channel mean impact vanishes. As SNR $\to \infty$, the power allocation approaches equi-power. These effects require that either the SNR is kept constant as K approaches infinity, or K is constant as the SNR approaches infinity. If both the K factor and the SNR increase, then there exists a K factor threshold for single-mode beamforming, which increases with the SNR. An example of this threshold is given in Section 3.6.

3.5.6 *Discussion*

Several observations can be drawn from the precoding solutions presented. First, the precoding matrix for all criteria in all CSIT scenarios has the same right eigenvectors. These vectors form the input-shaping matrix, matched to the covariance of the precoder input signal, independent of the CSIT and the SNR. Except for the PEP per-distance criterion with general statistical CSIT, the precoders also have the same left eigenvectors. These vectors are the beam directions, matched to the channel according to the CSIT. In most cases, the beam directions are also independent of the SNR. Second, the main difference among the precoding solutions for different criteria is the power allocation. This allocation follows the water-filling principle for all criteria, where more power is allocated to stronger modes, and weak modes are dropped, depending on the SNR. The selectivity in the power allocation, however, varies according to the criterion; more selective schemes tend to drop more modes at low SNRs. The capacity criterion produces the least selective power allocation. As the SNR increases, all the power allocation schemes, except for the PEP per-distance with perfect CSIT, approach equi-power, but at different rates. A more selective scheme approaches equi-power more slowly. In summary, the precoder optimally collects the input signal power and spatially re-distributes this power into the channel according to the criterion and the CSIT.

The water-filling type power allocation leads to mode-dropping at low SNRs. Therefore, care should be taken in the system design to ensure that the employed STBC functions in such situations, especially for high-rate codes. In most cases, the input-shaping matrix, i.e., the precoder right singular vectors, combines the STBC output, such that all symbols are transmitted even with mode-dropping. When the codeword covariance is white, $\mathbf{Q} = \mathbf{I}$, even though this precoder input shaping matrix can theoretically be omitted, some rotation matrix may still be necessary for practical constellations, to ensure the transmission of all distinct symbols. A similar rotation matrix may be necessary with spatial multiplexing to mix the spatial symbol streams before transmitting on each beam (mode), so that all streams are still transmitted in the event of mode-dropping. An initial study of the mixing effect can be found in [33]. The precoder input shaping matrix thus helps to prevent the adverse effects of mode-dropping on the system performance.

3.6 Precoder performance results and discussion

Using the system structure in Figure 3.16, the precoder designs are evaluated for system error performance in different CSIT scenarios via simulations. We generate i.i.d. random bit streams, encode these data with a convolutional code, interleave and map the coded bits into symbols, before encoding with an STBC and precoding for transmission. The signal is then sent through a randomly generated channel and white Gaussian noise is added. At the receiver, we detect and decode the signal, and measure both the uncoded and the coded error rate performance, respectively.

Specific simulation parameters

The simulation system has four transmit and two receive antennas. We employ the following STBC:

$$\mathbf{C} = \begin{pmatrix} c_1 & c_2 & c_3 & c_4 \\ -c_2^* & c_1^* & -c_4^* & c_3^* \\ c_3 & c_4 & c_1 & c_2 \\ -c_4^* & c_3^* & -c_2^* & c_1^* \end{pmatrix}.$$

This is a quasi-orthogonal code [26, 51] with second-order diversity, a spatial rate of 1, and allowing simple joint two-symbol detection. Note that although a 4×2 system can support up to a spatial rate of 2, we only simulate the spatial rate 1 case. For this STBC, the precoder input shaping matrix is the identity matrix and it is omitted. We implement the [133, 171] convolutional code with rate $\frac{1}{2}$, used in the IEEE 802.11a wireless LAN standard [23]; a block interleaver; and the quadrature phase-shift keying (QPSK) modulation. At the receiver, we use maximum-likelihood (ML) detection and a soft-input soft-output Viterbi decoder.

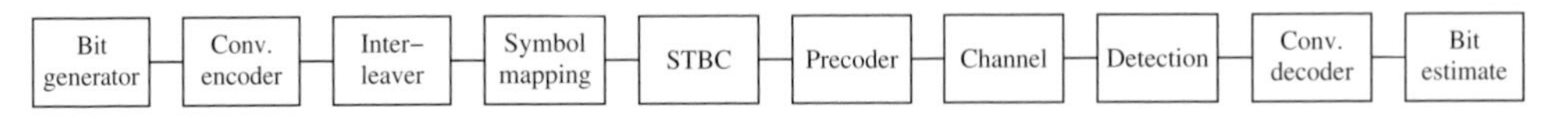

Figure 3.16. Simulation system configuration.

3.6.1 *Performance results*

We present the system performance for several CSIT scenarios: the perfect CSIT, the correlation CSIT, and the channel estimate CSIT (3.12) involving both mean and correlation CSI. We assume quasi-static block-fading channels. The data block-length used for the perfect and correlation CSIT is 96 bits, and for the channel estimate CSIT is 48 bits. We consider the performance without precoding and with precoding, using four criteria: capacity, error exponent (with $\alpha = 0.5$ in (3.30)), average PEP, and minimum-distance PEP. The MSE precoder design is not included due to ML detection.

Perfect CSIT

For perfect CSIT, the channel is assumed to be i.i.d. Rayleigh fading (i.e., $\mathbf{H}_m = 0$ and $\mathbf{R}_h = \mathbf{I}$). The error performance is shown in Figure 3.17. All four precoder designs achieve substantial gains, measured in both uncoded and coded error rates, with a larger gain in the latter case (up to 6 dB at 10^{-4} coded bit error rate). Such a gain is consistent with the predicted capacity gain in (3.16). Since the quasi-orthogonal STBC (QSTBC) provides only partial diversity, some additional diversity gain is obtained by the precoder in the uncoded systems, evident through the higher slopes of the precoded error curves. In both uncoded and coded systems, however, most of the precoding gain is array gain. The array gain is attributed to the optimal beam directions and the water-filling type power allocation. To differentiate the gains due to each of these effects, a two-beam precoder with optimal directions (3.42) but with equal power allocation is also studied. Results show that optimal beam directions alone achieve a significant portion of the precoding gain with perfect CSIT. A water-filling type power allocation further improves the precoding gain, especially at low SNRs. Thus, both the precoder beam directions and the power allocation contribute to the performance gain.

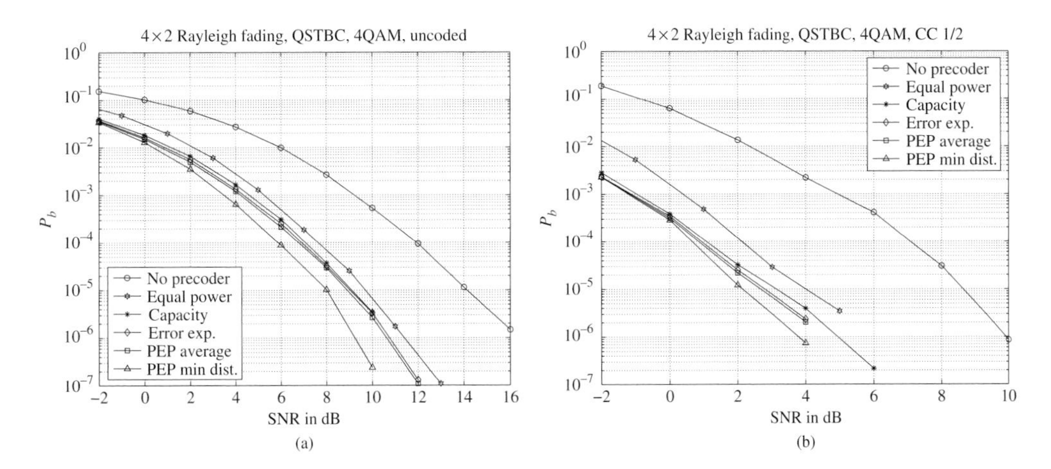

Figure 3.17. System performance with perfect CSIT: (a) uncoded; (b) coded.

These results also reveal only minor performance differences among precoder designs according to the four criteria. The minimum-distance PEP precoder for perfect CSIT is a single-mode beamformer; it achieves the best gain, attributed to the small number of receive antennas. The three precoders based on the capacity, the error exponent, and the average PEP criteria perform similarly. Note that this relative performance order is dependent on the CSIT, the number of antennas, the channel coding, and the STBC; thus, the order may change for a different system configuration.

Correlation CSIT

For correlation CSIT (3.48), we use the transmit correlation matrix (3.66) listed in the Appendix 3.1. This matrix has the eigenvalues [2.717, 0.997, 0.237, 0.049] and a condition number of 55.5. Thus, the transmit antennas are quite strongly correlated; we chose this correlation to emphasize the correlation CSIT gain.

The performance results are shown in Figure 3.18. Again, all four precoders achieve significant gains in both uncoded and coded domains (approximately 3 dB at 10^{-4} coded bit error rate). Note that the minimum-distance PEP precoder for correlation CSIT has moved away from the single-beam solution. A single-beam precoder matched to the dominant eigenvector of the correlation matrix is also included for comparison. In contrast to the perfect CSIT case, this single-beam scheme performs poorly; it has a diversity order of 1 and performs worse than no precoding at high SNRs. Several other observations can be made. First, the precoding gains depend strongly on the CSIT. A more complete CSIT will improve the precoding gain – the perfect CSIT provides the best gain. Second, for statistical CSIT, no precoding diversity gain is present. Both the coded and uncoded precoding error rate curves have the same diversity order of 2 as that of the QSTBC. Third, we again observe similar performance among the precoders for all four criteria. Note that the precoding gain also depends on the transmit correlation. A more correlated channel will result in a higher precoding gain.

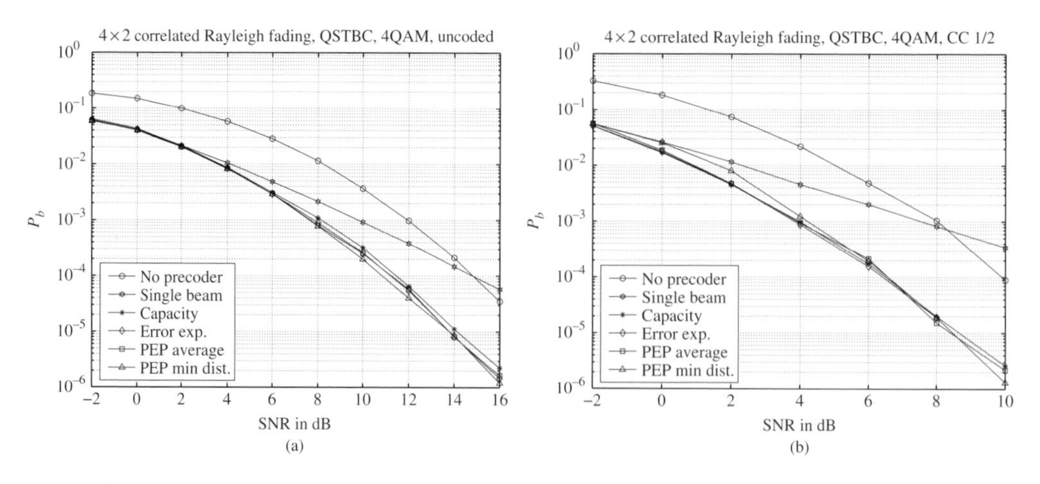

Figure 3.18. System performance with correlation CSIT: (a) uncoded; (b) coded.

Mean and correlation CSIT

We now examine the general CSIT framework (both mean and covariance CSIT) that involves channel estimates (3.12). For this CSIT scenario, we choose a representative precoder design based on the minimum-distance PEP criterion, for which an optimal precoder is known. Simulation results in the last two sections suggest that precoders based on the capacity, the error exponent, and the average PEP criteria have a similar performance.

For the channel statistics, we use the same transmit correlation (3.66), the channel mean (3.67), and set $K = 0.1$. We simulate the system performance with different values of the estimate quality ρ, using initial channel measurements $\mathbf{H}(0)$ randomly drawn from the channel distribution. The error rates are averaged over multiple initial measurements and multiple channel estimates, given each initial measurement.

The performances with and without precoding for $\rho = [0,\ 0.7,\ 0.8,\ 0.9,\ 0.96,\ 0.99]$ are given in Figure 3.19. Several observations follow. First, the precoding gain increases with a higher estimate quality. Depending on ρ, the gain ranges between statistical CSIT gain and perfect CSIT gain. Second, the initial channel measurement $\mathbf{H}(0)$ helps to increase the precoding gain over the statistical CSIT gain only when its correlation with the current channel is sufficiently strong, $\rho \geq 0.6$. When $\rho < 0.6$, precoding on the channel statistics can extract most of the available gain. Third, the simulations exhibit some differences in the slopes of the bit error rate curves at low-to-medium SNRs for large ρ values ($\rho \geq 0.9$). However, analyses show that the asymptotic bit error rate slope at high SNRs is independent of ρ for $\rho < 1$ [60], for which the system transmit diversity is determined by the code – the STBC for the uncoded system, and the combination of the STBC and the convolutional code for the coded one. Only when $\rho = 1$, corresponding to perfect CSIT, does the precoder deliver the maximum transmit diversity gain of order M_T. Thus, the precoder primarily offers an array gain; when the CSIT is perfect, it delivers additional diversity gain.

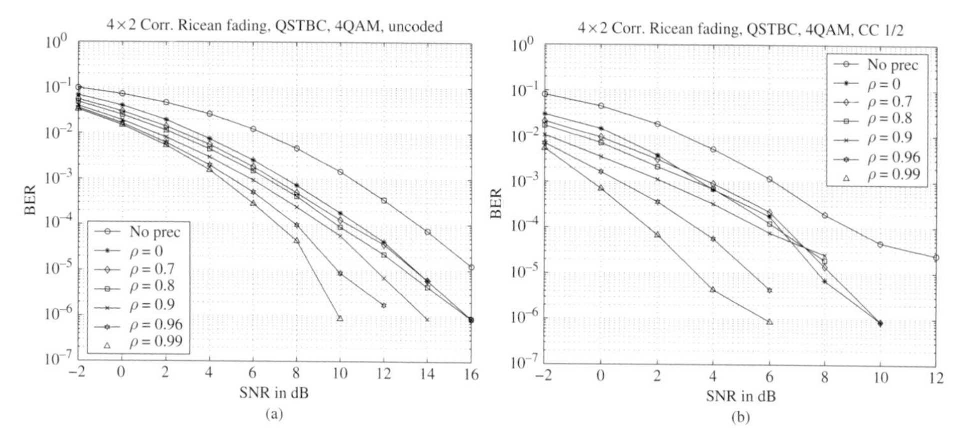

Figure 3.19. System performance with the channel estimate CSIT (3.12) using the minimum-distance PEP precoder: (a) uncoded; (b) coded.

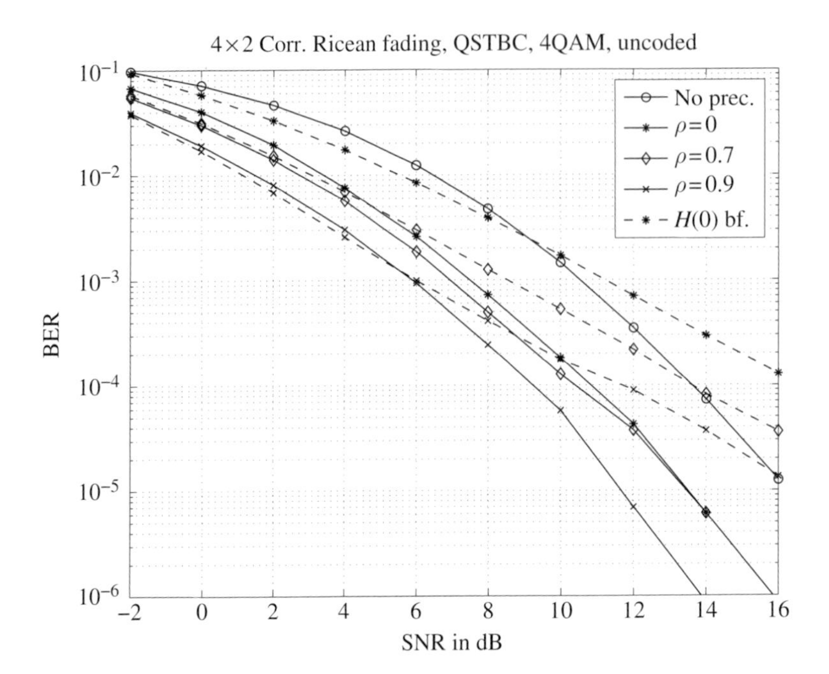

Figure 3.20. Comparison between the minimum-distance PEP precoder (solid lines) and a single-beam scheme (dashed lines) performance.

For comparison, we study a single-beam scheme that relies only on the initial channel measurement, shown in Figure 3.20. This scheme coincides with the optimal minimum-distance PEP precoder for a perfect CSIT ($\rho = 1$). For other ρ values, however, this scheme performs poorly. It loses all transmit diversity regardless of the STBC and only achieves second-order receive diversity due to ML detection with two receive antennas; this scheme performs worse than no precoding at high SNRs. The precoder exploiting the channel estimate CSIT (3.12), on the other hand, provides gains at all SNRs for all ρ. This result demonstrates the robustness of this CSIT framework.

Finally, we plot the number of beams of the minimum-distance PEP precoder as a function of the channel K factor and the SNR in Figure 3.21. A higher K factor leads to fewer beams; whereas a higher SNR leads to more beams. Note that other design criteria may lead to different precoding beam regions.

3.6.2 Discussion

We have presented numerical results on the precoding performance for different CSIT scenarios. The precoding gains are significant in both uncoded and coded domains, with higher gains in the latter. The gain depends on the CSIT, the number of antennas, the system configuration, and the SNR. The channel estimate CSIT model (3.12) is a robust framework ranging from statistical CSIT to perfect CSIT. The precoding gain usually increases with more antennas. For the simulated system configuration, we observe similar performance among precoders based on different criteria: capacity, error exponent,

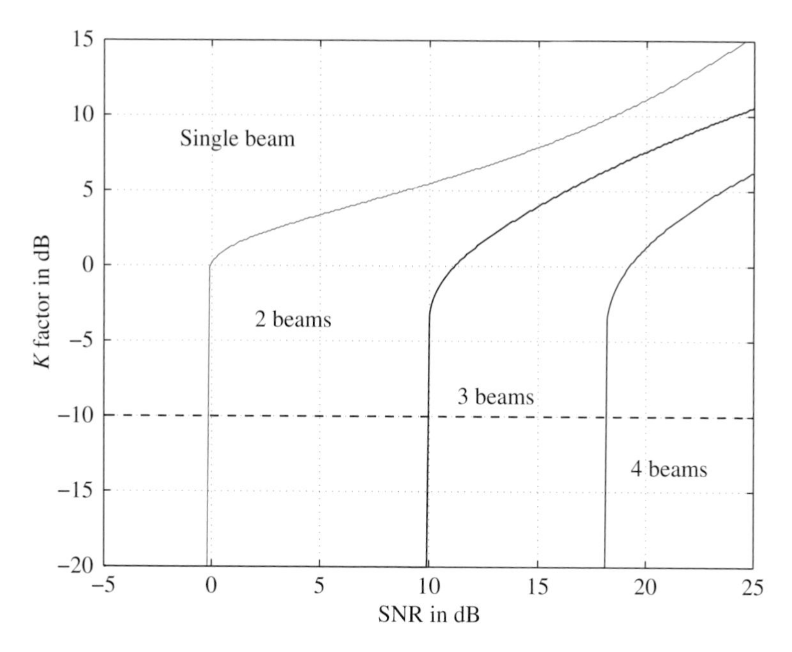

Figure 3.21. Numbers of beams used in the minimum-distance PEP precoder.

average PEP, and minimum-distance PEP. The precoding gain is composed of two parts: an array gain obtained by the optimal beam directions and a water-filling gain obtained by the power allocation; both result in an SNR advantage. When the CSIT is perfect, the precoders also deliver an additional diversity gain.

3.7 Applications in practical systems

This section focuses on practical issues in precoding. We first discuss how channel information is acquired by the transmitter using open- and closed-loop acquisition techniques. We then discuss codebook design in closed-loop systems to compress channel information efficiently. Finally, we give an overview of the precoding status in emerging wireless standards.

3.7.1 Channel acquisition methods

From Section 3.1, we know that the two principles for obtaining CSIT are reciprocity and feedback. Practical channel acquisition techniques based on these principles are categorized accordingly as open- and closed-loop methods. Open-loop methods are applicable to TDD systems that support duplex communications, where the time lag between the reverse and forward transmissions is relatively small compared to the channel coherence time (3.6). In FDD systems, however, due to the large frequency offset between the forward and reverse link (normally 5% of the carrier frequency), channel reciprocity generally does not hold; thus, open-loop channel acquisition cannot be used. On the other hand, closed-loop methods are applicable to both TDD and FDD systems. We discuss each method in more detail below, assuming CSIT is needed at a base node in a multi-user communication network. Thus, we refer to the transmission from the base node to a user as the forward-link, and from the user to the base as the reverse-link. Similar methods can be used for obtaining CSIT at a user node.

Open-loop methods

Open-loop acquisition methods obtain CSIT based on the channel reciprocity principle. Consider a base node that is a transmitter for the forward link and is a receiver for the reverse link. The node measures the reverse channel during receive and uses this measurement as the forward channel CSIT. In voice applications, the forward and reverse links to all the users operate in back-to-back time slots. Therefore, the reverse channel measurements can be made regularly using the embedded pilots in these transmissions. These measurements periodically refresh the CSIT. In data communications, the forward and reverse links may not operate back-to-back; hence, specially scheduled reverse link transmissions for channel measurements, known as channel sounding, are used. The subset of the users, for whom CSIT are required, are scheduled to send a sounding transmission. The sounding signals are orthogonal among simultaneously scheduled users. In OFDM systems, for example, different users may be assigned non-overlapping, interlaced tones that span the channel bandwidth; or they are assigned orthogonal pilot code sequences that overlap on the entire frequency band, similar to CDMA. Channel sounding is efficient for systems with many antennas at the base node.

The reciprocity principle is applicable for the "over the air" (i.e. antenna to antenna) segment of the forward and reverse links. In practical systems, however, signal processing is performed at the baseband, i.e., the channel is estimated at the receiver baseband section after the signal has passed through the receive RF chain. The transmit signal uses a different RF chain, which has a different transfer function from that of the receive chain, as depicted in Figure 3.22. Therefore, the reciprocity principle can only be applied after the transmit (or receive) RF chain is equalized to make the two chains identical. This equalization requires a calibration process, wherein the difference between the two chains is identified. Calibration is expensive and has made open-loop methods less attractive in practice.

Transmit–receive chain calibration and equalization have been widely studied. Let $H_1(f)$ and $H_2(f)$ be the transmit- and receive-chain transfer functions, respectively. One technique involves first finding $H_1(f)H_2(f)$ using the transmit and receive chains in a loop-back mode. Then, $H_2(f)$ is determined by injecting a calibration signal at the antenna. In both steps, thermal-noise-induced errors can be averaged out by taking multiple measurements [6]. When both $H_1(f)$ and $H_2(f)$ are available, it is easy to compute a digital equalizer incorporated into the baseband section to make the two chains effectively identical. This equalizer usually requires high numerical precision and accuracy. For a flat channel, e.g. per OFDM tone, the equalizer reduces to a complex

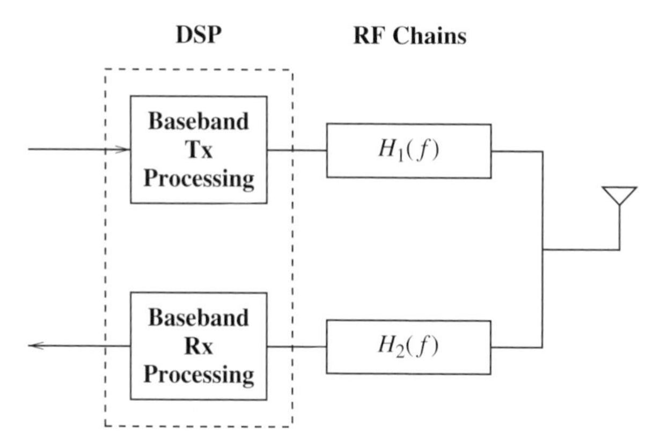

Figure 3.22. A transceiver front-end with transmit and receive RF chains.

scalar coefficient per antenna branch. Calibration must be performed periodically to track the slow time variations of the RF chains.

Closed-loop methods

Closed-loop acquisition methods use feedback from the receiver to send the channel information to the transmitter. The forward-link transmission from a base node, received by all the active users, includes pilot signals. Thus, these users can measure their respective receive channels. The required users then send these channel measurements on a reverse link back to the base node for use as the CSIT. The feedback communication can either be scheduled or piggybacked on on-going transmissions. Again, in data communications, the CSIT may be needed for only a subset of users; these users are scheduled to transmit their channel measurements. Closed-loop methods do not require the transmit–receive calibration that is necessary in open-loop techniques.

Nevertheless, feedback information is susceptible to errors, due to delay in the feedback loop. The usefulness of feedback depends on this time lag and the channel Doppler spread. For a time-varying channel in mobile communications, feedback techniques are usually effective up to a certain mobile speed, depending on the carrier frequency, the transmission frame length, and the feedback turn-around time. The effects of feedback delays and errors have been analyzed for various techniques in 3GPP [22, 34], revealing potentially severe performance degradation. Therefore, the optimal use of feedback information must account for the feedback quality, as in the framework (3.12).

Overhead in MIMO CSIT acquisition

In a closed-loop method, if multiple antennas are used for transmit, then additional pilot resources, proportional to the number of antennas, are needed. This training overhead on the forward link from a base node is independent of the number of users. The overhead for the feedback, however, is proportional to the number of designated users on the reverse link, multiplied with the size of the channel matrix, which is a product of the number of transmit antennas at the base node and the number of receive antennas at each user.

The overhead in an open-loop system is the product of the number of training pilots on the reverse link, which is proportional to the number of antennas at each user, and the number of users participating in the reverse channel sounding.

The overhead related to CSIT acquisition remains a major concern in multiple antenna systems. The overhead comparison in open- versus closed-loop systems typically favors open-loop systems. However, when the number of receive antennas on the forward link is much larger than the number of transmit antennas, closed-loop systems may have an advantage.

3.7.2 *Codebook design in closed-loop systems*

In closed-loop methods, the feedback requires precious transmission (bandwidth, time) resources. Thus, compressing, or quantizing, the feedback data is of great interest. Quantized channel feedback information can be modeled as the mean CSIT with a known error covariance [28, 38], similarly to (3.12). The precoder can then be designed at the transmitter using a suitable algorithm. Another approach is to design the precoder at the receiver and to send back the quantized precoding matrix, rather than the CSI. For example, a unitary precoder is designed based on the instantaneous channel measurement at the receiver, assuming that the CSI remains valid over the feedback duration [34, 35]. The choice between these two approaches depends on the system setup, the feedback overhead in each method, and the complexity requirement of the transmitter and the receiver. The ease of adapting the feedback information to channel variation is another factor to consider. In some systems, scheduling delays may not be known to the receiver; hence, having the CSI at the transmitter can make it easier to estimate the channel for precoding use.

The quantization method in closed-loop systems is a rich research topic. Several techniques have been explored. One technique is incremental encoding, in which only the relevant change in the channel is transmitted. Another technique is to design codebooks that efficiently encode important channel information. Since the channel eigen-directions are particularly important, several authors have studied sphere packing on the Grassmannian manifold as a way of designing compact codebooks [20, 34]. Codebook design depends on the channel statistics. For example, the design for an i.i.d. Rayleigh fading MIMO channel corresponds to finding subspaces in a Grassmannian manifold that are uniformly spaced, such that the distances between them, suitably defined, are approximately equal. There are still open questions regarding codebook designs for non-i.i.d. channels, particularly structures that are invariant to channel variations, and regarding the complexity in mapping the matrix channel or the precoder to a codeword.

3.7.3 *The role of channel information at the receiver*

In this chapter, we have assumed perfect channel knowledge at the receiver and focused on variable quality of CSIT alone. We now briefly discuss the role of receive channel side information (CSIR).

Early wireless systems did, in fact, assume very little or no CSIR and there has been a rich body of techniques for dealing with this problem. The typical solution was to assume that the channel is constant over two or more symbol periods and to encode the information in the difference of transmit symbols. The channel invariance between adjacent symbols can then be exploited to extract the desired information. No knowledge of the actual CSIR is needed. This technique of course comes with a capacity penalty, typically a 3 dB loss in SNR compared to coherent detection. Later, wireless systems were designed to guarantee good CSIR, alleviating the penalty.

In emerging mobile networks, however, good CSIR is increasingly difficult to obtain due to low SINR at the receiver and the frequency selectivity of the channel. We first discuss the extreme case of no CSIR and then comment on partial CSIR. Information-theoretic results for MIMO systems have shown that without either CSIR or CSIT, the channel coherence time T plays an important role. The high-SNR capacity then grows as $M^\star(1-M^\star/T)$ bps/Hz for every 3 dB increase in SNR, where $M^\star = \min\{M_T, M_R, \lfloor T/2 \rfloor\}$ and T is measured in symbol periods; whereas with perfect CSIR, it grows as $\min\{M_T, M_R\}$ [65]. Notice that the loss in capacity vanishes as T increases. Without CSIR, having more transmit than receive antennas provides no capacity gain at high SNR. Increasing the number of transmit antennas beyond T also provides no increase in capacity [37]. Furthermore, the optimal signal consists of mutually orthogonal vectors, of which the directions are isotropically distributed and statistically independent of the magnitudes. Thus any transmit precoding should be restricted to power loading alone. When $T \geq \min\{M_T, M_R\} + M_R$, these optimal vectors at high SNR have equi-power, implying that there is no need for precoding. Practical schemes for MIMO systems include differential space–time modulation techniques and non-coherent matrix designs. The differential techniques have received greater attention.

With partial CSIR, the channel estimate at the receiver is often modeled as the actual channel plus a zero-mean Gaussian noise. Recent results suggest designing a space–time constellation based on a metric combining between coherent and non-coherent criteria, depending on the CSIR quality [14]. Note also that such imperfect CSIR results in no diversity loss in space–time decoding [47].

3.7.4 *Precoding in emerging wireless standards*

Precoding has been successfully incorporated in the IEEE 802.16e standard for broadband wireless metropolitan networks. In the closed-loop approach, the precoder is based on either an initial channel measurement or the channel statistics. The users measure the channel using the forward-link preambles or pilots. A codebook technique is then used to feed back the best unitary fit of the channel measurement, along with a time-to-live parameter. The precoder uses the unitary fit until the time-to-live expires. Thereafter, the precoder relies on the channel statistics information, which is updated at a much slower rate and, therefore, is always valid. In an open-loop technique, a subset of users are scheduled to transmit a sounding signal. The base-station then estimates the channels for these users and determines the CSIT after the transmit–receive RF calibration.

MIMO is included in the IEEE 802.11n standard for wireless local area networks (WLANs). Space–time coding and spatial multiplexing are both supported. The current precoding proposals use an open-loop method. The reciprocity principle implies that the best beam on receive must be the best beam for transmit. The access point uses pre-formed beams for receive and transmit and records the beam(s) with the best signal strength on receive for each user, then uses the same beam(s) during transmit.

The 3GPP standard uses a closed-loop beamforming technique, based on feedback of the channel phase and amplitude information. Precoding is under discussion in high-speed downlink packet access (HSDPA) for mobile communication. Channel-sounding techniques appear to be the preferred approach.

3.8 Conclusion

Before summarizing this chapter, we briefly discuss other CSIT types and open problems in the rich area of exploiting CSIT.

3.8.1 *Other types of CSIT*

This chapter has mainly focused on the CSIT in the form of an estimate of the entire channel with its error covariance. However, there are other types of less complete CSIT. One example is the knowledge of the channel K factor and the channel phase distribution [57], where the solution requires beamforming on the average channel phase with variable antenna power allocation. Another CSIT example is a known channel condition number [19], where the solution suggests adapting the transmission spatial rate. Nonlinear precoding for MIMO channels with inter-symbol interference (ISI), using the Tomlinson–Harashima precoder, has also been studied [12].

3.8.2 *Open problems in exploiting CSIT*

Exploiting CSIT is a current research area that still has many open problems. For example, an analytical solution for the optimal precoder with statistical CSIT (both mean and covariance) for the capacity criterion remains unsolved. Precoding with a more general channel covariance structure (for example, a non-Kronecker model) has hardly been studied. Codebook design and its application in compressing channel information efficiently for closed-loop systems is also a rich research area. Finally, exploiting variable CSIT in multi-user systems is an important open area with promising practical applications.

3.8.3 *Summary*

This chapter has provided an overview of CSIT acquisition and linear precoding techniques exploiting CSIT in MIMO wireless systems. Principles and methods for acquiring transmit channel information are discussed, including open- and closed-loop techniques, and related issues, such as the sources of error, system overhead, and complexity. The definition of a CSIT form is given as an estimate of the channel at the transmit time with an associated error covariance. Such CSIT can be obtained using a possibly outdated channel measurement, along with the first- and second-order channel statistics, and the channel temporal correlation factor. Information-theoretic foundation confirms the optimality of a linear precoder in exploiting the CSIT. A linear precoder, in essence, is an input shaper together with a multi-mode beamformer with defined per-beam power.

The chapter provides linear precoder solutions for several CSIT scenarios, involving a channel mean (or channel estimate) and a channel covariance (or error covariance), according to different performance criteria: the ergodic capacity, the error exponent, the pairwise error probability, and the detection mean-squared error. Simulation examples, using a spatial rate 1 STBC transmission, demonstrate that precoding improves the error performance at all SNRs. For higher spatial rate transmissions (such as spatial multiplexing), although not illustrated, precoding also improves the capacity and error performance at all SNRs for systems with more transmit than receive antennas, and at low SNRs for equal or more receive antennas than transmit antennas.

The essential value of precoding is that it exploits the CSIT to add an array gain, resulting in increased SNR. This gain is achieved by the optimal eigen-beam directions (patterns) and the spatial water-filling type power allocation across these beams. Both of these features can be used to increase the transmission rate (the system capacity) or reduce the error rate, or both. If the CSIT is perfect, precoding can also deliver a diversity gain; in addition, it helps to reduce receiver complexity for higher spatial-rates by allowing parallel channel transmissions.

Having discussed precoding, the last block in the transmit processing chain, the next chapter will discuss space–time coding, which appears immediately prior to the precoding block.

3.9 Bibliographical notes

An overview of transmit precoding can be found in 'Introduction to Space–Time Wireless Communications' by Paulraj, Nabar, and Gore [39]. A key reference for the information-theoretic foundations for channels with causal CSIT is the paper by Caire and Shamai (Shitz) [8], with extension to MIMO systems by Skoglund and Jöngren [46]. References for discussion on beamforming and its use in the presence of partial CSIT from a capacity viewpoint include Narula, Lopez, Trott, and Wornell [38], Visotsky and Madhow [54], Jafar and Goldsmith [25], and Jorswieck and Boche [29]. Discussions of statistical CSIT-related precoding to enhance error performance can be found in papers by Jöngren, Skoglund, and Ottersten [28], Sampath and Paulraj [41], Zhou and Giannakis [67, 68], and Vu and Paulraj [61].

Appendix 3.1

For simulations, all channels are normalized for a constant average power gain of $M_T M_R$ (the product of the numbers of transmit and receive antennas). The normalization ensures the same average power gain in all channels for fair performance comparison. The channel parameters, mean (3.1) and covariance (3.2), can be written as

$$\mathbf{H}_m = \sqrt{\frac{K}{K+1}}\mathbf{H}_0 \qquad (3.65)$$
$$\mathbf{R}_h = \frac{1}{K+1}\mathbf{R}_0,$$

where $\mathbf{H}_0$ and $\mathbf{R}_0$ are the normalized channel mean and covariance, such that

$$||\mathbf{H}_0||_F^2 = M_T M_R$$
$$\mathrm{tr}(\mathbf{R}_0) = M_T M_R.$$

For the Kronecker antenna correlation model (3.4), the normalized channel covariance can be written as

$$\mathbf{R}_0 = \mathbf{R}_{t,0}^T \otimes \mathbf{R}_{r,0},$$

where

$$\mathrm{tr}(\mathbf{R}_{t,0}) = M_T$$
$$\mathrm{tr}(\mathbf{R}_{r,0}) = M_R.$$

When only transmit correlation exists, this correlation becomes

$$\mathbf{R}_t = \frac{1}{K+1}\mathbf{R}_{t,0}.$$

Most simulations in this chapter are performed for four transmit antenna channels, with two receive antennas for the error performance in Section 3.6, and two or four receive antennas for the capacity in Figure 3.15. These simulations use $K = 0.1$, except for Figure 3.21. Other channel parameters are listed below.

The transmit correlation matrix is

$$\mathbf{R}_{t,0} = \begin{bmatrix} 0.8758 & -0.0993-0.0877i & -0.6648-0.0087i & 0.5256-0.4355i \\ -0.0993+0.0877i & 0.9318 & 0.0926+0.3776i & -0.5061-0.3478i \\ -0.6648+0.0087i & 0.0926-0.3776i & 1.0544 & -0.6219+0.5966i \\ 0.5256+0.4355i & -0.5061+0.3478i & -0.6219-0.5966i & 1.1379 \end{bmatrix}. \qquad (3.66)$$

The mean for the 4×2 channel is

$$\mathbf{H}_0 = \begin{bmatrix} 0.0749 - 0.1438i & 0.0208 + 0.3040i & -0.3356 + 0.0489i & 0.2573 - 0.0792i \\ 0.0173 - 0.2796i & -0.2336 - 0.2586i & 0.3157 + 0.4079i & 0.1183 + 0.1158i \end{bmatrix}. \tag{3.67}$$

The mean for the 4×4 channel is

$$\mathbf{H}_0 = \begin{bmatrix} 0.2976 + 0.1177i & 0.1423 + 0.4518i & -0.0190 + 0.1650i & -0.0029 + 0.0634i \\ -0.1688 - 0.0012i & -0.0609 - 0.1267i & 0.2156 - 0.5733i & 0.2214 + 0.2942i \\ 0.0018 - 0.0670i & 0.1164 + 0.0251i & 0.5599 + 0.2400i & 0.0136 - 0.0666i \\ -0.1898 + 0.3095i & 0.1620 - 0.1958i & 0.1272 + 0.0531i & -0.2684 - 0.0323i \end{bmatrix}. \tag{3.68}$$

References

[1] S. Alamouti, "A simple transmit diversity technique for wireless communications," *IEEE J. Select. Areas Commun.*, vol. 16, pp. 1451–1458, Oct. 1998.

[2] P. Balaban and J. Salz, "Optimum diversity combining and equalization in digital data transmission with applications to cellular mobile radio – Part I: Theoretical considerations," *IEEE Trans. Commun.*, vol. 40, no. 5, pp. 885–894, May 1992.

[3] J.-C. Belfiore, G. Rekaya, and E. Viterbo, "The Golden code: a 2×2 full-rate space–time code with nonvanishing determinants," *IEEE Trans. Inform. Theory*, vol. 51, no. 4, pp. 1432–1436, Apr. 2005.

[4] M. Bengtsson and B. Ottersten, "Optimal and suboptimal transmit beamforming," in *Handbook of Antennas in Wireless Communications*. CRC Press, 2001.

[5] D. Bliss, A. Chan, and N. Chang, "MIMO wireless communication channel phenomenology," *IEEE Trans. Antennas Propagation*, vol. 52, no. 8, pp. 2073–2082, Aug. 2004.

[6] A. Bourdoux, B. Come, and N. Khaled, "Non-reciprocal transceivers in OFDM/SDMA systems: impact and mitigation," *Proc. Radio and Wireless Conf.*, pp. 183–186, Aug. 2003.

[7] S. Boyd and L. Vandenberghe, *Convex Optimization*. Cambridge University Press, 2003. [Online]. Available: http://www.stanford.edu/~boyd/cvxbook.html

[8] G. Caire and S. S. Shamai, "On the capacity of some channels with channel state information," *IEEE Trans. Inform. Theory*, vol. 45, no. 6, pp. 2007–2019, Sept. 1999.

[9] T. Cover and J. Thomas, *Elements of Information Theory*. John Wiley & Sons, 1991.

[10] H. El Gamal, G. Caire, and M. Damen, "Lattice coding and decoding achieve the optimal diversity–multiplexing tradeoff of MIMO channels," *IEEE Trans. Inform. Theory*, vol. 50, no. 6, pp. 968–985, June 2004.

[11] H. El Gamal and M. Damen, "Universal space–time coding," *IEEE Trans. Inform. Theory*, vol. 49, pp. 1097–1119, May 2003.

[12] R. Fischer, C. Stierstorfer, and J. Huber, "Precoding for point-to-multipoint transmission over MIMO ISI channels," *Proc. Int. Zurich Sem. Commun.*, pp. 208–211, Feb. 2004.

[13] R. Gallager, *Information Theory and Reliable Communication*. Wiley & Sons, 1968.

[14] J. Giese and M. Skoglund, "Space–time constellation design for partial CSI at the receiver," *Proc. IEEE Int. Symp. on Inform. Theory*, pp. 2213–2217, Sept. 2005.

[15] D. Goeckel, "Adaptive coding for time-varying channels using outdated fading estimates," *IEEE Trans. Commun.*, vol. 47, no. 6, pp. 844–855, June 1999.

[16] A. Goldsmith and P. Varaiya, "Capacity of fading channels with channel side information," *IEEE Trans. Inform. Theory*, vol. 43, no. 6, pp. 1986–1992, Nov. 1997.

[17] A. Graham, *Kronecker Products and Matrix Calculus with Application*. Ellis Horwood, 1981.

[18] T. Haustein and H. Boche, "Optimal power allocation for MSE and bit-loading in MIMO systems and the impact of correlation," *Proc. IEEE Int. Conf. on Acoustics, Speech, and Signal Processing*, vol. 4, pp. 405–408, Apr. 2003.

[19] R.W. Heath Jr. and A. Paulraj, "Switching between diversity and multiplexing in MIMO systems," *IEEE Trans. Commun.*, vol. 53, pp. 962–968, June 2005.

[20] B. Hochwald, T. Marzetta, T. Richardson, W. Sweldens, and R. Urbanke, "Systematic design of unitary space–time constellations," *IEEE Trans. Inform. Theory*, vol. 46, pp. 1962–1973, Sept. 2000.

[21] D. Hösli and A. Lapidoth, "The capacity of a MIMO Ricean channel is monotonic in the singular values of the mean," *Proc. 5th Int. ITG Conf. on Source and Channel Coding*, Jan. 2004.

[22] A. Hottinen, O. Tirkkonen, and R. Wichman, *Multi-antenna Transceiver Techniques for 3G and Beyond*. Wiley & Sons, 2003.

[23] I.S. 802.11a, "Part 11: Wireless LAN medium access control (MAC) and physical layer (PHY) specifications high-speed physical layer in the 5 GHz band," *IEEE Standards*, June 1999.

[24] I.S. 802.16e, "Part 16: Air interface for fixed and mobile broadband wireless access systems," *IEEE Standards*, Oct. 2005.

[25] S. Jafar and A. Goldsmith, "Transmitter optimization and optimality of beamforming for multiple antenna systems," *IEEE Trans. Wireless Commun.*, vol. 3, no. 4, pp. 1165–1175, July 2004.

[26] H. Jafarkhani, "A quasi-orthogonal space time block code," *IEEE Trans. Commun.*, vol. 49, no. 1, pp. 1–4, Jan. 2001.

[27] W. Jakes, *Microwave Mobile Communications*. IEEE Press, 1994.

[28] G. Jöngren, M. Skoglund, and B. Ottersten, "Combining beamforming and orthogonal space–time block coding," *IEEE Trans. Inform. Theory*, vol. 48, no. 3, pp. 611–627, Mar. 2002.

[29] E. Jorswieck and H. Boche, "Channel capacity and capacity-range of beamforming in MIMO wireless systems under correlated fading with covariance feedback," *IEEE Trans. Wireless Commun.*, vol. 3, no. 5, pp. 1543–1553, Sept. 2004.

[30] E. Jorswieck, A. Sezgin, H. Boche, and E. Costa, "Optimal transmit strategies in MIMO Ricean channels with MMSE receiver," *Proc. Vehicular Tech. Conf.*, Sept. 2004.

[31] T. Kailath, A. Sayed, and B. Hassibi, *Linear Estimation*. Prentice-Hall, 2000.

[32] J. Kermoal, L. Schumacher, K. Pedersen, P. Mogensen, and F. Frederiksen, "A stochastic MIMO radio channel model with experimental validation," *IEEE J. Select. Areas Commun.*, vol. 20, no. 6, pp. 1211–1226, Aug. 2002.

[33] T. Kim, G. Jöngren, and M. Skoglund, "Weighted space–time bit-interleaved coded modulation," *Proc. IEEE Inform. Theory Workshop*, pp. 375–380, Oct. 2004.

[34] D. Love and R. Heath, Jr., "Limited feedback unitary precoding for orthogonal space–time block codes," *IEEE Trans. Signal Processing*, pp. 64–73, Jan. 2005.

[35] D. Love and R. Heath, Jr., "Limited feedback unitary precoding for spatial multiplexing systems," *IEEE Trans. Inform. Theory*, vol. 51, pp. 2967–2976, Aug. 2005.

[36] A. Marshall and I. Olkin, *Inequalities: Theory of Majorization and its Applications*. Academic Press, 1979.

[37] T. Marzetta and B. Hochwald, "Capacity of a mobile multiple-antenna communication link in Rayleigh flat-fading," *IEEE Trans. Inform. Theory*, vol. 45, no. 1, pp. 139–157, Jan. 1999.

[38] A. Narula, M. Lopez, M. Trott, and G. Wornell, "Efficient use of side information in multiple-antenna data transmission over fading channels," *IEEE J. Select. Areas Commun.*, vol. 16, no. 8, pp. 1423–1436, Oct. 1998.

[39] A. Paulraj, R. Nabar, and D. Gore, *Introduction to Space–Time Wireless Communications*. Cambridge University Press, 2003.

[40] T. Rappaport, *Wireless Communications: Principles and Practice*. Prentice-Hall, 1996.

[41] H. Sampath and A. Paulraj, "Linear precoding for space–time coded systems with known fading correlations," *IEEE Commun. Lett.*, vol. 6, no. 6, pp. 239–241, June 2002.

[42] A. Sayeed, "Deconstructing multiantenna fading channels," *IEEE Trans. Signal Processing*, vol. 50, no. 10, pp. 2563–2579, Oct. 2002.

[43] B. Sethuraman, B. Sundar Rajan, and V. Shashidhar, "Full-diversity, high-rate space–time block codes from division algebras," *IEEE Trans. Inform. Theory*, vol. 49, pp. 2596–2616, Oct. 2003.

[44] C. Shannon, "Channels with side information at the transmitter," *IBM J. Res. Devel.*, vol. 2, no. 4, pp. 289–293, Oct. 1958.

[45] D. Shiu, G. Foschini, M. Gans, and J. Kahn, "Fading correlation and its effect on the capacity of multielement antenna systems," *IEEE Trans. Commun.*, vol. 48, no. 3, pp. 502–513, Mar. 2000.

[46] M. Skoglund and G. Jöngren, "On the capacity of a multiple-antenna communication link with channel side information," *IEEE J. Select. Areas Commun.*, vol. 21, no. 3, pp. 395–405, Apr. 2003.

[47] G. Taricco and E. Biglieri, "Space–time decoding with imperfect channel estimation," *IEEE Trans. Wireless Commun.*, vol. 4, no. 4, pp. 1874–1888, July 2005.

[48] V. Tarokh, H. Jafarkhani, and R. Calderbank, "Space–time block codes from orthogonal designs," *IEEE Trans. Inform. Theory*, vol. 45, no. 5, pp. 1456–1467, July 1999.

[49] V. Tarokh, N. Seshadri, and R. Calderbank, "Space–time codes for high data rate wireless communication: performance criterion and code construction," *IEEE Trans. Inform. Theory*, vol. 44, no. 2, pp. 744–765, Mar. 1998.

[50] I. Telatar, "Capacity of multi-antenna Gaussian channels," *Bell Laboratories Technical Memorandum*, Oct. 1995. Available: http://mars.bell-labs.com/papers/proof/

[51] O. Tirkkonen, A. Boariu, and A. Hottinen, "Minimal non-orthogonality rate 1 space–time block code for 3+ tx antennas," *Proc. IEEE ISSSTA2000*, vol. 2, pp. 429–432, Sep. 2000.

[52] G. Turin, "On optimal diversity reception, II," *IRE Trans. Commun. Systems*, vol. 10, no. 1, pp. 22–31, Mar. 1962.

[53] S. Venkatesan, S. Simon, and R. Valenzuela, "Capacity of a Gaussian MIMO channel with nonzero mean," *Proc. IEEE Vehicular Tech. Conf.*, vol. 3, pp. 1767–1771, Oct. 2003.

[54] E. Visotsky and U. Madhow, "Space–time transmit precoding with imperfect feedback," *IEEE Trans. Inform. Theory*, vol. 47, no. 6, pp. 2632–2639, Sept. 2001.

[55] M. Vu, *Exploiting Transmit Channel Side Information in MIMO Wireless Systems.* Stanford University PhD Dissertation, 2006.

[56] M. Vu and A. Paulraj, "Some asymptotic capacity results for MIMO wireless with and without channel knowledge at the transmitter," *Proc. 37th Asilomar Conf. Sig., Sys. and Comp.*, vol. 1, pp. 258–262, Nov. 2003.

[57] M. Vu and A. Paulraj, "Optimum space–time transmission for a high K factor wireless channel with partial channel knowledge," *Wiley J. Wireless Commun. Mobile Computing*, vol. 4, pp. 807–816, Nov. 2004.

[58] M. Vu and A. Paulraj, "Characterizing the capacity for MIMO wireless channels with non-zero mean and transmit covariance," *Proc. 43rd Allerton Conf. on Communications, Control, and Computing*, Sept. 2005.

[59] M. Vu and A. Paulraj, "Capacity optimization for Rician correlated MIMO wireless channels," *Proc. 39th Asilomar Conf. Sig., Sys. and Comp.*, Nov. 2005.

[60] M. Vu and A. Paulraj, "A robust transmit CSI framework with applications in MIMO wireless precoding," *Proc. 39th Asilomar Conf. Sig., Sys. and Comp.*, pp. 623–627, Nov. 2005.

[61] M. Vu and A. Paulraj, "Optimal linear precoders for MIMO wireless correlated channels with non-zero mean in space–time coded systems," *IEEE Trans. Signal Processing*, vol. 54, pp. 2318–2322, June 2006.

[62] W. Weichselberger, M. Herdin, H. Özcelik, and E. Bonek, "A stochastic MIMO channel model with joint correlation of both link ends," *IEEE Trans. Wireless Commun.*, vol. 5, no. 1, pp. 90–100, Jan. 2006.

[63] H. Yao and G. Wornell, "Structured space–time block codes with optimal diversity–multiplexing tradeoff and minimum delay," *Proc. IEEE Global Telecom. Conf.*, vol. 4, pp. 1941–1945, Dec. 2003.

[64] K. Yu, M. Bengtsson, B. Ottersten, D. McNamara, P. Karlsson, and M. Beach, "Second order statistics of NLOS indoor MIMO channels based on 5.2 GHz measurements," *Proc. IEEE Global Telecomm. Conf.*, vol. 1, pp. 25–29, Nov. 2001.

[65] L. Zheng and D. Tse, "Communication on the Grassmann manifold: a geometric approach to the noncoherent multiple-antenna channel," *IEEE Trans. Inform. Theory*, vol. 48, no. 2, pp. 359–383, Feb. 2002.

[66] L. Zheng and D. Tse, "Diversity and multiplexing: a fundamental trade-off in multiple-antenna channels," *IEEE Trans. Inform. Theory*, vol. 49, no. 5, pp. 1073–1096, May 2003.

[67] S. Zhou and G. Giannakis, "Optimal transmitter eigen-beamforming and space–time block coding based on channel mean feedback," *IEEE Trans. Signal Processing*, vol. 50, no. 10, pp. 2599–2613, Oct. 2002.

[68] S. Zhou and G. Giannakis, "Optimal transmitter eigen-beamforming and space–time block coding based on channel correlations," *IEEE Trans. Inform. Theory*, vol. 49, no. 7, pp. 1673–1690, July 2003.

4 Space–time coding for wireless communications: principles and applications

4.1 Introduction

The essential feature of wireless transmission is the randomness of the communication channel which leads to random fluctuations in the received signal commonly known as fading. This randomness can be exploited to enhance performance through *diversity*. We broadly define diversity as the method of conveying information through multiple independent instantiations of these random fades. There are several forms of diversity; our focus in this chapter will be on *spatial* diversity through multiple independent transmit/receive antennas. Information theory has been used to show that multiple antennas have the potential to dramatically increase achievable bit rates [76], thus converting wireless channels from narrow to wide data pipes.

The earliest form of spatial transmit diversity is the delay diversity scheme proposed in [81, 84] where a signal is transmitted from one antenna, then delayed one time slot, and transmitted from the other antenna. Signal processing is used at the receiver to decode the superposition of the original and time-delayed signals. By viewing multiple-antenna diversity as independent information streams, more sophisticated transmission (coding) schemes can be designed to get closer to theoretical performance limits. Using this approach, we focus on space–time coding (STC) schemes defined by Tarokh *et al.* [74] and Alamouti [5], which introduce temporal and spatial correlation into the signals transmitted from different antennas without increasing the total transmitted power or the transmission bandwidth. There is, in fact, a diversity gain that results from multiple paths between the base-station and the user terminal, and a coding gain that results from how symbols are correlated across transmit antennas. Significant performance improvements are possible with only two antennas at the base-station and one or two antennas at the user terminal, and with simple receiver structures. The second antenna at the user terminal can be used to further increase system capacity through interference suppression.

In only a few years, space–time codes have progressed from invention to adoption in the major wireless standards. For wideband code-division multiple access (WCDMA) where short spreading sequences are used, transmit diversity provided by space–time codes represents the difference between data rates of 100 and 384 kb/s. Our emphasis is on solutions that include channel estimation, joint decoding and equalization, and where the complexity of signal processing is practical. The new world of multiple transmit and receive antennas requires significant modification of techniques developed for single-transmit single-receive communication. Since receiver cost and complexity is an important consideration, our treatment of innovation in signal processing is grounded in systems with one, two or four transmit antennas and one or two receive antennas. For example, the interference cancellation techniques presented in Section 4 enable transmission of 1 Mb/s over a 200 kHz GSM/EDGE channel using four transmit and two receive antennas. Hence, our limitation on numbers of antennas does not significantly dampen user expectations.

Initial STC research efforts focused on narrowband flat-fading channels [5, 62, 74]. Successful implementation of STC over multi-user broadband frequency-selective channels requires the development of novel, practical, and high-performance signal processing algorithms for channel estimation, joint equalization/decoding, and interference suppression. This task is quite challenging due to the long delay spread of broadband channels which increases the number of channel parameters to be estimated and the number of trellis states in joint equalization/decoding, especially with multiple transmit antennas. This, in turn, places significant additional computational and power consumption loads on user terminals. On the other hand, development and implementation of such advanced algorithms for broadband wireless channels promises even more significant performance gains than those reported for narrowband channels [5, 62, 74] due to availability of multi-path (in addition to spatial) diversity gains that can be realized. By virtue of their design, space–time-coded signals enjoy rich algebraic structure that can (and should!) be exploited to develop near-optimum reduced-complexity modem signal processing algorithms.

The organization of this chapter is as follows. We start in Section 4.2 with background material where we set up the broadband wireless channel model assumed, followed by a discussion of transmit diversity and the concept of diversity order. Section 4.3 describes STC design criteria and discusses representative examples with both the trellis and block structure. We also give some recent developments in space–time codes. Section 4.4 shows through concrete examples from signal processing, coding theory, and networking, how the STC algebraic structure can be exploited to enhance system performance and reduce implementation complexity. The chapter concludes in Section 4.5 with a summary and a discussion of several future challenges.

4.2 Background

4.2.1 *Broadband wireless channel model*

A typical outdoor wireless propagation environment is represented in Figure 4.1 where the mobile wireless terminal is communicating with a wireless access point (base-station). The signal transmitted from the mobile may reach the access point directly (line-of-sight) or through multiple reflections on local scatterers (buildings, mountains, etc.). As a result,

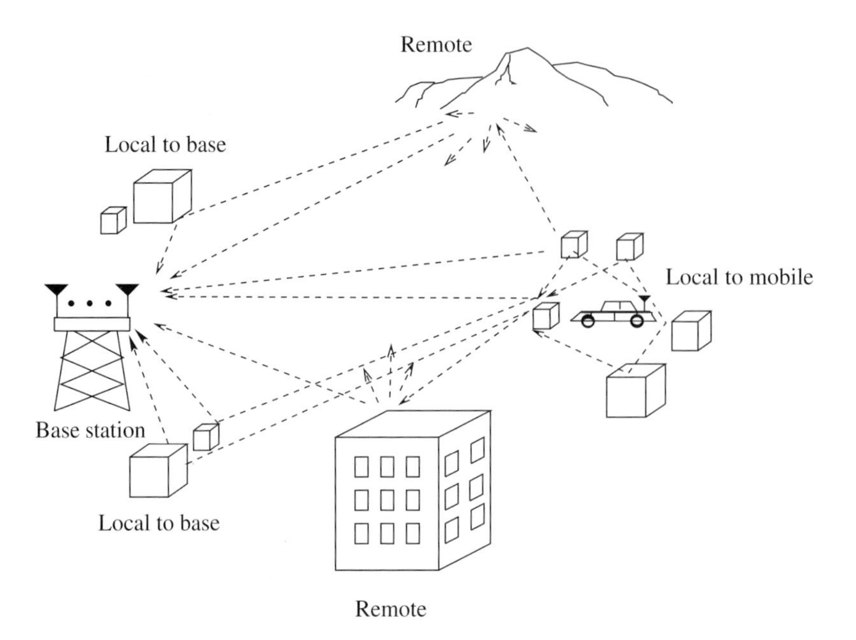

Figure 4.1. Radio propagation environment.

the received signal is affected by multiple random attenuations and delays. Moreover, the mobility of either the nodes or the scattering environment may cause these random fluctuations to vary with time. Furthermore, a shared wireless environment may cause undesirable interference to the transmitted signal. This combination of factors makes wireless a challenging communication environment. For a transmitted signal $s(t)$, the continuous-time received signal $y_c(t)$ can be expressed as

$$y_c(t) = \int h_c(t;\tau)\, s(t-\tau)\, d\tau + z(t)\,, \tag{4.1}$$

where $h_c(t;\tau)$ is the response of the time-varying channel[1] if an impulse is sent at time $t-\tau$, and $z(t)$ is the additive Gaussian noise. To collect discrete-time sufficient statistics we need to sample (4.1) faster than the Nyquist rate. That is, we sample (4.1) at a rate larger than $2(W_I + W_s)$, where W_I is the input bandwidth and W_s is the bandwidth of the channel time-variation. In this chapter, we assume that this criterion is met and therefore we focus on the following discrete-time model:

$$y(k) = y_c(kT_s) = \sum_{l=0}^{\nu} h(k;l)x(k-l) + z(k)\,, \tag{4.2}$$

where $y(k)$, $x(k)$, and $z(k)$ are the output, input, and noise samples at sampling instant k, respectively, and $h(k;l)$ represents the sampled time-varying channel impulse response

[1] Including the effects of transmit/receive filters.

(CIR) of finite memory ν. Any loss in modeling the channel as having a finite-duration impulse response can be made small by appropriately selecting ν.

Three key characteristics of broadband mobile wireless channels are time-selectivity, frequency-selectivity, and space-selectivity. Time-selectivity arises from mobility, frequency-selectivity arises from broadband transmission, and space-selectivity arises from the spatial interference patterns of the radio waves. The corresponding key parameters in the characterization of mobile broadband wireless channels are *coherence time*, *coherence bandwidth*, and *coherence distance*. The coherence time is the time duration over which each CIR tap can be assumed constant. It is approximately equal to the inverse of the Doppler frequency.[2] The channel is said to be *time-selective* if the symbol period is longer than the channel coherence time. The coherence bandwidth is the frequency duration over which the channel frequency response can be assumed flat. It is approximately equal to the inverse of the channel delay spread.[3] The channel is said to be *frequency-selective* if the symbol period is smaller than the delay spread of the channel. Likewise, the coherence distance is the maximum spatial separation over which the channel response can be assumed constant. This can be related to the behavior of arrival directions of the reflected radio waves and is characterized by the *angular spread* of the multiple paths [50, 65]. The channel is said to be space-selective between two antennas if their separation is larger than the coherence distance.

The channel memory causes interference among successive transmitted symbols that results in significant performance degradation unless corrective measures (known as equalization) are implemented. In this chapter, we shall use the terms *frequency-selective channel*, *broadband channel*, and *intersymbol interference (ISI) channel* interchangeably. The introduction of M_t transmit and M_r receive antennas leads to the following generalization of the basic channel model:

$$\mathbf{y}(k) = \sum_{l=0}^{\nu} \mathbf{H}(k; l)\,\mathbf{x}(k-l) + \mathbf{z}(k)\,, \tag{4.3}$$

where the $M_r \times M_t$ complex matrix $\mathbf{H}(k; l)$ represents the lth tap of the channel matrix response with $\mathbf{x} \in \mathbb{C}^{M_t}$ as the input and $\mathbf{y} \in \mathbb{C}^{M_r}$ as the output. The input vector may have independent entries to achieve high throughput (e.g. through spatial multiplexing) or correlated entries through coding or filtering to achieve high reliability (better distance properties, higher diversity, spectral shaping, or desirable spatial profile). Throughout this chapter, the input is assumed to be zero mean and to satisfy an average power constraint, i.e. $\mathbb{E}[||\mathbf{x}(k)||^2] \leq P$. The vector $\mathbf{z} \in \mathbb{C}^{M_r}$ models the effects of noise and interference.[4] It is assumed to be independent of the input and is modeled as a complex additive circularly-symmetric Gaussian vector with $\mathbf{z} \sim \mathcal{CN}(0, \mathbf{R}_{zz})$, i.e. a complex Gaussian vector with

[2] The Doppler frequency is a measure of the frequency spread experienced by a pure sinusoid transmitted over the channel. It is equal to the ratio of the mobile speed to the carrier wavelength.

[3] The channel delay spread is a measure of the time spread experienced by a pure impulse transmitted over the channel.

[4] Including co-channel interference, adjacent channel interference, and multi-user interference.

mean **0** and covariance $\mathbf{R}_{zz}$. Finally, we modify the basic channel model to accommodate a block or frame of N consecutive symbols. Now, (4.3) can be expressed in matrix notation as follows:

$$\mathbf{y} = \mathbf{H}\mathbf{x} + \mathbf{z}\,, \tag{4.4}$$

where $\mathbf{y}, \mathbf{z} \in \mathbb{C}^{N.M_r}$, $\mathbf{x} \in \mathbb{C}^{M_t(N+\nu)}$, and $\mathbf{H} \in \mathbb{C}^{N.M_r \times M_t(N+\nu)}$. In each input block, we insert a guard sequence of length equal to the channel memory ν to eliminate inter-block interference (IBI). In practice, the most common choices for the guard sequence are the all-zeros sequence (also known as *zero stuffing*) and the *cyclic prefix* (CP). When the channel is known at the transmitter, it is possible to increase throughput by optimizing the choice of the guard sequence.

The channel model in (4.4) includes several popular special cases. First, the quasi-static channel model follows by assuming the channel to be time-invariant within the transmission block. In this case, using the cyclic prefix makes the channel matrix **H** *block-circulant*, hence diagonalizable using the fast Fourier transform (FFT). Second, the flat-fading channel model follows by setting $\nu = 0$ which renders the channel matrix **H** a *block diagonal* matrix. Third, the channel model for single-antenna transmission, reception, or both follows directly by setting M_t, M_r, or both equal to 1, respectively.

4.2.2 Transmit diversity

Transmit diversity is more challenging to provision and realize than receive diversity because it involves the design of multiple correlated signals from a single information signal without utilizing CSI (typically not available accurately at the transmitter). Furthermore, transmit diversity must be coupled with effective receiver signal processing techniques that can extract the desired information signal from the distorted and noisy received signal. Transmit diversity is more practical than receive diversity for enhancing the downlink (which is the bottleneck in broadband asymmetric applications such as Internet browsing and downloading) to preserve the small size and low power consumption features of the user terminal. A common attribute of transmit and receive diversity is that both experience "diminishing returns" (i.e. diminishing SNR gains at a given error probability) as the number of antennas increases [50]. This makes them effective, from a performance–complexity trade-off point of view, for small numbers of antennas (typically less than four). This is in contrast with spatial multiplexing gains where the rate continues to increase linearly with the number of antennas (assumed equal at both ends).

There are two main classes of multiple-antenna transmitter techniques: closed-loop and open-loop. The former uses a feedback channel to send CSI acquired at the receiver back to the transmitter to be used in signal design while the latter does not require CSI. Assuming availability of ideal (i.e. error-free and instantaneous) CSI at the transmitter, closed-loop techniques have an SNR advantage of $10\log_{10}(M_t)$ dB over open-loop techniques due to the "array gain" factor [5]. However, several practical factors degrade the performance of closed-loop techniques including channel estimation errors at the receiver, errors in the feedback link (due to noise, interference, and quantization effects), and feedback delay which causes a mismatch between the available and the actual CSI. All of these factors combined with the extra bandwidth and system complexity resources needed for the feedback link make open-loop techniques more attractive as a robust means for improving downlink performance for high-mobility applications while closed-loop techniques (such as beamforming) become attractive under low-mobility conditions. Our focus in this chapter will be exclusively on open-loop spatial transmit diversity techniques due to their applicability to both scenarios[5]. Beamforming techniques are discussed extensively in several tutorial papers such as [38, 39].

The simplest example of open-loop spatial transmit diversity techniques is *delay diversity* [81, 84], where the signal transmitted at sampling instant k from the ith antenna is $x_i(k) = x(k - l_i)$ for $2 \leq i \leq M_t$ and $x_1(k) = x(k)$, where l_i denotes the time delay (in symbol periods) on the ith transmit antenna. Assuming a single receive antenna, the D-transform[6] of the received signal is given by

$$y(D) = x(D)\left(h_1(D) + \sum_{i=2}^{M_t} D^{l_i} h_i(D)\right) + z(D)\,. \tag{4.5}$$

It is clear from (4.5) that delay diversity transforms spatial diversity into multi-path diversity that can be realized through equalization [67]. For flat-fading channels, we can set $l_i = (i - 1)$ and achieve full (i.e. order-M_t) spatial diversity using a maximum-likelihood (ML) equalizer with $(2^b)^{M_t - 1}$ states, where 2^b is the input alphabet size. However, for frequency-selective channels, a delay of at least $l_i = (i-1)(\nu+1)$ is needed to ensure that coefficients from the various spatial FIR channels do not interfere with each other causing a diversity loss. This, in turn, increases equalizer complexity to $(2^b)^{(M_t-1)(\nu+1)}$ states, which is prohibitive even for moderate b, M_t, and ν. In Section 4.3, we describe another family of open-loop spatial transmit diversity techniques known as space–time block codes that achieve full spatial diversity with practical complexity even for frequency-selective channels with a long delay spread.

4.2.3 *Diversity order*

Error probability is particularly important as a performance criterion when we are coding over a small number of blocks (low-delay) where the Shannon capacity is zero [63] and, therefore, we need to design for low error probability. By characterizing the error probability, we can also formulate design criteria for space–time codes in Section 4.3.

Since we are allowed to transmit a coded sequence, we are interested in the probability that an erroneous codeword $\mathbf{e}$ is mistaken for the transmitted codeword $\mathbf{x}$. This is called the *pairwise error probability* (PEP) and is then used to bound the error probability. This is analyzed under the condition that the receiver has perfect channel state information. However, a similar analysis can be performed when the receiver does not know the channel state information but has statistical knowledge of the channel.

---

[5] It is also possible to combine closed-loop and open-loop techniques as shown recently in Soni *et al.* (2002).

[6] The D-transform of a discrete-time sequence $\{x(k)\}_{k=0}^{N-1}$ is defined as $x(D) = \sum_{k=0}^{N-1} x(k)D^k$. It is derived from the Z-transform by replacing the unit delay Z^{-1} by D.

For simplicity, we shall first present the result for a flat-fading Rayleigh channel (where $\nu = 0$). In the case when the receiver has perfect channel state information, we can bound the PEP between $\mathbf{x}$ and $\mathbf{e}$ (denoted by $P(\mathbf{x} \to \mathbf{e})$) as follows [41, 74]:

$$P(\mathbf{x} \to \mathbf{e}) \leq \left[\frac{1}{\prod_{n=1}^{M_t} (1 + \frac{E_s}{4N_0} \lambda_n)} \right]^{M_r}, \tag{4.6}$$

where λ_n are the eigenvalues of the matrix $\mathbf{A}(\mathbf{x}, \mathbf{e}) = \mathbf{B}^*(\mathbf{x}, \mathbf{e})\mathbf{B}(\mathbf{x}, \mathbf{e})$ and

$$\mathbf{B}(\mathbf{x}, \mathbf{e}) = \begin{pmatrix} \mathbf{x}_1(1) - \mathbf{e}_1(1) & \dots & \mathbf{x}_{M_t}(0) - \mathbf{e}_{M_t}(0) \\ \vdots & \vdots & \vdots \\ \mathbf{x}_1(N-1) - \mathbf{e}_1(N-1) & \dots & \mathbf{x}_{M_t}(N-1) - \mathbf{e}_{M_t}(N-1) \end{pmatrix}. \tag{4.7}$$

If q denotes the rank of $\mathbf{A}(\mathbf{x}, \mathbf{e})$ (i.e. the number of non-zero eigenvalues) then we can rewrite (4.6) as

$$P(\mathbf{x} \to \mathbf{e}) \leq \left(\prod_{n=1}^{q} \lambda_n \right)^{-M_r} \left(\frac{E_s}{4N_0} \right)^{-qM_r}. \tag{4.8}$$

Thus, we define the notion of diversity order as follows.

DEFINITION 4.1 *A scheme which has an average error probability $\bar{P}_e(\mathrm{SNR})$ as a function of SNR that behaves as*

$$\lim_{SNR \to \infty} \frac{\log(\bar{P}_e(\mathrm{SNR}))}{\log(\mathrm{SNR})} = -d \tag{4.9}$$

is said to have a diversity order of d.

In words, a scheme with diversity order d has an error probability at high SNR behaving as $\bar{P}_e(\mathrm{SNR}) \approx \mathrm{SNR}^{-d}$. Given this definition, we can see that the diversity order in (4.8) is at most qM_r. Moreover, in (4.8) we obtain an additional coding gain of $(\prod_{n=1}^{q} \lambda_n)^{1/q}$.

Note that in order to obtain the average error probability, one can calculate a naive union bound using the pairwise error probability given in (4.8). However, this bound may not be tight and a more careful upper bound for the error probability can be derived [68, 89]. However, if we ensure that *every* pair of codewords satisfies the diversity order in (4.8), then clearly the average error probability satisfies it as well. This is true when the transmission rate is held constant with respect to SNR. Therefore, code design for the diversity order through the pairwise error probability is a sufficient condition, although more detailed criteria can be derived based on a more accurate expression for the average error probability.

The error probability analysis can easily be extended to the case where we have quasi-static ISI channels with channel taps modeled as i.i.d. zero mean complex Gaussian random variables (see, for example [90] and references therein). In this case, the PEP can be written as

$$P(\mathbf{x} \to \mathbf{e}) \leq \left[\frac{1}{\prod_{n=1}^{M_t \nu} (1 + \frac{E_s}{4N_0} \tilde{\lambda}_n)} \right]^{M_r}, \tag{4.10}$$

where the eigenvalues $\tilde{\lambda}_n$ are those of $\tilde{\mathbf{A}}(\mathbf{x}, \mathbf{e}) = \tilde{\mathbf{B}}^*(\mathbf{x}, \mathbf{e})\tilde{\mathbf{B}}(\mathbf{x}, \mathbf{e})$,

$$\tilde{\mathbf{B}}(\mathbf{x}, \mathbf{e}) = \begin{pmatrix} \tilde{\mathbf{x}}^T(0) - \tilde{\mathbf{e}}^T(0) \\ \vdots \\ \tilde{\mathbf{x}}^T(N-1) - \tilde{\mathbf{e}}^T(N-1) \end{pmatrix}, \tag{4.11}$$

and

$$\tilde{\mathbf{x}}(k) = [\mathbf{x}^T(k), \ldots, \mathbf{x}^T(k-\nu)]^T. \tag{4.12}$$

Since $\tilde{\mathbf{A}}(\mathbf{x}, \mathbf{e})$ is a square matrix of size $M_t \nu$, clearly the maximal diversity order achievable for quasi-static ISI channels is $M_r M_t \nu$.

Finally, if we have a time-varying ISI channel, we can generalize (4.10) to

$$P(\mathbf{x} \to \mathbf{e}) \leq \left[\frac{1}{|\mathbf{I}_{M_r N M_t \nu} + \frac{E_s}{4N_0} \mathbf{F}(\mathbf{R}_h \otimes \mathbf{I}_{M_r M_t \nu})|} \right], \tag{4.13}$$

where $\otimes$ denotes a Kronecker product, $\mathbf{R}_h$ is the $N \times N$ correlation matrix of the channel tap process, and $\mathbf{F} = diag\{\mathbf{C}(0), \ldots, \mathbf{C}(N-1)\}$ with

$$\mathbf{C}(k) = \left[\tilde{\mathbf{x}}^T(k) - \tilde{\mathbf{e}}^T(k)\right] \otimes \mathbf{I}_{M_r}. \tag{4.14}$$

Again, it is clear that the maximal diversity attainable is $M_r M_t \nu N$, but for a given channel tap process, N is replaced by the number of dominant eigenvalues N_{dom} of the fading correlation matrix. This parameter is related to the Doppler spread of the channel and the block duration.

4.2.4 Rate–diversity trade-off

A natural question that arises is how many codewords can we have which allow us to attain a certain diversity order. For a flat Rayleigh fading channel, this has been examined in [58, 74] and the following result was obtained.[7]

[7] A constellation size refers to the alphabet size of each transmitted symbol. For example, a quadrature phase-shift keying (QPSK) modulated transmission has a constellation size of 4.

THEOREM 4.2 *If we use a constellation of size* 2^b *and the diversity order of the system is* qM_r*, then the rate* R *that can be achieved is bounded as*

$$R \leq (M_t - q + 1)\log_2|\mathcal{S}| \tag{4.15}$$

in bits per transmission.

One consequence of this result is that for maximum (M_tM_r) diversity order we can transmit at most b bits/s/Hz.

An alternate viewpoint in terms of the rate–diversity trade-off has been explored in [89] from a Shannon-theoretic point of view. Here the authors are interested in the multiplexing rate of a transmission scheme.

DEFINITION 4.3 *A coding scheme which has a transmission rate of* $R(\text{SNR})$ *as a function of SNR is said to have a multiplexing rate* r *if*

$$\lim_{SNR\to\infty} \frac{R(\text{SNR})}{\log(\text{SNR})} = r. \tag{4.16}$$

Therefore, the system has a rate of $r\log(\text{SNR})$ at high SNR. One way to contrast it with the statement in Theorem 4.2, is that the constellation size is also allowed to become larger with SNR. However, note that the naive union bound of the pairwise error probability (4.6) has to be used with care if the constellation size is also increasing with SNR. There is a trade-off between the achievable diversity and the multiplexing rate, and $d^{opt}(r)$ is defined as the supremum of the diversity gain achievable by *any* scheme with multiplexing rate r. The main result in [89] is given as the following theorem.

THEOREM 4.4 *For* $N > M_t + M_r - 1$*, and* $K = \min(M_t, M_r)$*, the optimal trade-off curve* $d^{opt}(r)$ *is given by the piecewise linear function connecting points in* $(k, d^{opt}(k))$, $k = 0, \ldots, K$*, where*

$$d^{opt}(k) = (M_r - k)(M_t - k). \tag{4.17}$$

The interesting interpretation of this result is that one can get large rates which grow with SNR if we reduce the diversity order from the maximum achievable. This diversity–multiplexing trade-off implies that a high multiplexing rate comes at the price of decreased diversity gain and is a manifestation of a corresponding trade-off between the error probability and the rate.

A different question was proposed in [22, 23], where it was asked whether there exists a strategy that combines high-rate communications with high reliability (diversity). Clearly the overall code will still be governed by the rate–diversity trade-off, but the idea is to ensure the reliability (diversity) of at least part of the total information. This allows a form of communication where the high-rate code opportunistically takes advantage of good channel realizations whereas the embedded high-diversity code ensures that at least part of the information is received reliably. In this case, the interest was not in a single pair of multiplexing rate and diversity order (r, d), but in a tuple (r_a, d_a, r_b, d_b) where rate r_a and diversity order d_a were ensured for part of the information with the rate–diversity pair (r_b, d_b) guaranteed for the other part. A class of space–time codes with such desired characteristics will be discussed in Section 4.3.5.

From an information-theoretic point of view Diggavi and Tse [26, 27] focused on the case when there is one degree of freedom (i.e. $\min(M_t, M_r) = 1$). In that case if we consider $d_a \geq d_b$ without loss of generality, the following result was established in [26, 27].

THEOREM 4.5 *When* $\min(M_t, M_r) = 1$, *then the diversity–multiplexing trade-off curve is successively refinable, i.e. for any multiplexing rates* r_a *and* r_b *such that* $r_a + r_b \leq 1$, *the diversity orders* $d_a \geq d_b$,

$$d_a = d^{opt}(r_a), \quad d_b = d^{opt}(r_a + r_b) \tag{4.18}$$

are achievable, where $d^{opt}(r)$ *is the optimal diversity order given in Theorem 4.4.*

Since the overall code has to still be governed by the rate–diversity trade-off given in Theorem 4.4, it is clear that the trivial outer bound to the problem is that $d_a \leq d^{opt}(r_a)$ and $d_b \leq d^{opt}(r_a + r_b)$. Hence Theorem 4.5 shows that the best possible performance can be achieved. This means that for $\min(M_t, M_r) = 1$, we can design ideal *opportunistic* codes. This new direction of enquiry is being currently explored (see [25, 27]).

4.3 Space–time coding principles

Space–time coding has received considerable attention in academic and industrial circles [3, 4] due to its many advantages. First, it improves the downlink performance without the need for multiple receive antennas at the terminals. For example, for wideband code-division multiple access (WCDMA), STC techniques were shown in [64] to result in substantial capacity gains due to the resulting "smoother" fading which, in turn, makes power control more effective and reduces the transmitted power. Second, it can be elegantly combined with channel coding, as shown in [74], realizing a coding gain in addition to the spatial diversity gain. Third, it does not require channel state information (CSI) at the transmitter, i.e. it operates in open-loop mode, thus eliminating the need for an expensive and, in the case of rapid channel fading, unreliable reverse link. Finally, they have been shown to be robust against non-ideal operating conditions such as antenna correlation, channel estimation errors, and Doppler effects [62, 73]. There has been extensive work on the design of space–time codes since their introduction in [74]. The combination of the turbo principle [8, 9] with space–time codes has been explored (see, for example, [7] and [55] among several other references). In addition, the application of linear low-density parity check (LDPC) codes [36] to space–time coding has been explored (see, for example [57] and references therein). We focus our discussion on the basic principles of space–time codes and next describe the two main flavors: trellis and block codes.

4.3.1 *Space–time code design criteria*

In order to design practical codes that achieve a performance target we need to glean insights from the analysis to arrive at design criteria. For example, in the flat-fading case of (4.8) we can state the following rank and determinant design criteria [41, 74].

- *Rank criterion.* In order to achieve maximum diversity $M_t M_r$, the matrix $\mathbf{B}(\mathbf{x}, \mathbf{e})$ from (4.7) has to be full rank for any codewords $\mathbf{x}, \mathbf{e}$. If the minimum rank of $\mathbf{B}(\mathbf{x}, \mathbf{e})$ over all pairs of distinct codewords is q, then a diversity order of qM_r is achieved.
- *Determinant criterion.* For a given diversity order target of q, maximize $(\prod_{n=1}^{q} \lambda_n)^{1/q}$ over all pairs of distinct codewords.

A similar set of design criteria can be stated for the quasi-static ISI fading channel using the PEP given in (4.10) and the corresponding error matrix given in (4.11). Therefore, if we need to construct codes satisfying these design criteria, we can guarantee performance in terms of diversity order. The main problem in practice is to construct such codes that do not have a large decoding complexity. This sets up a familiar tension on the design in terms of satisfying the performance requirements and having low-complexity decoding.

If coherent detection is difficult or too costly, one can employ non-coherent detection for the multiple-antenna channel [46, 88]. Though it is demonstrated in [88] that a training-based technique achieves the same capacity–SNR slope as the optimal, there might be a situation where inexpensive receivers are needed because channel estimation cannot be accommodated. In such a case, differential techniques which satisfy the diversity order might be desirable. There has been significant work on differential transmission with non-coherent detection (see, for example [45, 47] and references therein) and this is a topic we discuss briefly in Section 4.4.1.

The rank and determinant design criteria given above are suitable for transmission when we have a fixed input alphabet. As mentioned in Section 4.2.4, the rate–diversity trade-off can also be explored in the context of the multiplexing rate (see Definition 4.3). Therefore, it is natural to ask for the code-design criteria in this context. For the diversity–multiplexing guarantees, it is not clear that the rank and determinant criterion is the correct one to use. In fact, in [29], it is shown that when designing codes with the multiplexing rate in mind, the determinant criterion is not necessary for *specific* fading distributions. However, it has been shown that the determinant criterion again arises as a sufficient condition when designing codes for the diversity–multiplexing rate trade-off for specific constructions (see [32, 86] and references therein). For these constructions, it is shown that the determinant of the codeword difference matrix plays a crucial role in the diversity–multiplexing optimality of the codes.

Another multiplexing rate context in which the codeword difference matrix plays a crucial role in the space–time code design is in the design of *approximately universal* codes [75]. Traditionally space–time codes are designed for a *particular* distribution of the channel. Universal codes are designed to give an error probability which decays *exponentially* in SNR for all channels that are not in outage. Therefore, this provides a robust design rule which gives performance guarantees over the worst-case channel, rather than averaging over the channel statistics. For the multiple-transmit single-receive (MISO) channel, the code design is related to maximizing the *smallest* singular value of the codeword difference matrix. This corresponds to a worst-case channel that aligns itself with the weakest direction of the codeword difference matrix. This is in contrast to the *average* case, where we are interested in maximizing the product of the singular values (i.e. the determinant). In fact for MIMO channels, in certain cases, the maximizing the determinant of the codeword difference matrix again arises as the universal code design criterion [75].

4.3.2 Space–time trellis codes (STTC)

The space–time trellis encoder maps the information bit stream into M_t streams of symbols (each belonging in a size-2^b signal constellation) that are transmitted simultaneously.[8] STTC design criteria are based on minimizing the PEP bound in Section 4.2.3.

As an example, we consider the eight-state 8-PSK STTC for two transmit antennas introduced in [74]; the trellis description is given in Figure 4.2, where the edge label c_1c_2 means that symbol c_1 is transmitted from the first antenna and symbol c_2 from the second

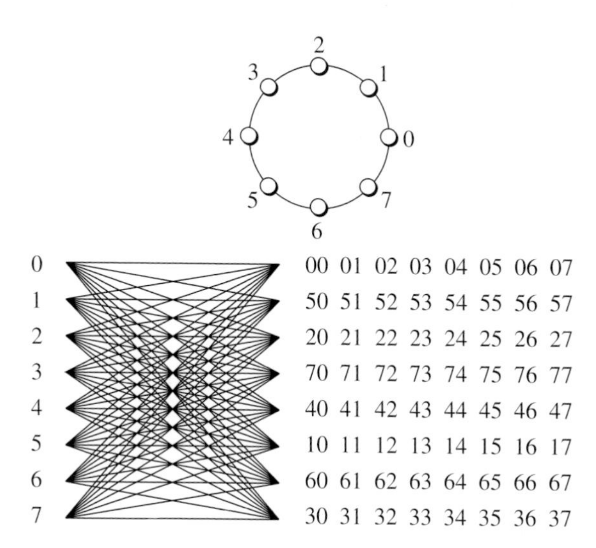

Figure 4.2. Eight-state 8-PSK space–time trellis code with two transmit antennas and a spectral efficiency of 3 bits/sec/Hz.

[8] The total transmitted power is divided equally among the M_t transmit antennas.

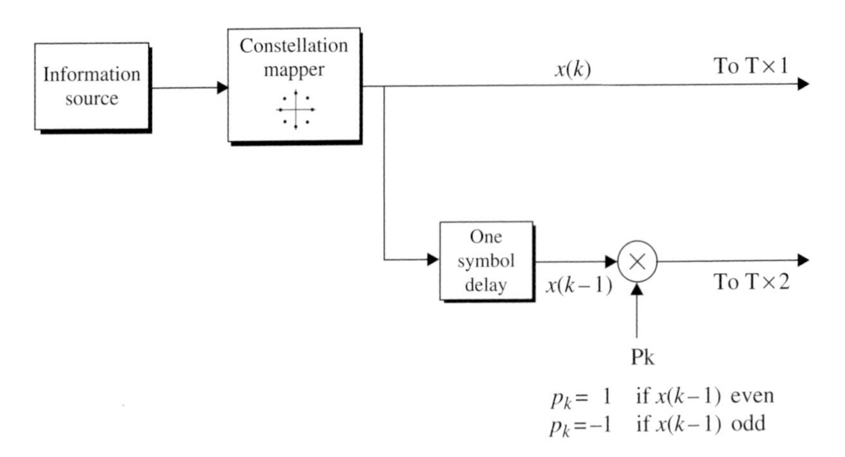

Figure 4.3. Equivalent encoder model for an eight-state 8-PSK space–time trellis code with two transmit antennas.

antenna. The different symbol pairs in a given row label the transitions out of a given state, in order, from top to bottom. An equivalent and convenient (for reasons to become clear shortly) implementation of the eight-state 8-PSK STTC encoder is depicted in Figure 4.3. This equivalent implementation clearly shows that the eight-state 8-PSK STTC is identical to classical delay diversity transmission [67] *except* that the delayed symbol from the second antenna is multiplied by -1 if it is an odd symbol, i.e. $\in \{1, 3, 5, 7\}$. This slight modification results in an additional coding gain over a flat-fading channel. We emphasize that this STTC does not achieve the maximum possible diversity gains (spatial and multi-path) on frequency-selective channels; however, its performance is near optimum for practical ranges of SNR on wireless links [34].[9] Furthermore, when implementing the eight-state 8-PSK STTC described above on a frequency-selective channel, its structure can be exploited to reduce the complexity of joint equalization and decoding. This is achieved by embedding the space–time encoder in Figure 4.3 in the two channels $h_1(D)$ and $h_2(D)$, resulting in an equivalent single-input single-output (SISO) data-dependent CIR with memory $(\nu + 1)$ whose D-transform is given by

$$h_{eqv}^{STTC}(k, D) = h_1(D) + p_k D h_2(D) , \tag{4.19}$$

where $p_k = \pm 1$ is data-dependent. Therefore, trellis-based joint space–time equalization and decoding with $8^{\nu+1}$ states can be performed on this equivalent channel. Without exploiting the STTC structure, trellis equalization requires $8^{2\nu}$ states and STTC decoding requires eight states.

The discussion in this section just illustrates one STTC example. Several other full-rate full-diversity STTCs for different signal constellations and different numbers of antennas were presented in [74].

[9] For examples of STTC designs for frequency-selective channels see, for example [54].

4.3.3 Space–time block codes (STBC)

The decoding complexity of STTC (measured by the number of trellis states at the decoder) increases *exponentially* as a function of the diversity level and the transmission rate [74]. In addressing the issue of decoding complexity, Alamouti [5] discovered an ingenious space–time block coding scheme for transmission with two antennas. According to this scheme (see also Appendix 4.1), input symbols are grouped in pairs where symbols x_k and x_{k+1} are transmitted at time k from the first and second antennas, respectively. Then, at time $k+1$, symbol $-x_{k+1}^*$ is transmitted from the first antenna and symbol x_k^* is transmitted from the second antenna, where $*$ denotes the complex conjugate transpose (cf. Figure 4.4). This imposes an orthogonal spatio-temporal structure on the transmitted symbols. Alamouti's STBC has been adopted in several wireless standards such as WCDMA [77] and CDMA2000 [78] due to the following attractive features. First, it achieves full-diversity at full transmission rate for any (real or complex) signal constellation. Second, it does not require CSI at the transmitter (i.e. open-loop). Third, maximum-likelihood decoding involves only *linear* processing at the receiver (due to the orthogonal code structure).

The Alamouti STBC has been extended to the case of more than two transmit antennas [72] using the theory of orthogonal designs. There it was shown that, in general, full-rate orthogonal designs exist for all real constellations for two, four, and eight transmit antennas only, while for all complex constellations they exist only for two transmit antennas (the Alamouti scheme). However, for particular constellations, it might be possible to construct full-rate orthogonal designs for larger numbers of transmit antennas. Moreover, if a rate loss is acceptable, then orthogonal designs exist for an arbitrary number of transmit antennas [72].

The advantage of orthogonal design is the simplicity of the decoder. However, using a sphere decoder, space–time codes that are not orthogonal, but are *linear* over the complex field can also be decoded efficiently. A class of space–time codes known as *linear dispersion codes* (LDC) was introduced in [43] where the orthogonality constraint is relaxed to achieve a higher rate while still enjoying (expected) polynomial decoding complexity for a wide SNR range by using the sphere decoder. This comes at the expense

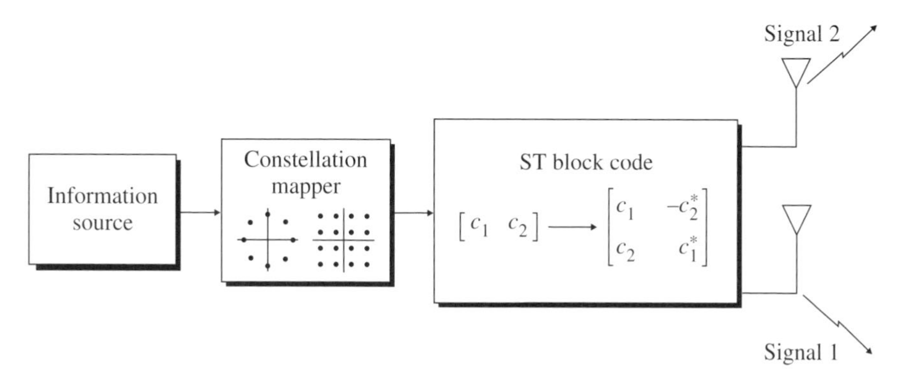

Figure 4.4. Spatial transmit diversity with Alamouti's space–time block code.

of signal constellation expansion and not guaranteeing maximum diversity gains (as in orthogonal designs). With M_t transmit antennas and a channel coherence time of T, the $T \times M_t$ transmitted signal space–time matrix $\mathbf{X}$ in LDC schemes has the form

$$\mathbf{X} = \sum_{q=1}^{Q} \alpha_q \mathbf{A}_q + j\beta_q \mathbf{B}_q , \tag{4.20}$$

where the real scalars α_q and β_q are related to the Q information symbols x_q (that belong to a size-2^b complex signal constellation) by $x_q = \alpha_q + j\beta_q$ for $q = 1, 2, \ldots, Q$. This LDC has a rate of $(Q/T) \log_2 M$. Several LDC designs were presented in [43] based on a judicious choice of the parameters T,Q and the so-called dispersion matrices $\mathbf{A}_q$ and $\mathbf{B}_q$ to maximize the mutual information between the transmitted and received signals.

An alternate way to attain diversity is to build the diversity into the modulation through constellation rotations. This basic idea was proposed by Boulle and Belfiore [10] and Kerpez [51], and developed for higher-dimensional lattices by Boutros and Viterbo ([11] and references therein). Therefore, one can construct modulation schemes with built-in diversity, with the caveat that the constellation size is actually increasing. The point to note here is that Theorem 4.2 refers to a rate versus diversity trade-off for a given constellation size. Therefore, in order to consider the efficiency of coding schemes based on the rotated constellations, one needs to take into account the expansion in the constellation size. As an alternative to alphabet constraints, other constellation constraints have been studied in order to produce codes with maximal rate as well as maximal diversity order (see [30] and references therein). Therefore, constellation rotations without alphabet constraints can yield the maximal performance of *both* rate (in terms of information constellation size) and diversity order. Note that in these cases there is a difference between the *information* constellation size and the constellation size of the transmitted codeword, which could be much larger.

Therefore, in this sense, the rotated codes are actually more suitable in the context of the *diversity–multiplexing* trade-off discussed in Section 4.2.4, where there are no transmit alphabet constraints. In fact, using such rotation-based codes, several diversity–multiplexing rate optimal codes have been constructed (see [32, 75, 86] and references therein).

Recently, STBCs have been extended to the frequency-selective channel case by implementing the Alamouti orthogonal signaling scheme at the *block* level instead of the *symbol* level. Depending on whether the implementation is done in the time or the frequency domain, three STBC structures for frequency-selective channels have been proposed: time-reversal (TR)-STBC [53], OFDM-STBC [56], and frequency-domain-equalized (FDE)-STBC [1]. As an illustration, next we present the space–time encoding scheme for FDE-STBC. Denote the nth symbol of the kth transmitted block from antenna i by $x_i^{(k)}$. At times $k = 0, 2, 4, \ldots$ pairs of length-N blocks $\mathbf{x}_1^{(k)}(n)$ and $\mathbf{x}_2^{(k)}(n)$ (for $0 \leq$

$n \leq N - 1$) are generated by the mobile user. Inspired by Alamouti's STBC, we encode the information symbols as follows [1]:

$$\mathbf{x}_1^{(k+1)}(n) = -\mathbf{x}_2^{*(k)}((-n)_N) \quad \text{and} \quad \mathbf{x}_x^{(k+1)}(n) = \mathbf{x}_1^{*(k)}((-n)_N)$$
$$\text{for } n = 0, 1, \ldots, N-1 \text{ and } k = 0, 2, 4, \ldots \tag{4.21}$$

where $(\cdot)_N$ denotes the modulo-N operation. In addition, a cyclic prefix of length ν (the maximum order of the FIR wireless channel) is added to each transmitted block to eliminate IBI and make all channel matrices *circulant*. We refer the reader to [2] for a detailed description and comparison of these schemes. The main point we would like to stress here is that these three STBC schemes can realize both spatial and multi-path diversity gains at practical complexity levels. For channels with a long delay spread, the frequency-domain implementation using a fast Fourier transform (FFT) either in a single-carrier or multi-carrier fashion becomes more advantageous from a complexity point of view.

4.3.4 A new non-linear maximum-diversity quaternionic code

In this section, we show how the STC algebraic structure inspires new code designs with desirable rate–diversity characteristics and low decoding complexity.

The only full-rate complex orthogonal design is the 2×2 Alamouti code [5], and as the number of transmit antennas increases, the available rate becomes unattractive. For example, for four transmit antennas, orthogonal STBC designs with rates of $\frac{1}{2}$ and $\frac{3}{4}$ were presented in [72]. This rate limitation of orthogonal designs caused a recent shift of research focus to non-orthogonal code design. These include a quasi-orthogonal design [48] for four transmit antennas that has rate 1 but achieves only second-order diversity. Full-diversity can be achieved by including signal rotations which expand the constellation. Another approach is the design of non-orthogonal but linear codes [18] for which decoding is efficient albeit not linear in complexity. In this project, we revisit the problem of designing orthogonal STBC for 4 TX. Another reason for our interest in orthogonal designs is that they limit the SNR loss incurred by differential decoding to its minimum of 3 dB from coherent decoding. The proposed code in this proposal is constructed by means of 2×2 arrays over the quaternions, thus resulting in a 4×4 array over the complex field. The proposed code is rate-1, full-diversity (for any M-PSK constellation), orthogonal over the complex field, but is *not* linear. For QPSK, the code does not suffer constellation expansion and enjoys a simple maximum-likelihood decoding algorithm. Consider the 4×4 space–time block code

$$\mathbf{X} = \begin{bmatrix} p & q \\ -q^* & \frac{q^* p^* q}{\|q\|^2} \end{bmatrix} \tag{4.22}$$

where each block entry is a quaternion. There is an isomorphism between quaternions $q = q_0 + iq_1 + jq_2 + kq_3$ and 2×2 complex matrices as follows:

$$q \leftrightarrow \begin{bmatrix} q^c(0) & q^c(1) \\ -q^{*c}(1) & q^{*c}(0) \end{bmatrix} = \mathbf{Q}, \tag{4.23}$$

where $q^c(0) = q_0 + iq_1$, $q^c(1) = q_2 + iq_3$. Therefore, we may replace the quaternions p and q by the corresponding 2×2 complex matrices to obtain a 4×4 STBC with complex entries. There is a classical correspondence between unit quaternions and rotations in $\mathbf{R}^3$ given by $q \longrightarrow \mathbf{T}_q : p \longrightarrow q^* pq$ (details of the transformation $\mathbf{T}_q$ are given in [14]). For QPSK, maximum-likelihood decoding requires a size-256 search. Through linear combining operations and appropriate application of the transformation $\mathbf{T}_q$, we showed how to exploit the quaternionic structure of this code to reduce the complexity of ML decoding to a size-16 search without loss of optimality.

In Figure 4.5, the significant performance gains achieved by the code in (4.22) in the IEEE 802.16 WiMAX environment [83] as compared to single-antenna transmission/reception translate to a 1.5- and 2.6-fold increase in the cell coverage area at 10^{-3} bit error rate when used with one and two receive antenna(s), respectively. We also compare in Figure 4.5 the effective throughput of our proposed quaternionic code with the rate-$\frac{3}{4}$ full-diversity Octonion code given in (A4.1) assuming QPSK modulation and an outer RS(15, 11) code for both. We observe that our proposed code achieves a throughput level of 1.46 bits per channel use, whereas the achievable throughput for the Octonion code is 1.1 bits per channel use (33% increase).

4.3.5 *Diversity-embedded space–time codes*

The trade-off between rate and diversity was explored within the framework of fixed alphabets by Tarokh, Seshadri, and Calderbank [74] and by Zheng and Tse [89] within an information-theoretic framework. Common to both is the observation that to achieve a high transmission rate, one must sacrifice diversity and vice versa. Consequently a large body of literature has mainly emphasized the design of codes that achieve a certain level of diversity (typically maximal diversity order), and a corresponding rate associated with it, i.e. a particular point on this rate–diversity trade-off (see [31, 59] and references therein).

As explained at the end of Section 4.2.4, a different point of view was proposed by Diggavi *et al.* [22, 23], where the code was designed to achieve a high rate but has embedded within it a higher-diversity (lower-rate) code (see Figure 4.6). Moreover, in this work it was argued that diversity can be viewed as a systems resource that can be allocated judiciously to achieve a desirable rate–diversity trade-off in wireless communications. In particular, it was argued that if one designs the overall system for a fixed rate–diversity operating point, we might be over-provisioning a resource which could be flexibly allocated to different applications. For example, real-time applications need lower delay and therefore higher reliability (diversity) compared to non-real-time applications. By giving flexibility in the diversity allocation, one can simultaneously accommodate multiple applications with disparate rate–diversity requirements [23].

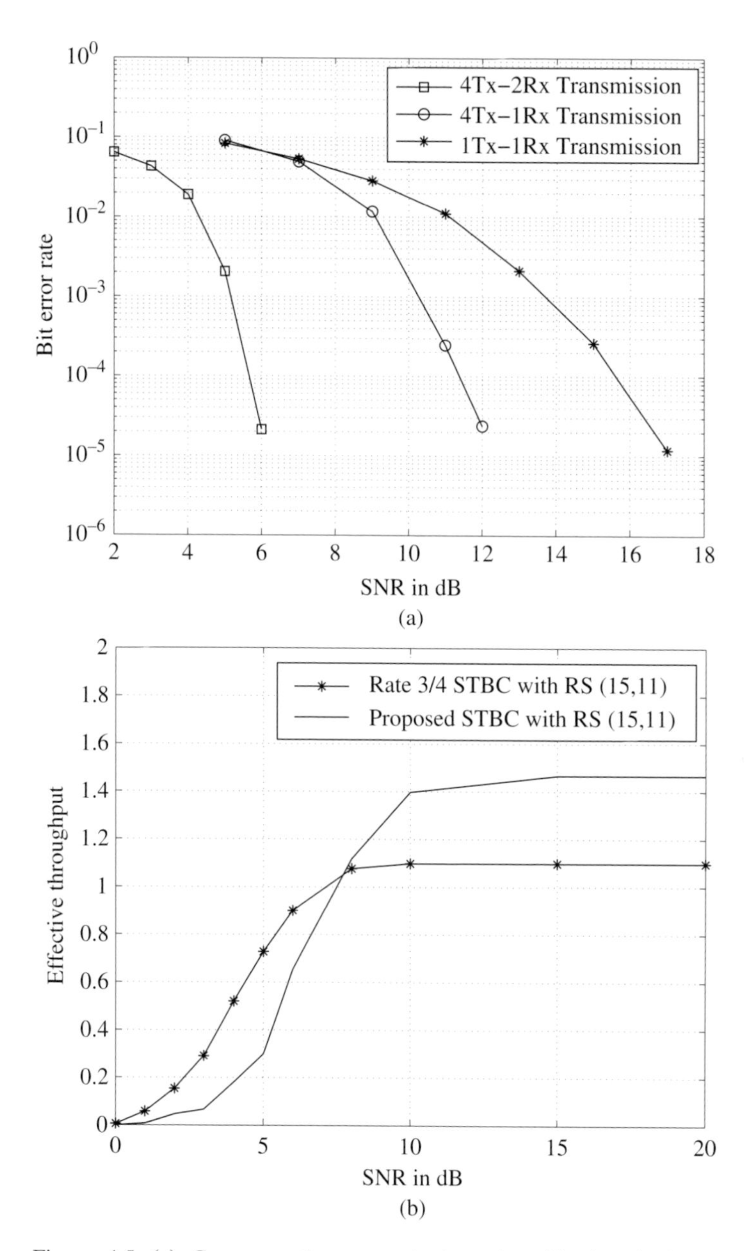

Figure 4.5. (a) Compares the quaternionic code with the single-antenna case for WiMAX. (b) Compares the effective throughput of the quaternionic code with an octonion in a quasi-static channel.

Figure 4.6. Embedded codebook.

Let $\mathcal{A}$ denote the message set from the first information stream and $\mathcal{B}$ denote that from the second information stream. The rates for the two message sets are, respectively, $R(\mathcal{A})$ and $R(\mathcal{B})$. The decoder jointly decodes the two message sets with average error probabilities, $\bar{P}_e(\mathcal{A})$ and $\bar{P}_e(\mathcal{B})$, respectively. We design the code $\mathbf{X}(\mathbf{a}, \mathbf{b})$, such that a certain tuple (R_a, D_a, R_b, D_b) of rates and diversities are achievable, where $R_a = R(\mathcal{A}) = \log(|\mathcal{A}|)/T$, $R_b = R(\mathcal{B}) = \log(|\mathcal{B}|)/T$ and analogously to [89] we define

$$D_a = \lim_{SNR\to\infty} \frac{\log \bar{P}_e(\mathcal{A})}{\log(\mathrm{SNR})}, \qquad D_b = \lim_{SNR\to\infty} \frac{\log \bar{P}_e(\mathcal{B})}{\log(\mathrm{SNR})}. \tag{4.24}$$

For *fixed-rate* codes it has been shown in [23] that to guarantee the diversity orders D_a, D_b we need to design codes such that

$$\min_{\mathbf{a}_1 \neq \mathbf{a}_2 \in \mathcal{A}} \min_{\mathbf{b}_1, \mathbf{b}_2 \in \mathcal{B}} rank(\mathbf{B}(\mathbf{x}_{\mathbf{a}_1, \mathbf{b}_1}, \mathbf{x}_{\mathbf{a}_2, \mathbf{b}_2})) \geq D_a/M_r \tag{4.25}$$

$$\min_{\mathbf{b}_1 \neq \mathbf{b}_2 \in \mathcal{B}} \min_{\mathbf{a}_1, \mathbf{a}_2 \in \mathcal{A}} rank(\mathbf{B}(\mathbf{x}_{\mathbf{a}_1, \mathbf{b}_1}, \mathbf{x}_{\mathbf{a}_2, \mathbf{b}_2})) \geq D_b/M_r, \tag{4.26}$$

where $\mathbf{B}$ is the codeword difference matrix as defined in (4.7). Basically, this implies that if we transmit a particular message $\mathbf{a} \in \mathcal{A}$, regardless of which message is chosen in message set $\mathcal{B}$, we are ensured a diversity level of D_a for this message set. A similar argument holds for message set $\mathcal{B}$. Using this criterion several diversity-embedded codes have been constructed and will be discussed in the following. This design rule is a generalization of the design rule for traditional space–time codes given in Section 4.3.1.

Linear diversity-embedded codes

In [23], linear constructions of diversity-embedded codes were given. These code designs are linear over the complex field in order to be able to decode them efficiently using the *sphere decoder* algorithm [19], which has an average complexity that is only polynomial (not exponential) in the rate, making it an attractive choice for decoding high-rate codes. Another constraint that we impose in our code designs is to *not expand the transmitted signal constellation*, in contrast with other designs based on constellation rotations. For illustration we focus on one code example given in [23].

Code example

Let $\mathcal{A}$ come from the message set $\{a(0), a(1)\} \in \mathcal{S}$ and $\mathcal{B}$ come from $\{b(0), b(1), b(2), b(3)\} \in \mathcal{S}$. Hence, $R_a = \frac{1}{2}\log|\mathcal{S}|$ and $R_b = \log|\mathcal{S}|$, leading to a total rate of $R_a + R_b = \frac{3}{2}\log|\mathcal{S}|$.

$$\mathbf{X} = \mathbf{X_a} + \mathbf{X_b} = \begin{bmatrix} a_1 & a_2 & b_3 & b_4 \\ -a_2^* & a_1^* & b_4^* & -b_3^* \\ b_1 & b_2 & a_1^* & -a_2 \\ -b_2^* & b_1^* & a_2^* & a_1 \end{bmatrix}, \tag{4.27}$$

where $\mathbf{X_a}$ is a function of variables a_1, a_2 and $\mathbf{X_b}$ is a function of variables b_1, b_2, b_3, b_4. This code is linear over the complex field so that it can be decoded using the sphere

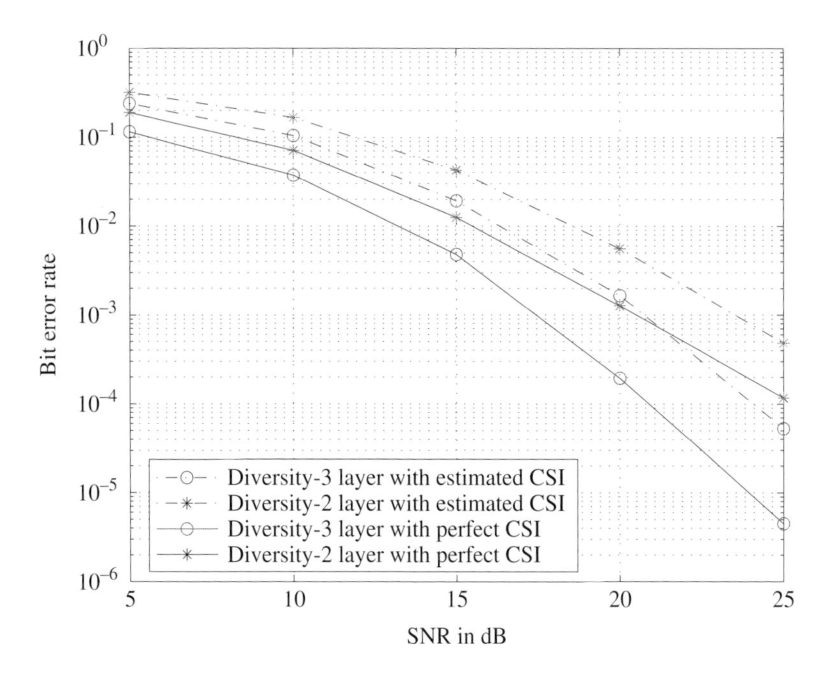

Figure 4.7. Performance of a diversity-embedding space–time block code with perfect and estimated CSI.

decoder [20], where the average complexity is polynomial rather than exponential in the rate. The proof that this code achieves diversity 3 for variables a_1, a_2 and diversity 2 for variables b_1, b_2, b_3, b_4 makes essential use of quaternion arithmetic. The code does not require channel knowledge at the transmitter and it outperforms time-sharing schemes [23]. The performance of this code with perfect and estimated CSI over a quasi-static flat-fading Rayleigh channel is depicted in Figure 4.7.

Non-linear diversity-embedded codes

Constructions of a class of non-linear diversity-embedded codes were given in [12, 25], and we explain the principles behind these constructions here. The basic idea of this class of non-linear codes is to use rank properties of binary matrices to construct codes in the complex domain with the desired diversity-embedding property. Given two message sets $\mathcal{A}, \mathcal{B}$, they are mapped to the space–time codeword $\mathbf{X}$ as shown below:

$$\mathcal{A}, \mathcal{B} \xrightarrow{f_1} \mathbf{K} = \begin{bmatrix} K(1,1) & \dots & K(1,T) \\ \vdots & \vdots & \vdots \\ K(M_t,1) & \dots & K(M_t,T) \end{bmatrix}$$

$$\xrightarrow{f_2} \mathbf{X} = \begin{bmatrix} x(1,1) & \dots & x(1,T) \\ \vdots & \vdots & \vdots \\ x(M_t,1) & \dots & x(M_t,T) \end{bmatrix},$$

where $K(m, n) \in \{0, 1\}^{\log(|\mathcal{S}|)}$, i.e. a binary string and $x(m, n) \in \mathcal{S}$. This construction is illustrated later in Figure 4.9 for a constellation size of L bits. The basic idea is that we choose sets $\mathcal{K}_1, \ldots, \mathcal{K}_L$ from which the sequence of binary matrices which encode the constellations are chosen. These sets of binary matrices are chosen so that they have given rank guarantees which reflect the ultimate diversity orders required for each message set. Given the diversity order requirements we can choose these sets appropriately. For example, if we desire a single diversity order (i.e. no diversity embedding) then we can choose all the sets of binary matrices to be identical. At the other extreme all the sets could be different, yielding L different levels of diversity embedding. Given the message set, we first choose the matrices $\mathbf{K}_1, \ldots, \mathbf{K}_L$. The first mapping f_1 is obtained by taking matrices and constructing the matrix $\mathbf{K} \in \mathbb{C}^{M_t \times T}$ each of whose entries is constructed by concatenating the bits from the corresponding entries in the matrices $\mathbf{K}_1, \ldots, \mathbf{K}_L$ into an L-length bit-string.

This matrix is then mapped to the space–time codeword through a constellation mapper f_2. This can be done by using an L-level binary partition of a quadrature amplitude modulation (QAM) or PSK signal constellation (see Figure 4.8).

As shown above, this structure can be used for one to L levels of diversity order. However, for simplicity, we restrict our attention to two levels of diversity (as shown in Figure 4.6). For concreteness consider a 4-QAM constellation, with two levels of diversity order, D_a, D_b. Given that $L = 2$, we then assign layer 1 to diversity order D_a and layer 2 to D_b with $D_a \geq D_b$. We choose sets of binary matrices $\mathcal{K}_1, \mathcal{K}_2$ with rank guarantees $D_a/M_r, D_b/M_r$ respectively. Let the set sizes be $|\mathcal{K}_1| = 2^{TR_a}, |\mathcal{K}_2| = 2^{TR_b}$, yielding the appropriate rates R_a, R_b. Therefore, given a message $m_a \in \mathcal{A}, m_b \in \mathcal{B}$, we choose the matrices $\mathbf{K}_1 \in \mathcal{K}_1, \mathbf{K}_2 \in \mathcal{K}_2$ corresponding respectively to the messages m_a, m_b. Given $\mathbf{K}_1, \mathbf{K}_2$, we can construct the space–time code $\mathbf{X}$ as illustrated in Figure 4.9. If we have constellations of size 2^L with $L > 2$ and we still need two levels of diversity, we can assign layers $1, \ldots, L_a$ to choose matrices with the *same* binary set $\mathcal{K}_1$ with rank guarantee D_a/M_r and layers, $L_a + 1, \ldots, L$ to choose matrices from the binary set $\mathcal{K}_2$ with rank

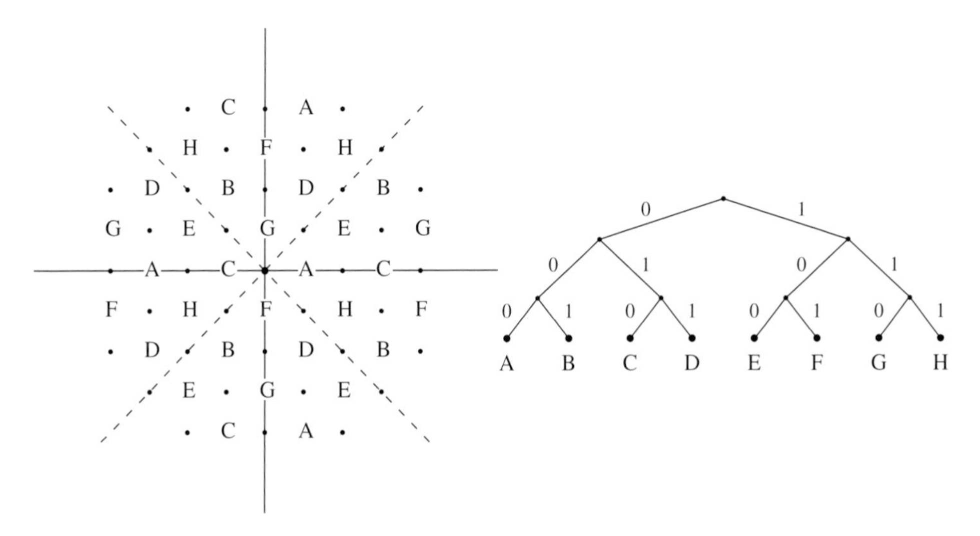

Figure 4.8. A binary partition of a 32-point QAM constellation.

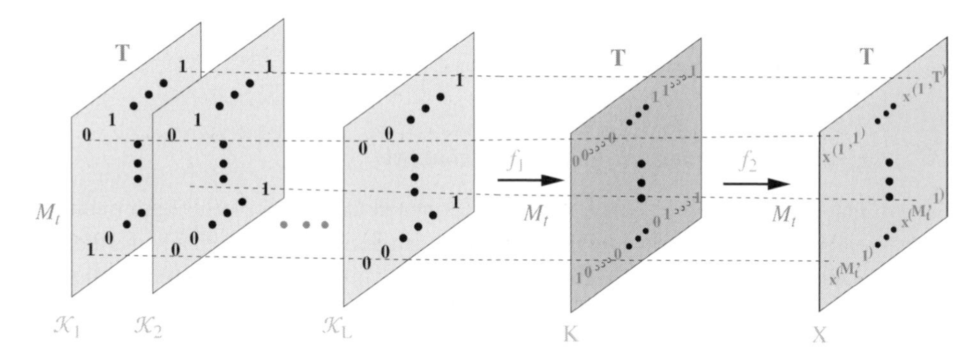

Figure 4.9. Schematic representation of the non-linear code construction.

guarantee D_b/M_r. By choosing set cardinalities as $|\mathcal{K}_1| = 2^{TR_a/L_a}$, $|\mathcal{K}_2| = 2^{TR_b/L_b}$, we get the corresponding rates R_a, R_b for the two diversity orders. Therefore as before given message $m_a \in \mathcal{A}$, $m_b \in \mathcal{B}$, we choose the sequence of matrices $\mathbf{K}_1, \ldots, \mathbf{K}_{L_a} \in \mathcal{K}_1$ based on m_a and matrices $\mathbf{K}_{L_a+1}, \ldots, \mathbf{K}_L \in \mathcal{K}_2$ based on m_b. Using this sequence of L matrices, we obtain the space–time codeword as seen in Figure 4.9.

In all this, the choice of the sets $\mathcal{K}_l, l = 1, \ldots, L$ has been unspecified. However in [58], sets of $M_t \times T$ binary matrices $\mathcal{P}(M_t, T, r)$ were constructed for $T \geq M_t$ such that the difference of any two matrices in the set had rank $\lfloor M_t - r \rfloor$ over the binary field. They showed that such sets had a cardinality of $|\mathcal{P}(M_t, T, r)| = 2^{T(r+1)}$. Therefore, these matrices yielded a rate of $r + 1$ bits/transmission. In our construction we use these matrices, with $d_l = \lfloor M_t - r_l \rfloor$, with $d_1 \geq d_2 \geq \cdots \geq d_L$. This yields a rate of $R_l = r_l + 1$ in each layer. In [25] it is shown that this construction for QAM constellations achieves the rate tuple $(R_1, M_r d_1, \ldots, R_L, M_r d_L)$, with the overall equivalent single-layer code achieving the rate–diversity point, $(\sum_l R_l, M_r d_L)$. As described above, we can have the desired number of layers by choosing several identical diversity/rate layers.

The optimal decoding is a maximum-likelihood decoder which jointly decodes the message sets. The performance of such a decoder is examined further in Section 4.4.2 along with applications of diversity-embedded codes.

4.4 Applications

In this section, we show how the STC algebraic structure can be exploited to enhance end-to-end system performance and reduce implementation complexity. This is illustrated through signal processing examples, new STC code constructions, and by examining interactions with the higher networking layers.

4.4.1 *Signal processing*

In this section, we demonstrate how the STC structure can be exploited to reduce the complexity of receiver signal processing algorithms for channel estimation, joint equalization and decoding (both channel-estimate-based and adaptive), and non-coherent detection.

Channel estimation for quasi-static channels

For quasi-static channels, CSI can be estimated at the receiver using a training sequence embedded in each transmission block. For single-transmit-antenna signaling, the training sequence is only required to have "good" (i.e. impulse-like) auto-correlation properties. However, for the M_t transmit-antenna scenarios, the M_t training sequences should, in addition, have "low" (ideally zero) cross-correlation. In addition, it is desirable (in order to avoid amplifier non-linear distortion) to use training sequences with constant amplitude. *Perfect root of unity sequences* (PRUS; see [16] have these ideal correlation and constant-amplitude properties. However, for a given training sequence length, PRUS do not always belong to standard signal constellations such as PSK. Additional challenges in channel estimation for multiple-transmit-antenna systems over the single-transmit-antenna case are the increased number of channel parameters to be estimated and the reduced transmit power (by a factor of M_t) for each transmit antenna.

In [35], it was proposed to encode a single training sequence by a space–time encoder to generate the M_t training sequences.[10] Strictly speaking, this approach is suboptimum since the M_t transmitted training sequences are cross-correlated by the space–time encoder which imposes a constraint on the possible generated training sequences. However, it turns out that, with proper design, the performance loss from optimal PRUS training is negligible [35]. Furthermore, this approach reduces the training sequence search space from $(2^b)^{M_t N_t}$ to $(2^b)^{N_t}$ (assuming equal input and output alphabet size of 2^b and length-N_t training sequences), making exhaustive searches more practical and thus facilitating the identification of good training sequences from standard signal constellations such as PSK.

The search space can be further reduced by exploiting special characteristics of the particular STC. As an example, consider the eight-state 8-PSK STTC for two transmit and one receive antennas whose equivalent CIR is given by (4.19). For a given transmission block (over which the two channels $h_1(D)$ and $h_2(D)$ are constant), the input sequence determines the equivalent channel. By transmitting only "even" training symbols from the sub-constellation $C_e = \{0, 2, 4, 6\}$, $p_k = +1$ and the equivalent channel is given by $h_e(D) = h_1(D) + Dh_2(D)$. On the other hand, transmitting only "odd" training symbols from the sub-constellation $C_0 = \{1, 3, 5, 7\}$, results in $p_k = -1$ and the equivalent channel $h_o(D) = h_1(D) - Dh_2(D)$. After estimating $h_e(D)$ and $h_o(D)$, we can compute

$$h_1(D) = \frac{h_e(D) + h_o(D)}{2} \quad \text{and} \quad h_2(D) = \frac{h_e(D) - h_o(D)}{2D}. \tag{4.28}$$

Consider a training sequence of the form $\mathbf{s} = [\mathbf{s}_e \;\; \mathbf{s}_o]$, where $\mathbf{s}_e$ has length $N_t/2$ and takes values in the C_e sub-constellation and $\mathbf{s}_o$ has length $N_t/2$ and takes values in the C_o

[10] We assume, for simplicity, the same space–time encoder for the training and the information symbols. However, they could be different in general.

sub-constellation. Note that if $\mathbf{s}_e$ is a good sequence in terms of MMSE for the estimation of $h_e(D)$, the sequence $\mathbf{s}_o$ created as $\mathbf{s}_o = a\,\mathbf{s}_e$ where $a = \exp(i\pi k/4)$ and any $k = 1, 3, 5, 7$ achieves the same MMSE for the estimation of $h_o(D)$. Thus, instead of searching over all possible 8^{N_t} sequences $\mathbf{s}$, we can further restrict the search space to the $4^{N_t}/2$ sequences $\mathbf{s}_e$. A reduced-size search can identify sequences $\mathbf{s}_e$ and $\mathbf{s}_o = a\,\mathbf{s}_e$ such that the channel estimation MMSE is achieved. We emphasize that similar reduced-complexity techniques can be developed for other STTCs by deriving their equivalent encoder models (as in Figure 4.3).

In summary, the special STC structure can be utilized to simplify training sequence design for multiple-antenna transmissions without sacrificing performance.

Integration of equalization and decoding

Our focus will be on Alamouti-type STBC with two transmit antennas. The treatment can be extended to more than two antennas using orthogonal designs [72] at the expense of some rate loss for complex signal constellations.

The main attractive feature of STBC is the quaternionic structure (see Appendix 4.1 for more discussions on quaternions) of the spatio-temporal channel matrix. This allows us to eliminate inter-antenna interference using a low-complexity linear combiner (which is a spatio-temporal matched filter and is also the maximum-likelihood detector in this case). Then, joint equalization and decoding for each antenna stream proceeds using any of well-known algorithms for the single-antenna case which can be implemented either in the time or frequency domains. For illustration purposes, we describe next a joint equalization and decoding algorithm for the single-carrier frequency-domain-equalizer (SC FDE)-STBC. A more detailed discussion and comparison is given in [2].

The SC FDE-STBC receiver block diagram is given in Figure 4.10. After analog-to-digital (A/D) conversion, the CP part of each received block is discarded. Mathematically, we can express the input–output relationship over the jth received block as follows:

$$\mathbf{y}^{(j)} = \mathbf{H}_1^{(j)}\mathbf{x}_1^{(j)} + \mathbf{H}_2^{(j)}\mathbf{x}_2^{(j)} + \mathbf{z}^{(j)}\,, \tag{4.29}$$

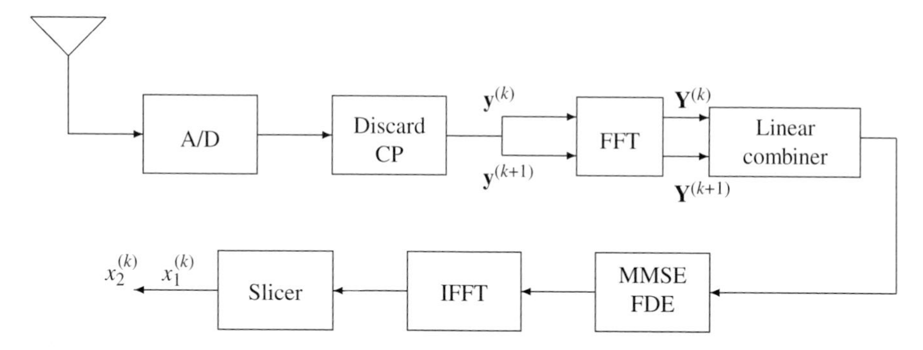

Figure 4.10. FDE-STBC receiver block diagram.

where $\mathbf{H}_1^{(j)}$ and $\mathbf{H}_2^{(j)}$ are $N \times N$ circulant matrices whose first columns are equal to $\mathbf{h}_1^{(j)}$ and $\mathbf{h}_2^{(j)}$, respectively, appended by $(N - \nu - 1)$ zeros and $\mathbf{z}^{(j)}$ is the noise vector. Since $\mathbf{H}_1^{(j)}$ and $\mathbf{H}_2^{(j)}$ are circulant matrices, they admit the eigen-decompositions

$$\mathbf{H}_1^{(j)} = \mathbf{Q}^* \mathbf{\Lambda}_1^{(j)} \mathbf{Q} \quad ; \quad \mathbf{H}_2^{(j)} = \mathbf{Q}^* \mathbf{\Lambda}_2^{(j)} \mathbf{Q} \,,$$

where $\mathbf{Q}$ is the orthonormal FFT matrix and $\mathbf{\Lambda}_1^{(j)}$ (resp. $\mathbf{\Lambda}_2^{(j)}$) is a diagonal matrix whose (n, n) entry is equal to the nth FFT coefficient of $\mathbf{h}_1^{(j)}$ (resp. $\mathbf{h}_2^{(j)}$). Therefore, applying the FFT to $\mathbf{y}^{(j)}$, we find (for $j = k, k+1$)

$$\mathbf{Y}^{(j)} = \mathbf{Q}\mathbf{y}^{(j)} = \mathbf{\Lambda}_1^{(j)} \mathbf{X}_1^{(j)} + \mathbf{\Lambda}_2^{(j)} \mathbf{X}_2^{(j)} + \mathbf{Z}^{(j)} \,.$$

The SC FDE-STBC encoding rule is given by Al-Dhahir [1]

$$\mathbf{X}_1^{(k+1)}(m) = \mathbf{X}_2^{*(k)}(m) \quad \text{and} \quad \mathbf{X}_2^{(k+1)}(m) = -\mathbf{X}_1^{*(k)}(m) \tag{4.30}$$

for $m = 0, 1, \ldots, N-1$ and $k = 0, 2, 4, \ldots$. The length-N blocks at the FFT output are then processed in pairs resulting in the two blocks (we drop the time index from the channel matrices since they are assumed fixed over the two blocks under consideration)

$$\underbrace{\begin{bmatrix} \mathbf{Y}^{(k)} \\ \mathbf{Y}^{*(k+1)} \end{bmatrix}}_{\mathbf{Y}} = \underbrace{\begin{bmatrix} \mathbf{\Lambda}_1 & \mathbf{\Lambda}_2 \\ -\mathbf{\Lambda}_2^* & \mathbf{\Lambda}_1^* \end{bmatrix}}_{\mathbf{\Lambda}} \underbrace{\begin{bmatrix} \mathbf{X}_1^{(k)} \\ \mathbf{X}_2^{(k)} \end{bmatrix}}_{\mathbf{X}} + \underbrace{\begin{bmatrix} \mathbf{Z}^{(k)} \\ \mathbf{Z}^{*(k+1)} \end{bmatrix}}_{\mathbf{Z}}, \tag{4.31}$$

where $\mathbf{X}_1^{(k)}$ and $\mathbf{X}_2^{(k)}$ are the FFTs of the information blocks $\mathbf{x}_1^{(k)}$ and $\mathbf{x}_2^{(k)}$, respectively, and $\mathbf{Z}$ is the noise vector. We used the encoding rule in (4.30) to arrive at (4.31). To eliminate *inter-antenna interference*, the linear combiner $\mathbf{\Lambda}^*$ is applied to $\mathbf{Y}$. Owing to the quaternionic structure of $\mathbf{\Lambda}$, a second-order diversity gain is achieved. Then, the two decoupled blocks at the output of the linear combiner are equalized separately using the MMSE FDE [66] which consists of N complex taps per block that mitigate *intersymbol interference*. Finally, the MMSE-FDE output is transformed back to the time domain using the inverse FFT where decisions are made.

Adaptive techniques

The coherent receiver techniques described up to now require CSI which is estimated and tracked using training sequences/pilot symbols inserted in each block and then used to compute the optimum joint equalizer/decoder settings. An alternative to this two-step channel-estimate-based approach is *adaptive* space–time equalization/decoding where CSI is not explicitly estimated at the receiver. Adaptive receivers still require a training overhead to converge to their optimum settings which, in the presence of channel variations, are adapted using previous decisions to *track* these variations. The celebrated least mean square (LMS) adaptive algorithm [44] is widely used in single-antenna communication systems today due to its low implementation complexity. However, it has been shown to exhibit slow convergence and suffer significant performance degradation (relative to performance achieved with the optimum settings) when applied to broadband MIMO channels due to the large number of parameters that need to be simultaneously adapted and the wide eigenvalue spread problems encountered on those channels. Faster convergence can be achieved by implementing a more sophisticated family of algorithms known as recursive least squares (RLS). However, their high computational complexity compared to LMS and their notorious behavior when implemented with finite precision limit their appeal in practice. It was shown in [87] that the orthogonal structure of STBC can be exploited to develop fast-converging RLS-type adaptive FDE-STBC at LMS-type complexity. A brief overview is given next.

Our starting point in deriving the adaptive algorithm is the relation

$$\begin{bmatrix} \hat{\mathbf{X}}_1^{(k)} \\ \hat{\mathbf{X}}_2^{(k)} \end{bmatrix} = \begin{bmatrix} \mathbf{A}_1 & \mathbf{A}_2 \\ \mathbf{A}_2^* & -\mathbf{A}_1^* \end{bmatrix} \mathbf{Y}, \tag{4.32}$$

where $\mathbf{Y}$ was defined in (4.31) and the diagonal matrices $\mathbf{A}_1$ and $\mathbf{A}_2$ are given by

$$\mathbf{A}_1 = \mathbf{\Lambda}_1^*.diag\left\{\frac{1}{\tilde{\mathbf{\Lambda}}(i,i)+\frac{1}{\mathrm{SNR}}}\right\}_{i=0}^{N-1} \; ; \; \mathbf{A}_2 = \mathbf{\Lambda}_2^*.diag\left\{\frac{1}{\tilde{\mathbf{\Lambda}}(i,i)+\frac{1}{\mathrm{SNR}}}\right\}_{i=0}^{N-1}, \tag{4.33}$$

with $\tilde{\mathbf{\Lambda}}(i,i) = |\mathbf{\Lambda}_1(i,i)|^2 + |\mathbf{\Lambda}_2(i,i)|^2$. Alternatively, we can write

$$\begin{bmatrix} \hat{\mathbf{X}}_1^{(k)} \\ \hat{\mathbf{X}}_2^{(k)} \end{bmatrix} = \begin{bmatrix} diag(\mathbf{Y}^{(k)}) & -diag(\mathbf{Y}^{*(k+1)}) \\ diag(\mathbf{Y}^{(k+1)}) & diag(\mathbf{Y}^{*(k)}) \end{bmatrix} \begin{bmatrix} \mathbf{W}_1^* \\ \mathbf{W}_2 \end{bmatrix} = \mathbf{U}_k \mathcal{W}, \tag{4.34}$$

where $\mathbf{W}_1^*$ and $\mathbf{W}_2$ are vectors containing the diagonal elements of $\mathbf{A}_1^*$ and $\mathbf{A}_2$, respectively, and $\mathcal{W}$ is a $2N \times 1$ vector containing the elements of $\mathbf{W}_1^*$ and $\mathbf{W}_2$. The $2N \times 2N$ quaternionic matrix $\mathbf{U}_k$ contains the received symbols for blocks k and $k+1$. Equation (4.34) can be used to develop a frequency-domain block-adaptive RLS algorithm for $\mathbf{W}$ which, using the special quaternionic structure of the problem, can be simplified to the following LMS-type recursions (see [87] for details of the derivation)

$$\mathcal{W}_{k+2} = \mathcal{W}_k + \begin{bmatrix} \mathbf{P}_{k+2} & \mathbf{0} \\ \mathbf{0} & \mathbf{P}_{k+2} \end{bmatrix} \mathbf{U}_{k+2}(\mathbf{D}_{k+2} - \mathbf{U}_{k+2}\mathcal{W}_k), \tag{4.35}$$

where $\mathbf{D}_{k+2} = \left[\mathbf{X}_1^{(k+2)} \; \mathbf{X}_2^{*(k+2)}\right]^T$ for the training mode and $\mathbf{D}_{k+2} = \left[\hat{\mathbf{X}}_1^{(k+2)} \; \hat{\mathbf{X}}_2^{*(k+2)}\right]^T$ for the decision-directed mode. The $N \times N$ diagonal matrix $\mathbf{P}_{k+2}$ is computed by the recursion

$$\mathbf{P}_{k+2} = \lambda^{-1}(\mathbf{P}_k - \lambda^{-1}\mathbf{P}_k\Gamma_{k+2}\mathbf{P}_k), \tag{4.36}$$

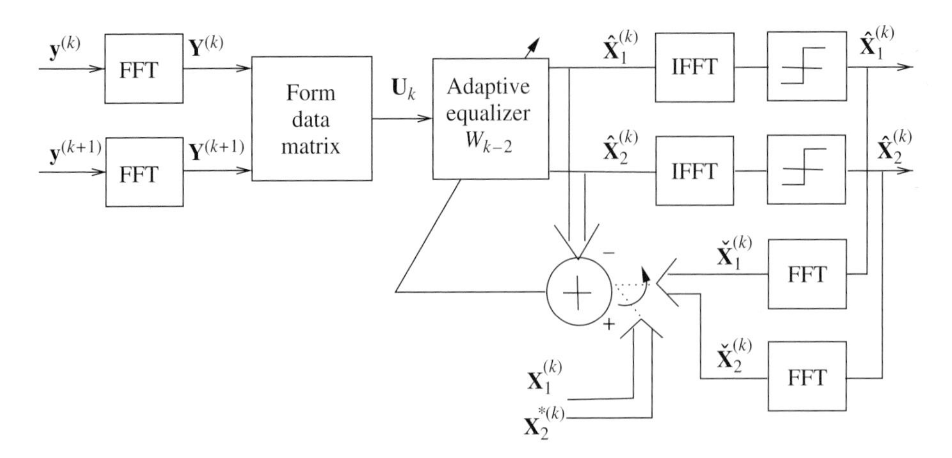

Figure 4.11. Block diagram of an adaptive FDE-STBC joint equalizer/decoder.

where the diagonal matrices Γ_{k+2} and Δ_{k+2} are computed from the recursions

$$\begin{aligned}\Gamma_{k+2} &= diag(\mathbf{Y}^{(k)})\Delta_{k+2} diag(\mathbf{Y}^{*(k)}) \\ &\quad + diag(\mathbf{Y}^{(k+1)})\Delta_{k+2} diag(\mathbf{Y}^{*(k+1)}) \\ \Delta_{k+2} &= (\mathbf{I}_N + \lambda^{-1}(diag(\mathbf{Y}^{(k)})\mathbf{P}_k diag(\mathbf{Y}^{*(k)}) \\ &\quad + diag(\mathbf{Y}^{(k+1)})\mathbf{P}_k diag(\mathbf{Y}^{*(k+1)})))^{-1} .\end{aligned}$$

The initial conditions are $\mathcal{W}_0 = \mathbf{0}$, $\mathbf{P}_0 = \delta \mathbf{I}_N$, where δ is a large number, and the forgetting factor λ is chosen to be close to 1.

The block diagram of the adaptive FDE-STBC is shown in Figure 4.11. Pairs of consecutive received blocks are transformed to the frequency domain using the FFT, then the data matrix in (4.34) is formed. The filter output (the product $\mathbf{U}_k \mathcal{W}_{k-2}$) is transformed back to the time domain using an inverse FFT (IFFT) and passed to a decision device to generate data estimates. The output of the adaptive equalizer is compared to the desired response to generate an error vector which is, in turn, used to update the equalizer coefficients according to the RLS recursions. The equalizer operates in a training mode until it converges, then it switches to a decision-directed mode where previous decisions are used for tracking. When operating over fast time-varying channels, retraining blocks can be transmitted periodically to prevent equalizer divergence (see [87]).

Non-coherent techniques

Non-coherent transmission schemes do not require channel estimation, hence eliminating the need for bandwidth-consuming training sequences and reducing terminal complexity. This becomes more significant for rapidly fading channels where frequent retraining is needed to track channel variations and for multiple-antenna broadband transmission scenarios where more channel parameters (several coefficients for each transmit–receive antenna pair) need to be estimated. One class of non-coherent techniques are blind identification and detection schemes. Here, the structure of the channel (finite impulse response), the input constellation (finite alphabet) and the output (cyclostationarity) are exploited to eliminate training symbols. Such techniques have a vast literature and we refer the interested reader to a good survey in [79]. Another class of non-coherent techniques is the generalized ML receiver in [82].

Several non-coherent space–time transmission schemes have been proposed for flat-fading channels including differential STBC schemes with two [71] or more [49] transmit antennas and group differential STC schemes (see, for example [47] and references therein). Here, we describe a differential space–time transmission scheme for frequency-selective channels we recently proposed in [21] that achieves full diversity (spatial and multi-path) at rate one[11] with two transmit antennas. This scheme is a differential form for the OFDM-STBC structure described in [56]. A time-domain differential space–time scheme with single-carrier transmission is presented in [21].

We consider two symbols $X_1(m)$ and $X_2(m)$ drawn from a PSK constellation which, in a conventional OFDM system, would be transmitted over two consecutive OFDM blocks on the same subcarrier m. Following the Alamouti encoding scheme, the two source symbols are mapped as

$$\mathbf{X}^{(1)}(m) = [X_1(m), X_2(m)]^T, \quad \mathbf{X}^{(2)}(m) = [-X_2^*(m), X_1^*(m)]^T, \tag{4.37}$$

where $\mathbf{X}^{(1)}$ represents the information-bearing vector for the first OFDM block and $\mathbf{X}^{(2)}$ corresponds to the second OFDM block.[12] Let N denote the FFT size. Then $\mathbf{X}^{(1)}$ and $\mathbf{X}^{(2)}$ are length-$2N$ vectors holding the symbols to be transmitted by the two transmit antennas. Consequently, after taking the FFT at the receiver, we have (at subcarrier m)

$$\begin{aligned} &\begin{pmatrix} Y_1(m) & Y_2(m) \\ -Y_2^*(m) & Y_1^*(m) \end{pmatrix} \\ &= \begin{pmatrix} H_1(m) & H_2(m) \\ -H_2^*(m) & H_1^*(m) \end{pmatrix} \begin{pmatrix} X_1(m) & -X_2^*(m) \\ X_2(m) & X_1^*(m) \end{pmatrix} + \text{noise}, \end{aligned} \tag{4.38}$$

where $H_1(m)$ and $H_2(m)$ are the frequency responses of the two channels at subcarrier m.

For block k and subcarrier m, denote the source symbols as $\mathbf{u}_m^{(k)} = \begin{bmatrix} u_{1,m}^{(k)} & u_{2,m}^{(k)} \end{bmatrix}^T$, the transmitted matrix as $\mathbf{X}_m^{(k)}$, and the received matrix as $\mathbf{Y}_m^{(k)}$. Then, in the absence of noise, (4.38) is written as $\mathbf{Y}_m^{(k)} = \mathbf{H}_m \mathbf{X}_m^{(k)}$, where we assume that the channel is fixed over two consecutive blocks. Using the quaternionic structure of $\mathbf{H}_m$, it follows that

$$\mathbf{Y}_m^{*(k-1)} \mathbf{Y}_m^{(k)} = (|H_1(m)|^2 + |H_2(m)|^2)\, \mathbf{X}_m^{*(k-1)} \mathbf{X}_m^{(k)}.$$

[11] This does not include the rate penalty incurred by concatenating OFDM-STBC with an outer code and interleaving across tones which is common to all OFDM systems (see, for example, [66] for more discussion).

[12] Intuitively, each OFDM subcarrier can be thought of as a flat-fading channel and the Alamouti code is applied to each of the OFDM subcarriers. As a result, the Alamouti code yields diversity gains at every subcarrier.

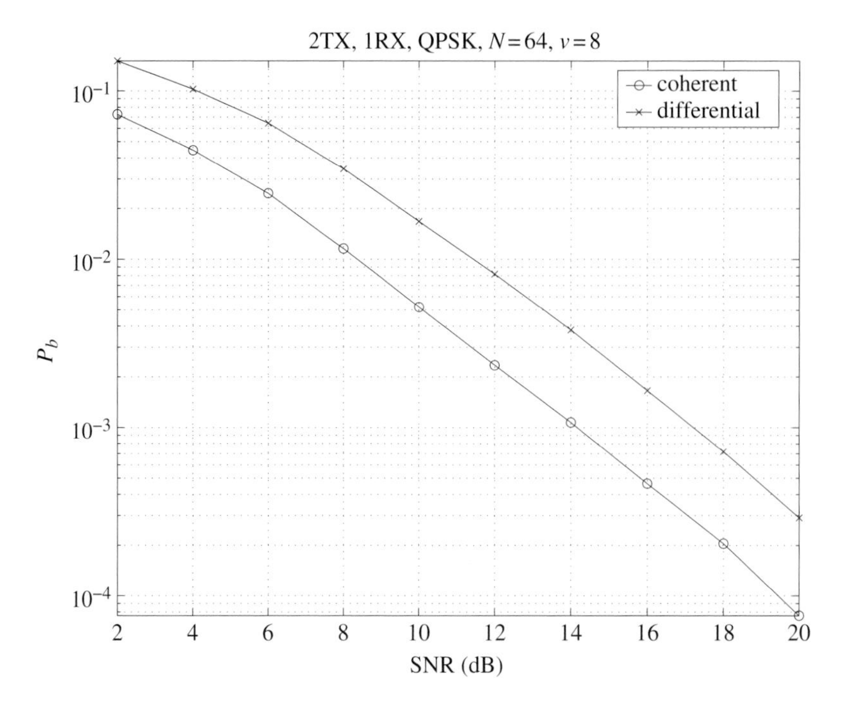

Figure 4.12. Performance comparison between coherent and differential OFDM-STBC with 2TX, 1RX, QPSK modulation, an FFT size of 64, $\nu = 8$.

Since we would like to estimate the source symbols contained in

$$\mathbf{U}_m^{(k)} \stackrel{def}{=} \begin{pmatrix} u_{1,m}^{(k)} & -u_{2,m}^{*(k)} \\ u_{2,m}^{(k)} & u_{1,m}^{*(k)} \end{pmatrix},$$

we define the differential transmission rule $\mathbf{X}_m^{(k)} = (\mathbf{X}_m^{*(k-1)})^{-1}\mathbf{U}_m^{(k)}$. Note that no inverse computation is needed in computing $(\mathbf{X}_m^{*(k-1)})^{-1}$ due to the quaternionic structure of $\mathbf{X}_m^{*(k-1)}$. Figure 4.12 illustrates the 3 dB SNR loss of differential OFDM-STBC relative to its coherent counterpart (with perfect CSI assumed) for an indoor wireless environment.

4.4.2 Applications of diversity-embedded codes

Given that we can construct diversity-embedded codes, we would like to examine how such codes can impact the wireless communication system design. We follow Diggavi *et al.* [25] in examining three applications of diversity-embedded codes: (i) A natural application would be for applications requiring unequal error protection (UEP). For example, image, audio, or video transmission might need multiple levels of error protection for sensitive and less sensitive parts of the message. (ii) A second application could be to improve the overall throughput by opportunistically using the good channel realizations without channel state feedback. (iii) A third application could be in reducing delay in packet transmission using the different diversity orders for prioritized scheduling.

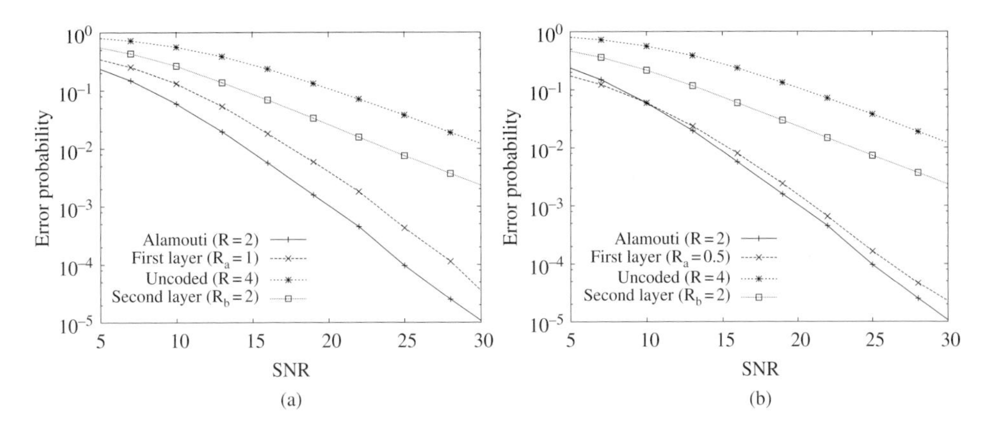

Figure 4.13. UEP performance of diversity-embedded codes.

In this section, we examine the impact of diversity embedding for each of these applications by comparing it to conventional single-layer codes. In all the numerical results we have $M_r = 1$. The numerical results given below are from [25].

In Figure 4.13, the performance of a diversity-embedded code which is designed for $M_t = 2$ and 4-QAM signal constellation is given. In Figure 4.13(a), the embedded code has $R_a = 1$ bit/transmission at diversity order $D_a = 2$ and $R_b = 2$ bits/transmission at diversity order $D_b = 1$. We compare this with a "full-rate," maximal diversity order code (Alamouti code with $R = 2$ bits/transmission and diversity order $D = 2$). We also plot the performance of uncoded transmission with rate $R = 4$ bits/transmission and diversity order $D = 1$. Qualitatively, we can see that the embedded code gives two levels of diversity. Note that as expected, we do pay a penalty in rate (or error performance) over a single-layer code designed for the specific diversity order, but the penalty can be made smaller by cutting the rate for one of the diversity layers as demonstrated in Figure 4.13(b).

Figure 4.14 illustrates the advantage of diversity-embedded codes in terms of opportunistically utilizing the channel conditions without feedback. In Figure 4.14(a), we see that for 4-QAM, and $M_t = 2$, the diversity-embedded code outperforms the Alamouti single-layer code designed for full diversity. However, at very high SNR, single-layer transmission designed at the lower diversity order outperforms a diversity-embedded code since it transmits at a higher rate. This illustrates that for moderate SNR regimes, diversity-embedded codes outperform single-layer codes in terms of average throughput. Figure 4.14(b) for 8-QAM shows that this regime increases with constellation size.

In the final application given in [25], the delay behavior if we combine a rudimentary ACK/NACK feedback about the transmitted information along with space–time codes, is examined. In the single-layer code the traditional automatic repeat request (ARQ) protocol is used wherein if a packet is in error, it is re-transmitted. In a diversity-embedded code, since different parts of the information get unequal error protection, we can envisage an alternative use of the ARQ. For two diversity levels, we assume that ACK/NACK is received separately for each diversity layer. The mechanism proposed in [25] is illustrated in Figure 4.15. The information is sent along two streams, one on the higher diversity level and the other on the lower diversity level. If the packet on the higher diversity level

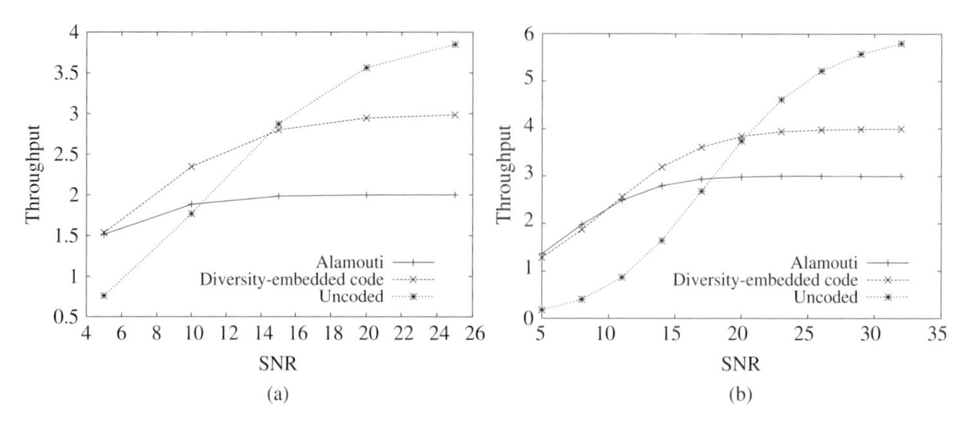

Figure 4.14. Throughput comparison of diversity-embedded codes with single-layer codes: (a) 4-QAM; (b) 8-QAM.

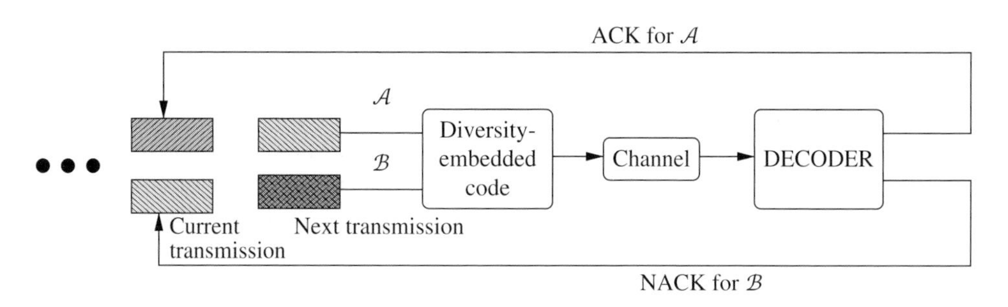

Figure 4.15. ARQ mechanism for diversity-embedded codes using prioritized scheduling.

goes through but the lower diversity level one fails, then in the next transmission, the failed packet is sent on the higher diversity level and therefore receives a higher "priority." Therefore the lower priority packet opportunistically rides along with the higher priority packet and thereby opportunistically uses the channel to reduce the delay.

In Figure 4.16, we examine the impact of the ARQ mechanism illustrated in Figure 4.15 with a comparison to single-layer schemes. In Figure 4.16(a) we transmit both the single-layer and the diversity-embedded code using the same transmit 4-QAM alphabet. We assume that the diversity-embedded code gets ACK/NACK feedback on both levels separately. Figure 4.16(b) illustrates the same principle, but with 8-QAM for the diversity-embedded code and 4-QAM for the single-layer code. The single-layer scheme at the lower diversity order ($D = 1$) has double the rate of the maximal diversity single-layer code. The comparison is made for the same packet size and therefore it gets individual ACK/NACK for its packets. Figure 4.16 shows that there is an SNR regime where diversity-embedded codes give lower average delay than the single-layer codes. Qualitatively this is similar to the throughput maximization of Figure 4.14.

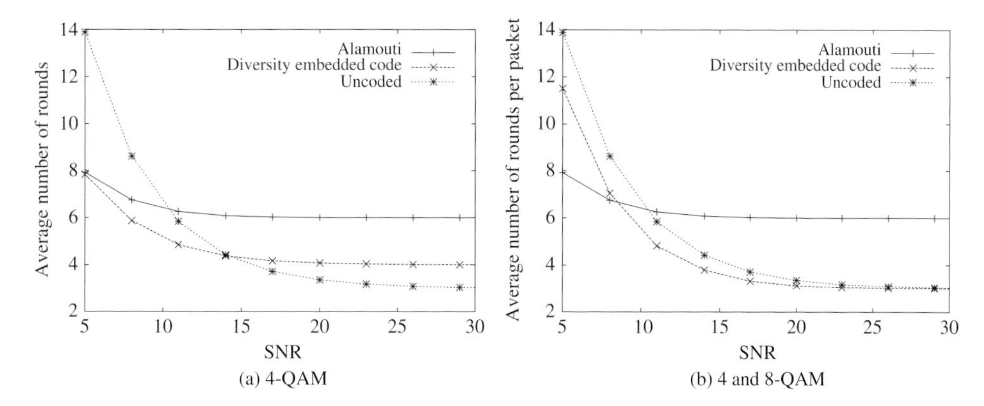

Figure 4.16. (a) Delay comparison of diversity-embedded codes with single-layer codes when used with ARQ. In (b) the single-layer code uses 4-QAM whereas the diversity-embedded code uses 8-QAM with a rate of 2 bits/transmission for each diversity level.

4.4.3 *Interactions with network layers*

Multiple access: interference cancellation

Spatial diversity implies that different users expect to see different channel conditions. We can double the number of STBC users (and hence network capacity) by adding a second receiver at the base-station and employing interference cancellation techniques. We can also deliver higher rates by multiplexing parallel data streams, and in previous work we have described how to use four antennas at the base-station and two antennas at the mobile to deliver twice the standard data rate on a GSM channel.

Our approach is to use algebraic structure to design a single-receiver architecture that cancels interference when it is present and delivers increased diversity gain when it is not. We illustrate this for the Alamouti code by showing that a second antenna at the receiver can separate two users, each employing the Alamouti code. Consider vectors $\mathbf{r}_1, \mathbf{r}_2$ where the entries of $\mathbf{r}_i$ are the signals received at antenna i over two consecutive time slots. If $\mathbf{c} = (c_1, c_2)$ and $\mathbf{s} = (s_1, s_2)$ are the codewords transmitted by the first and second users, then

$$\mathbf{r} = \begin{bmatrix} \mathbf{r}_1 \\ \mathbf{r}_2 \end{bmatrix} = \begin{bmatrix} \mathbf{H}_1 & \mathbf{G}_1 \\ \mathbf{H}_2 & \mathbf{G}_2 \end{bmatrix} \begin{bmatrix} \mathbf{c} \\ \mathbf{s} \end{bmatrix} + \begin{bmatrix} \mathbf{w}_1 \\ \mathbf{w}_2 \end{bmatrix},$$

where the vectors $\mathbf{w}_1$ and $\mathbf{w}_2$ are complex Gaussian random variables with zero mean and covariance $N_0\mathbf{I}_2$. The matrices $\mathbf{H}_1$ and $\mathbf{H}_2$ capture the path gains from the first user to the first and second receive antennas. The matrices $\mathbf{G}_1$ and $\mathbf{G}_2$ capture the path gains from the second user to the first and second receive antennas. What is important is that all these matrices share the Alamouti structure. Define

$$\mathbf{D} = \begin{bmatrix} \mathbf{I}_2 & -\mathbf{G}_1\mathbf{G}_2^{-1} \\ -\mathbf{H}_2\mathbf{H}_1^{-1} & \mathbf{I}_2 \end{bmatrix}$$

and observe that

$$\mathbf{Dr} = \begin{bmatrix} \mathbf{H} & \mathbf{0} \\ \mathbf{0} & \mathbf{G} \end{bmatrix} \begin{bmatrix} \mathbf{c} \\ \mathbf{s} \end{bmatrix} + \begin{bmatrix} \tilde{\mathbf{w}}_1 \\ \tilde{\mathbf{w}}_2 \end{bmatrix},$$

where $\mathbf{H} = \mathbf{H}_1 - \mathbf{G}_1\mathbf{G}_2^{-1}\mathbf{H}_2$ and $\mathbf{G} = \mathbf{G}_2 - \mathbf{H}_2\mathbf{H}_1^{-1}\mathbf{G}_1$.

The matrix $\mathbf{D}$ transforms the problem of joint detection of two co-channel users into the separate detection of two space–time users. It plays the role of the decorrelating detector in CDMA systems; detection of the codeword $\mathbf{c}$ is through projection onto the orthogonal complement of $[\mathbf{G}_1^T, \mathbf{G}_2^T]$. The algebraic structure of the Alamouti code (closure under addition, multiplication, and taking inverses) implies that the matrices $\mathbf{H}$ and $\mathbf{G}$ have the same structure as $\mathbf{H}_1, \mathbf{H}_2, \mathbf{G}_1$, and $\mathbf{G}_2$. Next, we show how the algebraic structure of the Alamouti code leads to a single-receiver structure that cancels interference when it is present and delivers increased diversity gain when it is not. The covariance matrix $\mathbf{M}$ of the received signal is given by

$$\mathbf{M} = E[\mathbf{rr}^*] = \underbrace{\begin{bmatrix} \mathbf{H}_1 \\ \mathbf{H}_2 \end{bmatrix} \begin{bmatrix} \mathbf{H}_1^* \ \mathbf{H}_2^* \end{bmatrix}}_{\substack{\text{orthogonal projection} \\ \text{on } \langle \mathbf{h}_1, \mathbf{h}_2 \rangle}} + \underbrace{\begin{bmatrix} \mathbf{G}_1 \\ \mathbf{G}_2 \end{bmatrix} \begin{bmatrix} \mathbf{G}_1^* \ \mathbf{G}_2^* \end{bmatrix}}_{\substack{\text{orthogonal projection} \\ \text{on } \langle \mathbf{g}_1, \mathbf{g}_2 \rangle}} + \frac{1}{\text{SNR}}\mathbf{I}_4$$

and if

$$\begin{bmatrix} \mathbf{H}_1 \\ \mathbf{H}_2 \end{bmatrix} = [\mathbf{h}_1 \ \mathbf{h}_2] \quad \text{and} \quad \begin{bmatrix} \mathbf{G}_1 \\ \mathbf{G}_2 \end{bmatrix} = [\mathbf{g}_1 \ \mathbf{g}_2]$$

then it can be shown that if $i \neq j$ then for all integers k, we have $\mathbf{h}_i\mathbf{M}^k\mathbf{h}_j^* = \mathbf{g}_i\mathbf{M}^k\mathbf{g}_j^* = 0$ (see Section 4 of [13]).

The MMSE receiver looks for a linear combination $\alpha^*\mathbf{r}$ of received signals that is close to some linear combination $\beta_1 c_1 + \beta_2 c_2$ of the codeword $\mathbf{c}$. The solution turns out to be

$$\alpha_1 = (\mathbf{M} - \mathbf{h}_2\mathbf{h}_2^*)^{-1}\mathbf{h}_1; \qquad \beta_1 = 1, \qquad \beta_2 = \frac{\mathbf{h}_2^*\mathbf{M}^{-1}\mathbf{h}_1}{1 - \mathbf{h}_2^*\mathbf{M}^{-1}\mathbf{h}_1}$$

$$\alpha_2 = (\mathbf{M} - \mathbf{h}_1\mathbf{h}_1^*)^{-1}\mathbf{h}_2; \qquad \beta_2 = 1, \qquad \beta_1 = \frac{\mathbf{h}_1^*\mathbf{M}^{-1}\mathbf{h}_2}{1 - \mathbf{h}_1^*\mathbf{M}^{-1}\mathbf{h}_2}.$$

Either $\beta_1 = 0$ and $\beta_2 = 1$, or $\beta_2 = 0$ and $\beta_1 = 1$! The MMSE interference canceller maintains the separate detection feature of space–time block codes; errors in decoding $\mathbf{c}_1$ do not influence the decoding of $\mathbf{c}_2$ and vice versa. Generalizations to the case of frequency-selected channels are described in [28].

Integration of physical, link, and transport layers

It is well-known that errors at the wireless physical layer reverberate across layers and have a negative impact on transport control protocol (TCP) performance (see [6] and references therein).

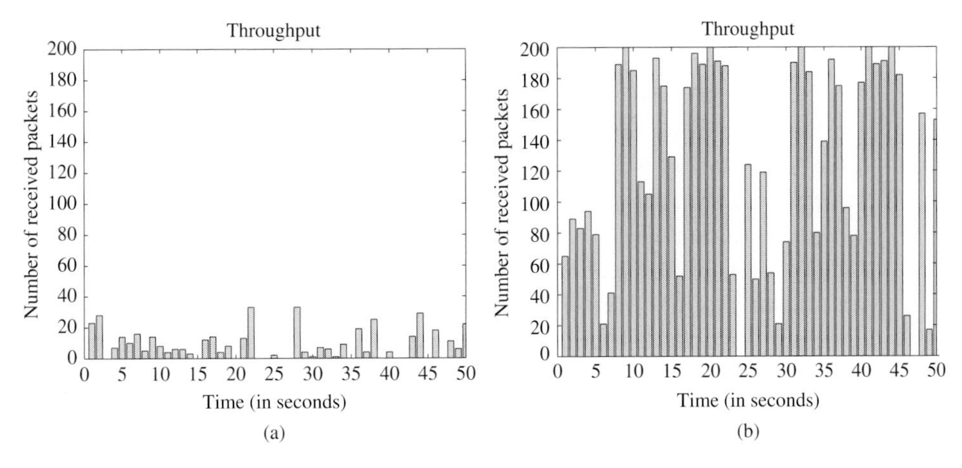

Figure 4.17. TCP throughput with one (a) and two (b) transmit antennas.

Roughly speaking, TCP interprets frame/packet losses as signs of network congestion, and cuts the transmission rate by half, whenever these error events occur. When the link layer does not hide frame errors, TCP times out and does not transmit anything for a significant amount of time. Figure 4.17 (see the discussion in [70]) shows the ability of space–time block codes (in this case the Alamouti code) to shift the SNR point at which TCP breaks down resulting in significant improvements in throughput. In this example, there are no frame retransmissions at the link-layer, and the operating point is in an SNR region where the bit-error rate (BER) performance with only one transmit antenna is below the TCP breaking-point threshold, and the BER with space–time codes is above this particular threshold.

Network utility maximization (NUM)

This brief section provides a glimpse into an emerging foundation for networking that is both mathematically rich and practically relevant, with a promising track record of impact on commercial systems (for a survey see [15]).

The layered architecture is one of the most fundamental and influential structures in network design. Each layer in the protocol stack hides the complexity of the layer below and provides a service to the layer above. While the general principle of layering is widely recognized as one of the key reasons for the enormous success of the Internet, there is little quantitative understanding as a systematic, rather than an ad hoc, process of designing a layered protocol stack for wired and wireless networks. One possible perspective to rigorously and holistically understand layering is to integrate the various protocol layers into a single coherent theory, by regarding them as carrying out an asynchronous distributed computation over the network to implicitly solve a global objective. Such a theory will expose the interconnection between protocol layers and can be used to study rigorously the performance trade-off in protocol layering, as different ways to distribute a centralized computation. Even though the design of a complex system will always be broken down into simpler modules, this theory will allow us to carry out this layering process systematically and to explicitly trade-off design objectives.

The approach of "protocol as a distributed solution" to some global optimization problem in the form of NUM has been successfully tested in trials for TCP. The key innovation from this line of work is to view the network as an optimization solver and the congestion control protocol as distributed algorithms solving a specified NUM. The framework of NUM has recently been substantially extended from an analytic tool of reverse-engineering TCP congestion control to a general approach for understanding interactions across layers. Application needs form the objective function, i.e. the network utility to be maximized, and the restrictions in the communication infrastructure are translated into many constraints of a generalized NUM problem. Such problems may be very difficult non-linear, non-convex optimization with integer constraints. There are many different ways to decompose a given problem, each of which corresponds to a different layering scheme. These decomposition (i.e. layering) schemes have different trade-offs in efficiency, robustness, and asymmetry of information and control, thus some are "better" than others depending on the criteria set by the network users and managers.

The key idea in "layering as optimization decomposition" is as follows. Different *decompositions* of an optimization problem, in the form of a generalized NUM are mapped to different *layering* schemes in a communication network, and from functions of primal or Lagrange dual *variables* coordinating the subproblems to the *interfaces* among the layers. Since different decompositions correspond to different layer architectures, we can also tackle the question "how to and how not to layer" by investigating the pros and cons of decomposition techniques. Furthermore, by comparing the objective function values under various forms of optimal decompositions and suboptimal decompositions, we can seek "separation theorems" among layers: conditions under which strict layering incurs no loss of optimality. The robustness of these separation theorems can be further characterized by sensitivity analysis in optimization theory: how much will the differences in the objective value (between different layer schemes) fluctuate as constant parameters in utility maximization are perturbed.

4.5 Discussion and future challenges

A discussion about the issues and trade-offs involved in MIMO system design is now in order. These issues include the choice of key system parameters including the block length, the carrier frequency, and the number of transmit/receive antennas in addition to the operating environment conditions such as high versus low SNR, high versus low mobility, and strict versus relaxed delay constraints.

The length of the transmission block N (relative to the symbol period and the channel memory ν) is an important design parameter. Shorter blocks experience less channel time variation (which reduces the need for channel tracking within the block), incur smaller delay, and have smaller receiver complexity (typically, block-by-block signal processing algorithm complexity grows in a quadratic or cubic manner with the block size). On the other hand, smaller blocks could incur a significant throughput penalty due to overhead (needed for various functions including guard sequence, synchronization, training, etc.).

Concerning the carrier frequency f_c, the current trend is towards higher f_c where more radio-frequency (RF) bandwidth is available, the antenna size is smaller (at the same radiation efficiency), and the antenna spacing requirements (to ensure independent fading) are less stringent due to the decreased wavelength. On the other hand, the main challenges in migrating towards higher f_c are the higher costs of manufacturing reliable RF components, the increased propagation loss, and the increased sensitivity to Doppler effects.

When selecting the number of transmit/receive antennas, several practical considerations must be taken into account, as described next. Under strict delay constraints, achieving high diversity gains (i.e. high reliability) becomes critical in order to minimize the need for retransmissions. Since transmit/receive diversity gains experience diminishing returns as their numbers increase, complexity considerations dictate the use of small antenna arrays (typically no more than four antennas at each end). Current technology limitations favor using more antennas at the base-station than at the user terminal.

For delay-tolerant applications (such as data file transfers), achieving high throughput takes precedence over achieving high diversity and larger antenna arrays (of course still limited by cost and space constraints) can be used to achieve high spatial rate multiplexing gains. Likewise, high-mobility channel conditions substantially impact the choice of system parameters such as the use of shorter blocks, lower carrier frequencies, and non-coherent or adaptive receiver techniques.

STTC [74] use multiple transmit antennas to achieve diversity and coding gains. The first gain manifests itself as an increase in the slope of the BER versus SNR curve (on a log–log scale) at high SNR, while the latter gain manifests itself as a horizontal shift in that same curve. At low SNR, it becomes more important to maximize the coding gain while at high SNR diversity gains dominate performance. For SNR ranges typically encountered on broadband wireless terrestrial links, it might be wise to sacrifice some diversity gain in exchange for more coding gain. For example, using only two transmit and one receive antennas for a channel with a delay spread as high as 16 taps, the maximum (spatial and multi-path) diversity gain possible is $16 \times 2 \times 1 = 32$. For typical SNR levels in the 10–25 dB range, it suffices to design STCs that achieve a much smaller diversity level (e.g. up to 8) to limit the receiver complexity and to use the extra degrees of freedom in code design to achieve a higher coding gain. STC have also been shown to result in significant improvements in the networking throughput [70].

Wireless networks present an opportunity to re-examine functional abstractions of traditional network layer protocols. Cross-layer interactions in wireless networks can optimize throughput by making additional performance information visible between layers in the IP protocol stack. Spatial diversity is critical in improving data rates and reliability of individual links and leads to innovations in scheduling that optimize global throughput. Space–time codes designed for small numbers of transmit and receive antennas have been shown to significantly improve link capacity, and also system capacity through resource allocation. This coding technology can be integrated with sophisticated signal processing to provide a complete receiver that has computational complexity essentially implementable on current chip technology. This bounding of signal processing complexity is important given the energy constraints at the mobile terminal.

Many challenges still exist at the physical layer on the road to achieving high-rate and reliability wireless transmission. We conclude this chapter by enumerating some of these challenges.

- **Signal processing**. While effective and practical joint equalization and decoding schemes that exploit the multi-path diversity available in frequency-selective channels have been developed, the full exploitation of time diversity in fast time-varying channels remains elusive. The main challenge here is the development of practical adaptive algorithms that can track the rapid variations of the large number of taps in MIMO channels and/or equalizers. While some encouraging steps have been made in this direction [52, 87], the allowable Doppler rates (which depend on the mobile speed and carrier frequency) for high performance are still quite limited.

  Another signal processing challenge is the design of MIMO training sequences that are resilient to practical impairments such as receiver synchronization errors (for example, residual frequency offsets in OFDM). It is also of practical importance to construct training sequences with low peak-to-average power ratio to extend battery life by improving efficiency of the transmit power amplifier. Some of our recent work in this area is described in [60, 61].
- **Code design**. One challenging code design problem that has attracted significant interest recently is the design of practical space–time codes that achieve the optimal rate–diversity trade-off [88] and have a practical decoding complexity. As mentioned in Section 4.3, some progress has been made in [32, 75, 86]. Another challenging problem is the design of non-coherent encoding/decoding schemes for the family of diversity-embedding codes.
- **Networking**. The interference cancellation techniques described in Section 4.3 can be extended in several directions. Given the importance of mobile ad hoc networks and their lack of fixed infrastructure and centralized control, it would be interesting to drop the assumption of time-synchronous users and explore the asynchronous case. Furthermore, given the commercial interest in wireless systems with four transmit antennas, it is important to explore interference cancellation based on the octonion space–time block code, and on the non-linear quaternionic code described in Section 4.3.4.

  Another challenging problem is the investigation of cross-layer interactions between embedded-diversity coding and link-layer ARQ protocols (which come in several hybrid and selective forms). In particular, the reliability that is lost when spatial diversity is traded for rate can be recovered by the time diversity gained through ARQ retransmission. Conversely, when rate is traded for spatial diversity, it would be interesting to quantify the value of the reduced latency in terms of throughput, delay, and power consumption.

4.6 Bibliographical notes

The past decade has witnessed significant progress in the understanding and design of space–time codes. The information-theoretic underpinnings of space–time codes were given in [76] and [33]. These authors established that multiple antennas can make wireless communication a high data-rate pipe. Though the rudiments of transmit spatial diversity were proposed in [81, 84], the basis of modern space–time coding was given in [5, 74]. Since then, space–time codes have been extended through linear (see [30, 43, 72] and references therein) and non-linear designs (see [42, 59] and references therein). In another line of work, non-coherent space–time codes, their design, analysis and information-theoretic properties have been studied by Hochwald and Marzetta [46], Hochwald and Sweldens [45], Hughes [47], Zheng and Tse [89] and references therein.

The diversity–multiplexing trade-off was first established in [74] for transmit alphabet size. The informat in the context of the *fixed* theoretic question was posed and answered in [89]. Since then, there has been a significant effort in designing codes achieving the diversity–multiplexing trade-off (see [32, 75, 86] and references therein). The idea of diversity-embedded space–time codes was first proposed in [23] where design criteria and some constructions were given. Diversity embedding from an information-theoretic viewpoint was examined in [27].

There has been extensive work in the area of signal processing techniques for space–time codes (see, for example, the papers in the special issue [4] and references therein). A more extensive survey of developments in diversity communications can also be found in [24]. Several recent textbooks [40, 80] give excellent introductions to modern wireless communications.

Appendix 4.1 Algebraic structure: quadratic forms

The simplest form of transmit diversity is the delay diversity scheme proposed by Wittneben [84] for two transmit antennas, where a signal is transmitted from the second antenna, then delayed one time slot and transmitted from the first antenna. Orthogonal designs [72] are a class of space–time block codes that achieve maximal diversity with decoding complexity that is linear in the size of the constellation. The most famous example was discovered by Alamouti [5], and is described by a 2×2 matrix where the columns represent different time slots, the rows represent different antennas, and the entries are the symbols to be transmitted. The encoding rule is

$$\begin{bmatrix} c_1 & c_2 \end{bmatrix} \rightarrow \begin{bmatrix} c_1 & c_2 \\ -c_2^* & c_1^* \end{bmatrix}.$$

Assuming a quasi-static flat-fading channel, the signals r_1, r_2 received over two consecutive time slots are given by

$$\begin{bmatrix} r_1 \\ -r_2^* \end{bmatrix} = \begin{bmatrix} h_1 & h_2 \\ -h_2^* & h_1^* \end{bmatrix} \begin{bmatrix} c_1 \\ -c_2^* \end{bmatrix} + \begin{bmatrix} w_1 \\ -w_2^* \end{bmatrix},$$

where h_1, h_2 are the path gains from the two transmit antennas to the mobile, and the noise samples w_1, w_2 are independent samples of a zero-mean complex Gaussian random variable with noise energy N_0 per complex dimension. Thus $\mathbf{r} = \mathbf{H}\mathbf{c} + \mathbf{w}$, where the matrix $\mathbf{H}$ is orthogonal. The reason for broad commercial interest in the Alamouti code is that both coherent and non-coherent detection are remarkably simple. If the path gains are known at the mobile (typically this is accomplished at some sacrifice in rate by inserting pilot tones into the data frame for channel estimation) then the receiver is able to form

$$\mathbf{H}^*\mathbf{r} = \|\mathbf{h}\|^2\mathbf{c} + \mathbf{w}'.$$

The new noise term $\mathbf{w}'$ is still white, so that c_1, c_2 can be decoded separately rather than jointly, which is far more complex.

Let $u_0, u_1, \ldots, u_{s-1}$ be positive integers, and let $x_0, x_1, \ldots, x_{s-1}$ be commuting indeterminates. A *real orthogonal design* of type $(u_0, u_1, \ldots, u_{s-1})$ and size N is an $N \times N$ matrix $\mathbf{X}$ with entries $0, \pm x_0, \pm x_1, \ldots, \pm x_{s-1}$ satisfying

$$\mathbf{X}\mathbf{X}^T = \sum_{j=0}^{s-1} u_j x_j^2 \mathbf{I}_N.$$

There are s indeterminates and N time slots, so the rate of the orthogonal design is s/N.

$N = 2$. A real orthogonal design of type (1,1) and size $N = 2$ corresponds to the representation of the complex numbers $\mathbf{C}$ as a 2×2 matrix algebra over the real numbers $\mathbf{R}$. The complex number $x_0 + \mathbf{i}x_1$ corresponds to the matrix

$$\begin{bmatrix} x_0 & x_1 \\ -x_1 & x_0 \end{bmatrix}.$$

$N = 4$. A real orthogonal design of type (1,1,1,1) and size $N = 4$ corresponds to the representation of the quaternions $\mathbf{Q}$ as a 4×4 matrix algebra over the real numbers $\mathbf{R}$. The quaternion $x_0 + \mathbf{i}x_1 + \mathbf{j}x_2 + \mathbf{k}x_3$ corresponds to the matrix

$$\left[\begin{array}{cc|cc} x_0 & x_1 & x_2 & x_3 \\ -x_1 & x_0 & -x_3 & x_2 \\ \hline -x_2 & x_3 & x_0 & -x_1 \\ -x_3 & -x_2 & x_1 & x_0 \end{array}\right] = x_0\mathbf{I}_4 + x_1\left[\begin{array}{cc|cc} & 1 & & \\ -1 & & & \\ \hline & & & -1 \\ & & 1 & \end{array}\right]$$

$$+x_2\left[\begin{array}{cc|cc} & & 1 & \\ & & & 1 \\ \hline -1 & & & \\ & -1 & & \end{array}\right] + x_3\left[\begin{array}{cc|cc} & & & 1 \\ & & -1 & \\ \hline & 1 & & \\ -1 & & & \end{array}\right].$$

$N = 8$. A real orthogonal design of type $(1, 1, \ldots, 1)$ and size $N = 8$ corresponds to the representation of the octonions or Cayley numbers as an eight-dimensional algebra over the real numbers $\mathbf{R}$. This algebra is non-associative as well as non-commutative.

A *complex orthogonal design* of size N and type $(u_0, u_1, \ldots, u_{s-1}; v_1, v_2, \ldots, v_t)$ is a matrix $\mathbf{Z} = \mathbf{X} + \mathbf{i}\mathbf{Y}$, where $\mathbf{X}$ and $\mathbf{Y}$ are real orthogonal designs of type $(u_0, u_1, \ldots, u_{s-1})$ and $(v_1, v_2, \ldots, v_t)$ respectively, and where

$$\mathbf{Z}\mathbf{Z}^* = \sum_{j=0}^{s-1} u_j x_j^2 + \sum_{j=1}^{t} v_j y_j^2 \mathbf{I}_N.$$

Since

$$\mathbf{ZZ}^* = (\mathbf{X}+\mathbf{iY})(\mathbf{X}^T - \mathbf{iY}^T) = (\mathbf{XX}^T + \mathbf{YY}^T) + \mathbf{i}(\mathbf{YX}^T - \mathbf{XY}^T)$$

it follows that $\mathbf{YX}^T = \mathbf{XY}^T$. A pair of real orthogonal designs that is connected in this way is called an *amicable pair* (see [37] for more information). Note that if $t = s$, then the entries of $\mathbf{X}+\mathbf{iY}$ are linear combinations of the complex indeterminates $z_k = x_k + \mathbf{i}y_k$ and their complex conjugates $z_k^* = x_k - \mathbf{i}y_k$. In fact, the definition of a complex orthogonal design found in [72] is given in terms of these indeterminates. The rate of a complex orthogonal design is $(s+t)/2N$.

A complex design of size N with $t = s+1$ determines a real orthogonal design of size $2N$ through the substitution

$$x_0 + \mathbf{i}x_1 \rightarrow \begin{bmatrix} x_0 & x_1 \\ -x_1 & x_0 \end{bmatrix}.$$

$N = 2$. This is the Alamouti space–time block code. We may view quaternions as pairs of complex numbers, where the product of quaternions (a, b) and (c, d) is given by $(ac - bd^*, ad + bc^*)$. These are Hamilton's biquaternions, and if we associate the pair (a, b) with the 2×2 complex matrix

$$\begin{bmatrix} a & b \\ -b^* & a^* \end{bmatrix}$$

then we see that the rule for multiplying biquaternions coincides with the rule for matrix multiplication.

$N = 4$. The Alamouti space–time block code determines the full-rate 4×4 real orthogonal design via the above substitutions. However the full-rate 8×8 real orthogonal design cannot be obtained from a 4×4 complex design.

The representation of the octonions as 4-tuples of complex numbers provides an example of an extremal complex design. The product $\mathbf{c} = \mathbf{ab}$ of octonions $\mathbf{a} = (a_0, a_1, a_2, a_3)$ and $\mathbf{b} = (b_0, b_1, b_2, 0)$ is given by

$$\begin{aligned} c_0 &= a_0 b_0 - b_1^* a_1 - b_2^* a_2 - a_3^* b_3 \\ c_1 &= b_1 a_0 + a_1 b_0^* - a_3 b_2^* + b_3 a_2^* \\ c_2 &= b_2 a_0 - a_1^* b_3 + a_2 b_0^* + b_1^* a_3 \\ c_3 &= b_3 a_0^* + a_1 b_2 - b_1 a_2 + a_3 b_0. \end{aligned}$$

It follows that right multiplication of an octonion $\mathbf{a}$ by octonions of the form $\mathbf{b} = (b_0, b_1, b_2, 0)$ can be represented as $\mathbf{ab} = \mathbf{a}R(b_0, b_1, b_2, 0)$, where

$$R(b_0, b_1, b_2, 0) = \begin{bmatrix} b_0 & b_1 & b_2 & 0 \\ -b_1^* & b_0^* & 0 & b_2 \\ -b_2^* & 0 & b_0^* & -b_1 \\ 0 & -b_2^* & b_1^* & b_0 \end{bmatrix}. \qquad \text{(A4.1)}$$

The columns of this matrix are orthogonal; hence $R(b_0, b_1, b_2, 0)$ is a rate-$\frac{3}{4}$ complex orthogonal design.

Given t symmetric, anti-commuting orthogonal matrices of size N, let $\rho_t(N) - 1$ be the number of skew-symmetric, anti-commuting orthogonal matrices of size N that anti-commute with the initial set of t matrices. The next two theorems are proved using Clifford algebras [17] and are due to Wolfe [85].

THEOREM A4.1 *There exists an amicable pair* $\mathbf{X}, \mathbf{Y}$ *of real orthogonal designs of size* N, *where* $\mathbf{X}$ *has type* $(1, \ldots, 1)$ *on variables* $x_0, x_1, \ldots, x_{s-1}$ *and* $\mathbf{Y}$ *has type* $(1, \ldots, 1)$ *on variables* $y_1, y_2, \ldots, y_t$, *if and only if* $s \leq \rho_t(N) - 1$.

THEOREM A4.2 *Let* $\mathbf{X}, \mathbf{Y}$ *be an amicable pair of real orthogonal designs of size* $N = 2^h N_0$, *where* N_0 *is odd. Then the total number of real variables in* $\mathbf{X}$ *and* $\mathbf{Y}$ *is at most* $2h+2$, *and this bound is achieved by designs* $\mathbf{X}, \mathbf{Y}$ *that each involve* $h+1$ *variables.*

In fact, a group of Pauli matrices that appears in the construction of quantum error correcting codes can be used to construct pairs $\mathbf{X}, \mathbf{Y}$ where the entries of $\mathbf{X}$ are $0, \pm x_0, \ldots, \pm x_s$ and the entries of $\mathbf{Y}$ are $0, \pm y_1, \ldots, \pm y_t$ [13].

References

[1] N. Al-Dhahir, "Single-carrier frequency-domain equalization for space–time block-coded transmissions over frequency-selective fading channels," *IEEE Commun. Letters*, vol. 5, no. 7, pp. 304–306, July 2001.

[2] N. Al-Dhahir, "Overview and comparison of equalization schemes for space–time-coded signals with application to EDGE," *IEEE Trans. Signal Processing,* vol. 50, no. 10, pp. 2477–2488, Oct. 2002.

[3] N. Al-Dhahir, C. Fragouli, A. Stamoulis, Y. Younis, and A. R. Calderbank, "Space–time processing for broadband wireless access," *IEEE Commun. Magazine*, vol. 40, no. 9, pp. 136–142, Sept. 2002.

[4] N. Al-Dhahir, G. Giannakis, B. Hochwald, B. Hughes, and T. Marzetta, "Guest editorial on space–time coding," *IEEE Trans. Signal Processing,* vol. 50, no. 10, pp. 2381–2384, Oct. 2002.

[5] S. Alamouti, "A simple transmit diversity technique for wireless communications," *IEEE J. Select. Areas Commun.*, vol. 16, no. 8, pp. 1451–1458, Oct. 1998.

[6] H. Balakrishnan, V. N. Padmanabhan, S. Seshan, and R. H. Katz, "A comparison of mechanisms for improving TCP performance over wireless links," *IEEE/ACM Trans. Networking*, vol. 5, no. 6, pp. 756–769, Dec. 1997.

[7] G. Bauch and N. Al-Dhahir, "Reduced-complexity space–time turbo-equalization for frequency-selective MIMO channels," *IEEE Trans. Wireless Commun.*, vol. 1, no. 4, pp. 819–828, Oct. 2002.

[8] S. Benedetto and G. Montorsi, "Unveiling turbo codes: some results on parallel concatenated coding schemes," *IEEE Trans. Inform. Theory*, vol. 42, no. 2, pp. 409–428, March 1996.

[9] C. Berrou and A. Glavieux, "Near optimum error correcting coding and decoding: turbo-codes," *IEEE Trans. Commun.*, vol. 44, no. 10, pp. 1261–1271, Oct. 1996. Also see C. Berrou, A. Glavieux, and P. Thitimajshima, *Proceedings of ICC'93.*

[10] K. Boulle and J. C. Belfiore, "Modulation schemes designed for the Rayleigh channel," In *Proc. Conf. Inform. Sci. Syst. (CISS '92)*, pp. 288–293, March 1992.

[11] J. Boutros and E. Viterbo, "Signal space diversity: a power and bandwidth efficient diversity technique for the Rayleigh fading channel," *IEEE Trans. Inform. Theory*, vol. 44, no. 4, pp. 1453–1467, July 1998.

[12] A. R. Calderbank, S. N. Diggavi, and N. Al-Dhahir, "Space–time signaling based on Kerdock and Delsarte–Goethals codes," In *Proc. ICC*, pp. 483–487, June 2004.

[13] A. R. Calderbank and A. F. Naguib, "Orthogonal designs and third generation wireless communication." In J. W. P. Hirschfeld, ed., *Surveys in Combinatorics 2001, London Mathematical Society Lecture Note Series 288*, pp. 75–107. Cambridge University Press, 2001.

[14] A. R. Calderbank, S. Das, N. Al-Dhahir, and S. Diggavi, "Construction and analysis of a new quaternionic space–time code for 4 transmit antennas," *Commun. Inform. Syst.*, vol. 5, no. 1, pp. 97–121, 2005.

[15] M. Chiang, S. H. Low, J. C. Doyle, and A. R. Calderbank, "Layering as optimization decomposition," *Proceedings of the IEEE.* to appear, 2006.

[16] D. Chu, "Polyphase codes with good periodic correlation properties," *IEEE Trans. Inform. Theory*, vol. 18, pp. 531–532, July 1972.

[17] W. K. Clifford, "Applications of Grassman's extensive algebra," *Amer. J. Math.*, vol. 1, pp. 350–358, 1878.

[18] M. O. Damen, K. Abed-Meraim, and J.-C. Belfiore, "Diagonal algebraic space–time block codes," *IEEE Trans. Inform. Theory*, vol. 48, no. 3, pp. 628–636, March 2002.

[19] M. O. Damen, A. Chkeif, and J.-C. Belfiore, "Lattice codes decoder for space–time codes," *IEEE Commun. Letters*, vol. 4, pp. 161–163, May 2000.

[20] M. O. Damen, H. El-Gamal, and N. Beaulieu, "On optimal linear space–time constellations," *Intl. Conf. on Communications. (ICC)*, May 2003.

[21] S. Diggavi, N. Al-Dhahir, A. Stamoulis, and A. R. Calderbank, "Differential space–time coding for frequency-selective channels," *IEEE Commun. Letters*, vol. 6, no. 6, pp. 253–255, June 2002.

[22] S. N. Diggavi, N. Al-Dhahir, and A. R. Calderbank, "Diversity embedded space–time codes," In *IEEE Global Communications Conf. (GLOBECOM)*, pp. 1909–1914, Dec. 2003.

[23] S. N. Diggavi, N. Al-Dhahir, and A. R. Calderbank (2004a). "Diversity embedding in multiple antenna communications," In P. Gupta, G. Kramer, and A. J. van Wijngaarden, eds, *Network Information Theory*, pp. 285–302. *AMS Series on Discrete Mathematics and Theoretical Computer Science*, vol. 66. Appeared as a part of *DIMACS Workshop on Network Information Theory*, March 2003.

[24] S. N. Diggavi, N. Al-Dhahir, A. Stamoulis, and A. R. Calderbank, "Great expectations: the value of spatial diversity to wireless networks," *Proceedings of the IEEE*, vol. 92, pp. 217–270, Feb. 2004.

[25] S. N. Diggavi, S. Dusad, A. R. Calderbank, and N. Al-Dhahir, "On embedded diversity codes," In *Allerton Conf. on Communication, Control, and Computing*, 2005.

[26] S. N. Diggavi and D. N. C. Tse, "On successive refinement of diversity," In *Allerton Conf. on Communication, Control, and Computing*, 2004.

[27] S. N. Diggavi and D. N. C. Tse, "Fundamental limits of diversity-embedded codes over fading channels," In *IEEE Intl. Symp. on Information Theory (ISIT)*, pp. 510–514, Sept. 2005.

[28] S. N. Diggavi, N. Al-Dhahir, and A. R. Calderbank, "Algebraic properties of space–time block codes in intersymbol interference multiple access channels," *IEEE Trans. Inform. Theory*, vol. 49, no. 10, pp. 2403–2414, Oct. 2003.

[29] H. El-Gamal, G. Caire, and O. Damen, "Lattice coding and decoding achieve the optimal diversity–multiplexing of MIMO channels," *IEEE Trans. Inform. Theory*, vol. 50, no. 6, pp. 968–985, June 2004.

[30] H. El-Gamal and O. Damen, "Universal space–time coding," *IEEE Trans. Inform. Theory*, vol. 49, no. 5, pp. 1097–1119, May 2003.

[31] H. El-Gamal and A. R. Hammons, "A new approach to layered space–time coding and signal processing," *IEEE Trans. Inform. Theory*, vol. 47 no. 6, pp. 2321–2334, Sept. 2001.

[32] P. Elia, K. R. Kumar, S. A. Pawar, P. V. Kumar, and H.-F. Lu, "Explicit minimum-delay space–time codes achieving the diversity–multiplexing gain trade off," Submitted September 2004.

[33] G. J. Foschini, "Layered space–time architecture for wireless communication in a fading environment when using multi-element antennas," *Bell Labs Techn. J.*, vol. 1, no. 2, pp. 41–59, 1996.

[34] C. Fragouli, N. Al-Dhahir, and W. Turin, "Effect of spatio-temporal channel correlation on the performance of space–time codes," In *ICC*, vol. 2, pp. 826–830, 2002.

[35] C. Fragouli, N. Al-Dhahir, and W. Turin, "Training-based channel estimation for multiple-antenna broadband transmissions," *IEEE Trans. Wireless Commun.*, vol. 2, no. 2, pp. 384–391, March 2003.

[36] R. G. Gallager, *Low Density Parity Check Codes*, Cambridge, MA: MIT Press, 1963. Available at http://justice.mit.edu/people/gallager.html.

[37] A. V. Geramita and J. Seberry, "Orthogonal Designs, Quadratic Forms and Hadamard Matrices. Lecture Notes in Pure and Applied Mathematics," vol. 43. New York: Marcel Dekker, 1979.

[38] L. Godara, "Applications of antenna arrays to mobile communications. Part I. Performance improvement, feasibility, and system considerations," *Proceedings of the IEEE*, vol. 85, pp. 1031–1060, July 1997.

[39] L. Godara, "Applications of antenna arrays to mobile communications. Part II. Beamforming and direction-of-arrival considerations," *Proceedings of the IEEE*, vol. 85, pp. 1195–1245, 1997.

[40] A. Goldsmith, *Wireless Communications,* Cambridge: Cambridge University Press, 2005.

[41] J.-C. Guey, M. P. Fitz, M. R. Bell, and W.-Y. Kuo, "Signal design for transmitter diversity wireless communication systems over Rayleigh fading channels," *IEEE Trans. Commun.*, vol. 47, no. 4, pp. 527–537, April 1999.

[42] A.R. Hammons and H. El-Gamal, "On the theory of space–time codes for PSK modulation," *IEEE Trans. Inform. Theory*, vol. 46, no. 2, pp. 524–542, March 2000.

[43] B. Hassibi and B. Hochwald, "High-rate codes that are linear in space and time," *IEEE Trans. Inform. Theory*, vol. 48, no. 7, pp. 1804–1824, July 2002.

[44] S. Haykin, *Adaptive Filter Theory*, 2nd edn. Upper Saddle River, NJ: Prentice-Hall, 1991.

[45] B. Hochwald and W. Sweldens, "Differential unitary space–time modulation," *IEEE Trans. Commun.*, vol. 48, no. 12, pp. 2041–2052, Dec. 2000.

[46] B.M. Hochwald and T.L. Marzetta, "Capacity of a mobile multiple-antenna communication link in Rayleigh flat-fading," *IEEE Trans. Inform. Theory*, vol. 45, no. 1, pp. 139–157, Jan. 1999.

[47] B.L. Hughes "Differential space–time modulation," *IEEE Trans. Inform. Theory*, vol. 46, no. 7, pp. 2567–2578, Nov. 2000.

[48] H. Jafarkhani, "A quasi-orthogonal space–time block code," *IEEE Commun. Letters*, vol. 49, no. 1, pp. 1–4, Jan. 2001.

[49] H. Jafarkhani and V. Tarokh, "Multiple transmit antenna differential detection from generalized orthogonal designs," *IEEE Trans. Inform. Theory*, vol. 47, no. 6, pp. 2626–2631, Sept. 2001.

[50] W.C. Jakes, *Microwave Mobile Communications*. New York: IEEE Press, 1974.

[51] K.J. Kerpez, "Constellations for good diversity performance," *IEEE Trans. Commun.*, vol. 41, no. 9, pp. 1412–1421, Sept. 1993.

[52] C. Komninakis, C. Fragouli, A. Sayed, and R. Wesel, "Multi-input multi-output fading channel tracking and equalization using Kalman estimation," *IEEE Trans. Signal Processing*, vol. 50, no. 5, pp. 1065–1076, May 2002.

[53] E. Lindskog and A. Paulraj, "A transmit diversity scheme for delay spread channels," In *Intl. Conf. on Communications (ICC)*, pp. 307–311, 2000.

[54] Y. Liu, M. Fitz, and O. Takeshita, Space–time codes performance criteria and design for frequency selective fading channels. In *Intl. Conf. on Communications (ICC)*, vol. 9, pp. 2800–2804, 2001.

[55] Y. Liu, M.P. Fitz, and O.Y. Takeshita, "Full rate space–time turbo codes," *IEEE J. Select. Areas Commun.*, vol. 19, no. 5, pp. 969–980, May 2001.

[56] Z. Liu, G. Giannakis, A. Scaglione, and S. Barbarossa, "Decoding and equalization of unknown multipath channels based on block precoding and transmit-antenna diversity," In *Asilomar Conf. on Signals, Systems, and Computers*, pp. 1557–1561, 1999.

[57] B. Lu, X. Wang, and K.R. Narayanan, "LDPC-based space–time coded OFDM systems over correlated fading channels: performance analysis and receiver design," *IEEE Trans. Commun.*, vol. 50, no. 1, pp. 74–88, Jan. 2002

[58] H.F. Lu and P.V. Kumar, "Rate–diversity trade-off of space–time codes with fixed alphabet and optimal constructions for PSK modulation," *IEEE Trans. Inform. Theory*, vol. 49, no. 10, pp. 2747–2752, Oct. 2003.

[59] H.F. Lu and P.V. Kumar, "A unified construction of space–time codes with optimal rate–diversity trade-off," *IEEE Trans. Inform. Theory*, vol. 51, no. 5, pp. 1709–1730, May 2005.

[60] H. Minn and N. Al-Dhahir, "PAR-constrained training signal designs for MIMO OFDM channel estimation in the presence of frequency offsets," In *Vehicular Technology Conf.*, 2005.

[61] H. Minn and N. Al-Dhahir, "Training signal design for MIMO OFDM channel estimation in the presence of frequency offsets," In *Wireless Communications and Networking Conf.*, 2005.

[62] A. Naguib, V. Tarokh, N. Seshadri, and A. R. Calderbank, "A space–time coding modem for high-data-rate wireless communications," *IEEE J. Select. Areas in Commun.*, vol. 16, no. 8, pp. 1459–1477, Oct. 1998.

[63] L. H. Ozarow, S. Shamai, and A. D. Wyner, "Information theoretic considerations for cellular mobile radio," *IEEE Trans. Vehicular Technol.*, vol. 43, no. 2, pp. 359–378, May 1994.

[64] S. Parkvall, M. Karlsson, M. Samuelsson, L. Hedlund, and B. Goransson, "Transmit diversity in WCDMA: link and system level results," In *Vehicular Technology Conf.*, pp. 864–868, 2000.

[65] T. Rappaport, *Wireless Communications*, New York: IEEE Press, 1996.

[66] H. Sari, G. Karam, and I. Jeanclaude, "Transmission techniques for digital terrestrial TV broadcasting," *IEEE Commun. Magazine*, vol. 33, no. 2, pp. 100–109, 1995.

[67] N. Seshadri and J. Winters, "Two signaling schemes for improving the error performance of frequency-division-duplex (FDD) transmission systems using transmitter antenna diversity," In *Vehicular Technology Conf. (VTC)*, pp. 508–511, 1993.

[68] S. Siwamogsatham, M. P. Fitz, and J. H. Grimm, "A new view of performance analysis of transmit diversity schemes in correlated Rayleigh fading," *IEEE Trans. Inform. Theory*, vol. 48, no. 4, pp. 950–956, April 2002.

[69] R. Soni, M. Buehrer, and R. Benning, "Intelligent antenna system for cdma2000," *IEEE Signal Processing Magazine*, vol. 19, no. 4, pp. 54–67, Oct. 2004.

[70] A. Stamoulis and N. Al-Dhahir, "Impact of space–time block codes on 802.11 network throughput," *IEEE Trans. Wireless Commun.*, vol. 2, no. 5, pp. 1029–1039, Sept. 2003.

[71] V. Tarokh and H. Jafarkhani, "A differential detection scheme for transmit diversity," *IEEE J. Select. Areas Commun.*, vol. 18, no. 7, pp. 1169–1174, July 2000.

[72] V. Tarokh, H. Jafarkhani, and A. R. Calderbank, "Space–time block codes from orthogonal designs," *IEEE Trans. Inform. Theory*, vol. 45, no. 5, pp. 1456–1467, 1999.

[73] V. Tarokh, A. Naguib, N. Seshadri, and A. R. Calderbank, "Space–time codes for high data rate wireless communication: performance criteria in the presence of channel estimation errors, mobility, and multiple paths," *IEEE Trans. Commun.*, vol. 47, no. 2, pp. 199–207, Feb. 1999.

[74] V. Tarokh, N. Seshadri, and A. R. Calderbank, "Space–time codes for high data rate wireless communications: performance criterion and code construction," *IEEE Trans. Inform. Theory*, vol. 44, no. 2, pp. 744–765, Mar. 1998.

[75] S. Tavildar and P. Viswanath, "Approximately universal codes over slow-fading channels," *IEEE Trans. Inform. Theory,* vol. 52 no. 7 pp. 3233–3258, July 2006. See also http://www.ifp.uiuc.edu/~pramodv/pubs.html.

[76] I. Telatar, "Capacity of multi-antenna Gaussian channels," *Eur. Trans. Telecomm.*, vol. 10, no. 6, pp. 585–595, 1999.

[77] T.I. "Space–time block coded transmit antenna diversity for WCDMA," Texas Instruments SMG2 document 581/98, submitted October 1998.

[78] TIA, "The CDMA 2000 candidate submission," Draft of TIA 45.5 Subcommittee.

[79] L. Tong and S. Perreau, "Multichannel blind identification: from subspace to maximum likelihood methods," *Proceedings of the IEEE*, vol. 86, no. 10, pp. 1951–1968, Oct. 1998.

[80] D. N. C. Tse and P. Viswanath, *Fundamentals of Wireless Communication*, Cambridge: Cambridge University Press, 2005.

[81] J. Uddenfeldt and A. Raith, "Cellular digital mobile radio system and method of transmitting information in a digital cellular mobile radio system," US Patent no. 5,088,108, 1992.

[82] M. Uysal, N. Al-Dhahir, and C. N. Georghiades, "A space–time block-coded OFDM scheme for unknown frequency-selective fading channels," *IEEE Commun. Letters*, vol. 5, no. 10, pp. 393–395, Oct. 2001.

[83] S. Vaughan-Nichols, "Achieving wireless broadband with WiMAX," *IEEE Computer Magazine*, vol. 37, no. 6, pp. 10–13, June 2004.

[84] A. Wittneben, "A new bandwidth efficient transmit antenna modulation diversity scheme for linear digital modulation," In *Intl. Conf. on Communications (ICC)*, pp. 1630–1634, May 1993.

[85] W. Wolfe, "Amicable orthogonal designs – existence," *Canadian J. Math.*, vol. 28, pp. 1006–1020, 1976.

[86] H. Yao and G. Wornell, "Achieving the full MIMO diversity–multiplexing frontier with rotation based space–time codes," In *Allerton Conf. on Communication, Control, and Computing*, 2003.

[87] W. Younis, A. Sayed, and N. Al-Dhahir, "Efficient adaptive receivers for joint equalization and interference cancellation in multi-user space–time block-coded systems," *IEEE Trans. Signal Processing*, vol. 51, no. 11, pp. 2849–2862, Nov. 2003.

[88] L. Zheng and D. N. C. Tse, "Communication on the Grassmann manifold: a geometric approach to the noncoherent multiple-antenna channel," *IEEE Trans. Inform. Theory*, vol. 48, no. 2, pp. 359–383, Feb. 2002.

[89] L. Zheng and D. N. C. Tse, "Diversity and multiplexing: a fundamental trade off in multiple-antenna channels," *IEEE Trans. Inform. Theory*, vol. 49, no. 5, pp. 1073–1096, May 2003.

[90] A. Zhou and G. B. Giannakis, "Space–time coding with maximum diversity gains over frequency-selective fading channels," *IEEE Signal Processing Letters*, vol. 8, no. 10, pp. 269–272, Oct. 2001.

5 Fundamentals of receiver design

5.1 Introduction

This chapter is devoted to MIMO receivers, with special focus on single-user systems and frequency-flat channels (multi-user systems and more general channels will be the subject of the next chapter). We start with a brief discussion of uncoded MIMO systems, describing their optimum (maximum-likelihood, ML) receivers. Since these may exhibit a complexity that makes them unpractical, it is important to seek receivers that achieve a close-to-optimum performance while keeping a moderate complexity: these would remove the practical restriction to small signal constellations or few antennas. Linear receivers and receivers based on the sphere-detection algorithm are examined as possible solutions to the complexity problem. Next, we study iterative processing of received signals. We introduce here the idea of factor graphs. Their use offers a versatile tool, allowing one to categorize in a simple way the approximations on which MIMO receivers and their algorithms are based. In addition, they yield a "natural" way for the description of iterative (*turbo*) algorithms, and of their convergence properties through the use of EXIT-charts. Using factor graphs, we describe iterative algorithms for the reception of MIMO signals, along with some noniterative schemes that can be easily developed by using the factor-graph machinery.

A basic assumption in this chapter is that channel state information (i.e., the values taken on by all path gains) is available at the receiver, while the transmitter knows the channel distribution (i.e., the joint probability density function of the channel gains). In addition, the channel is *quasi-static* (i.e., it remains constant throughout the transmission of a whole data frame or codeword), and the transmitted signals are two-dimensional.

This chapter is organized as follows. Section 5.2 describes simple receivers for uncoded signals. Section 5.3 introduces factor graphs, the sum–product algorithm, and the turbo algorithms. The next two sections categorize MIMO receivers based on factor-graph concepts. We consider separately uncoded (Section 5.4) and coded (Section 5.5) MIMO systems. Finally, Section 5.6 provides additional details on some suboptimum receivers that have been presented in the literature.

5.2 Reception of uncoded signals

We consider first uncoded MIMO transmission, and the usual input–output relation (see Fig. 1.2)

$$\mathbf{y} = \mathbf{H}\mathbf{x} + \mathbf{n} \tag{5.1}$$

where $\mathbf{n}$ is a spatio-temporally white zero-mean circularly symmetric complex Gaussian (ZMCSCG) noise vector with variance N_0 per component, and we have omitted, for notational simplicity, the factor $\sqrt{E_s/M_T}$, as it is not relevant in our ensuing discussion. ML detection of $\mathbf{x}$ requires the minimization, with respect to $\mathbf{x}$, of a squared norm:

$$\widehat{\mathbf{x}} = \arg\min_{\mathbf{x}} \|\mathbf{y} - \mathbf{H}\mathbf{x}\|_F^2 = \arg\min_{\mathbf{x}} \sum_{i=1}^{M_R} \left| y_i - \sum_{j=1}^{M_T} h_{i,j} x_j \right|^2 \tag{5.2}$$

where $h_{i,j}$ is the (i, j)th component of the matrix $\mathbf{H}$. It is seen that exact calculation of the right-hand side of the above requires the summation of $M_R \times M_T$ terms for each value of $\mathbf{x}$. The $M_T \times 1$ vector $\mathbf{x}$ can take on 2^{QM_T} values, where 2^Q is the size of the modulation format used, and Q is the number of bits per signal. This is the number of computations of the norm in (5.2) necessary to the minimization if exhaustive search for the minimum is used. We see, in particular, that the complexity of an ML receiver grows exponentially with Q and M_T, which restricts its implementation to small signal constellations and small numbers of transmit antennas. We now describe briefly a number of detectors that, at the possible price of a loss in performance, reduce the receiver complexity. Next, we introduce an algorithm that allows one to achieve ML detection with a lower number of calculations.

5.2.1 *Linear receivers*

The basic idea here is to preprocess the received signal by transforming it linearly:

$$\tilde{\mathbf{y}} \triangleq \mathbf{A}\mathbf{y} = \mathbf{A}\mathbf{H}\mathbf{x} + \mathbf{A}\mathbf{n}$$

so that the transformed channel-matrix $\mathbf{AH}$ becomes close to a diagonal matrix. This allows the detection of each component of $\mathbf{x}$ to be performed separately. The preprocessing matrix $\mathbf{A}$ can be chosen to remove the off-diagonal elements of $\mathbf{AH}$ (zero-forcing receiver):

$$\tilde{\mathbf{y}} = \mathbf{H}^{\dagger}\mathbf{y} \tag{5.3}$$

($\mathbf{H}^{\dagger}$ is the Moore–Penrose pseudoinverse of $\mathbf{H}$), or to minimize the joint effects of off-diagonal elements of $\mathbf{AH}$ and of the filtered noise $\mathbf{An}$ (linear minimum mean-square error receiver, or LMMSE):

$$\tilde{\mathbf{y}} = \left(\mathbf{H}^H\mathbf{H} + \frac{N_0}{E}\mathbf{I} \right)^{-1} \mathbf{H}^H\mathbf{y} \tag{5.4}$$

where E is the average energy of one component of vector $\mathbf{x}$, N_0 is the noise variance, and $\mathbf{I}$ is the $M_T \times M_T$ identity matrix. It is expected that the simplification entailed by linear receivers comes at the price of a poorer performance, especially in the case of a number of receive antennas equal to that of transmit antennas (see, e.g., [11] and [29, p. 152 ff.]).

5.2.2 Decision-feedback receivers

The preprocessing step of decision-feedback detection entails the decomposition of the channel matrix $\mathbf{H}$ in the form of the product [24, p. 112]

$$\mathbf{H} = \mathbf{QR} \tag{5.5}$$

Assume for simplicity $M_R \geq M_T$ (the most general case is treated in [15]). Then, $\mathbf{Q}^H\mathbf{Q} = \mathbf{I}_{M_T}$, and the $M_T \times M_T$ matrix $\mathbf{R}$ is upper triangular. With this decomposition, the transformed observed vector $\tilde{\mathbf{y}} \triangleq \mathbf{Q}^H\mathbf{y}$ takes the form

$$\tilde{\mathbf{y}} = \mathbf{Rx} + \tilde{\mathbf{n}} \tag{5.6}$$

with the transformed noise vector $\tilde{\mathbf{n}} \triangleq \mathbf{Q}^H\mathbf{n}$ retaining the statistical properties of $\mathbf{n}$. It can be immediately seen that minimizing the metric $m(\mathbf{x}) \triangleq \|\mathbf{y} - \mathbf{Hx}\|_F^2$ is tantamount to minimizing

$$\tilde{m}(\mathbf{x}) \triangleq \|\tilde{\mathbf{y}} - \mathbf{Rx}\|_F^2 \tag{5.7}$$

The structure of $\mathbf{R}$ suggests the following detection technique: first detect x_{M_T} by minimizing $|\tilde{y}_{M_T} - r_{M_T,M_T}x_{M_T}|^2$; then, using the decision $\hat{x}_{M_T}$, detect x_{M_T-1} by minimizing $|\tilde{y}_{M_T-1} - r_{M_T-1,M_T-1}x_{M_T-1} - r_{M_T-1,M_T}\hat{x}_{M_T}|^2 + |\tilde{y}_{M_T} - r_{M_T,M_T}\hat{x}_{M_T}|^2$. This algorithm is prone to error propagation (in fact, a wrong detection of any component of $\mathbf{x}$ is likely to cause wrong decisions of the components detected after it). Notice that the performance of the decision-feedback receiver can be improved by optimizing the antenna labeling in some way, i.e., the order in which the signal components are detected.

5.2.3 Sphere detection

The sphere detection algorithm (SDA) achieves the performance of ML (or a close approximation to it) with lower complexity. The basic idea is to restrict the search for the optimum $\mathbf{x}$ to a smaller subset of potential candidates. Typically, the search is constrained to the interior of a hypersphere centered at $\mathbf{y}$ and having radius r:

$$\|\mathbf{y} - \mathbf{Hx}\|_F^2 \leq r^2 \tag{5.8}$$

Observe that, if $r = \infty$, the use of (5.8) does not entail any complexity reduction, as the problem is exactly the same as the ML detection problem. The complexity is actually reduced if one is able to choose an appropriate value for r, i.e., one that is small enough to decrease considerably the number of vectors $\mathbf{x}$ to be checked, but not so small that the hypersphere is empty.

There are several possible implementations of the basic idea of sphere detection (see, e.g., [2, 6, 15] and references therein). Here we describe a simple version of the SDA, and comment briefly on possible extensions. Consider a tree such that the bottom leaves correspond to all possible vectors $\mathbf{x}$, with the components of $\mathbf{x}$ labeling its branches from bottom to top. Figure 5.1 shows such a tree for the case of three transmit antennas and quaternary modulation. Assume further, without any loss in performance, that $\mathbf{y}$ is premultiplied by $\mathbf{Q}^H$ as in (5.6). With reference to Figure 5.1, ML detection can be viewed as traversing the tree by computing the metric $\tilde{m}(x) = \|\tilde{\mathbf{y}} - \mathbf{R}\mathbf{x}\|_F^2$ for all its branches, and retaining the minimum value found. SDA consists of reducing the number of branches to be checked, by suitably pruning the tree.

A simple algorithm runs as follows. Using decision feedback as described in Section 5.2.2, obtain a preliminary estimate $\widehat{\mathbf{x}}$ of $\mathbf{x}$, and compute the corresponding metric $\tilde{m}(\widehat{\mathbf{x}}) \triangleq \|\tilde{\mathbf{y}} - \mathbf{R}\widehat{\mathbf{x}}\|_F^2$. This value is chosen as the square radius of the sphere within which we look for the ML vector. The tree is now traversed, depth-first, from top to bottom, and the metric computed incrementally by adding one by one the terms of the summation

$$\|\tilde{\mathbf{y}} - \mathbf{R}\mathbf{x}\|_F^2 = \sum_{i=1}^{M_T} |\tilde{y}_i - (\mathbf{R}\mathbf{x})_i|^2 \tag{5.9}$$

Whenever it is discovered that at a node the partial sum accumulated is already bigger than or equal to $\tilde{m}(\widehat{\mathbf{x}})$, then there is no point in checking the leaves below that node. These are consequently pruned out, and hence removed from further consideration. If a new $\mathbf{x}$ is found whose metric is smaller than $\tilde{m}(\widehat{\mathbf{x}})$, then this takes the place of $\widehat{\mathbf{x}}$ in the rest of the algorithm (geometrically, this corresponds to shrinking the hypersphere in which the ML $\mathbf{x}$ is searched).

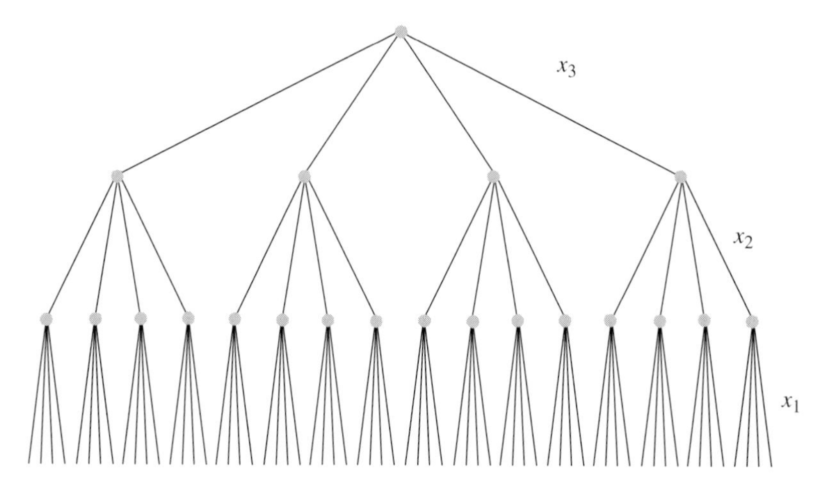

Fig. 5.1. A tree describing the implementation of the SDA for a MIMO system with $M_T = 3$ and quaternary modulation.

Several variations of this basic algorithm are possible. Among these,

(1) Instead of having a tree with depth M_T whose branches are labeled by complex signals, one may generate a tree with depth $2M_T$ with branches labeled by *real* signals obtained by separating real and imaginary parts of the components of $\mathbf{x}$. In this way, the number of terms in (5.9), and hence the expected number of nodes to be visited, increases, but the processing at a single node decreases. In [6, p. 1569, Section III.B] it is argued that, for very large scale integration (VLSI) implementation of the SDA, using complex signals may be more efficient.
(2) A breadth-first search, or an *L-best* search, can be implemented in lieu of the depth-first search described above. The L-best search consists of an approximation of the breadth-first search, whereby at each level of the tree only L nodes are kept, namely, those with the smallest partial metrics. This solution may not lead to the ML signal (and hence is suboptimum), but it has the advantage of providing a deterministic throughput.
(3) With branches labeled by real signals, and with a square signal constellation (e.g., 64-QAM), a version of the SDA generates, at each node, a real interval where the signals labeling the lower branches are searched for.

5.3 Factor graphs and iterative processing

We move now to the examination of iterative MIMO receivers. To motivate our study, based on factor graphs, let us consider *maximum a posteriori probability* (MAP) decisions. Specifically, assume a code in a signal space, that is, a set $\mathcal{C}$ of vectors $\mathbf{x} = (x_1, \ldots, x_n)$ whose components x_i are elements of the signal set $\mathcal{X}$. MAP decisions consist of observing the output $\mathbf{y} = (y_1, \ldots, y_n)$ of a channel whose input is $\mathbf{x}$, and of finding

$$\hat{x}_i = \arg\max_{x_i \in \mathcal{X}} f(x_i \mid \mathbf{y}), \qquad i = 1, \ldots, n \tag{5.10}$$

where the maximization is consistent with the code structure. MAP decisions minimize the symbol error probability. Notice also that, if all symbols are equally likely, MAP decisions are equivalent to maximum-likelihood decisions.

Generally, the maximization in (5.10) is easy to perform once $f(x_i \mid \mathbf{y})$ has been computed, as it is sufficient to compute this function for all possible values of x_i (usually, there is a relatively small number of them in $\mathcal{X}$). The complex part is the calculation of the functions $f(x_i \mid \mathbf{y})$, called *a posteriori probabilities* (APPs). In fact, this requires a *marginalization* operation, i.e., the computation of

$$f(x_i \mid \mathbf{y}) = \sum_{x_1}\sum_{x_2}\cdots\sum_{x_{i-1}}\sum_{x_{i+1}}\cdots\sum_{x_n} f(\mathbf{x} \mid \mathbf{y}) \tag{5.11}$$

where $f(\mathbf{x} \mid \mathbf{y})$ is known, and $\mathbf{x} \in \mathcal{C}$. It is convenient to introduce the compact notation $\sim x_i$ to denote the set of indices $x_1, \ldots, x_{i-1}, x_{i+1}, \ldots, x_n$ to be summed over, so that we can write

$$f(x_i \mid \mathbf{y}) = \sum_{\sim x_i} f(\mathbf{x} \mid \mathbf{y}) \tag{5.12}$$

In this framework, the MAP decision on symbol x_i, $i = 1, \ldots, n$, is made in two steps: marginalization (i.e., computation of $f(x_i \mid \mathbf{y})$) and hard decision (i.e., maximization of $f(x_i \mid \mathbf{y})$ with respect to x_i).

5.3.1 *Factor graphs*

With $x_i \in \mathcal{X}$, $i = 1, \ldots, n$, the complexity of marginalization, computed as in (5.11), grows exponentially with n. A simplification can be achieved when f, the function to be marginalized, can be factored as a product of functions, each with less than n arguments. Consider, for example, a function $f(x_1, x_2, x_3)$ that factors as follows:

$$f(x_1, x_2, x_3) = g_1(x_1, x_2) g_2(x_1, x_3) \tag{5.13}$$

Its marginal $f_1(x_1)$ can be computed as

$$\begin{aligned} f_1(x_1) &\triangleq \sum_{\sim x_1} f(x_1, x_2, x_3) \\ &= \sum_{x_2} \sum_{x_3} g_1(x_1, x_2) g_2(x_1, x_3) = \sum_{x_2} g_1(x_1, x_2) \cdot \sum_{x_3} g_2(x_1, x_3) \end{aligned}$$

where we see that this marginalization can be achieved by computing separately the two simpler marginals $\sum_{x_2} g_1(x_1, x_2)$ and $\sum_{x_3} g_2(x_1, x_3)$, and finally taking their product. This procedure can be represented in a graphical form by drawing a *factor graph*. This describes the fact that the function f factors as in (5.13), and is shown in Fig. 5.2. Each node here can be viewed as a processor that computes a function whose arguments label the incoming edges, and each edge as a channel along which these processors exchange data. We see that the first sum $\sum_{x_2} g_1(x_1, x_2)$ can be computed locally at the g_1 node, because x_1 and x_2 are available there; similarly, the second sum $\sum_{x_3} g_2(x_1, x_3)$ can be computed locally at the g_2 node, because x_1 and x_3 are available there.

Formally, we describe a (*normal*) factor graph as a set of *nodes*, *edges*, and *half-edges*. Every factor corresponds to a unique node, and every variable to a unique edge or half-edge. The node representing the function g is connected to the edge or half-edge representing the variable x if and only if x is an argument g. Half-edges are connected

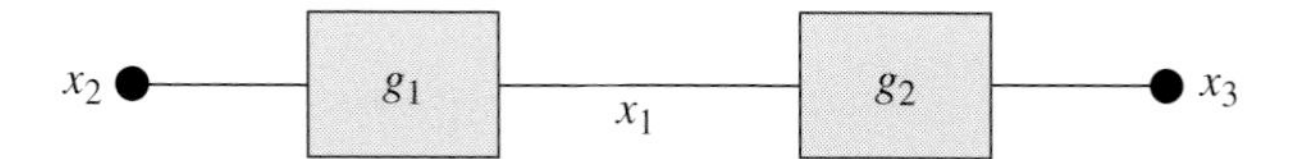

Fig. 5.2. Factor graph of the function $f(x_1, x_2, x_3) = g_1(x_1, x_2) g_2(x_1, x_3)$.

to only one node, and terminate in a filled circle •. In the example of Fig. 5.2 we have two nodes representing the factors g_1 and g_2, one edge, and two half-edges. An important feature of factor graphs is the presence or absence of *cycles*: we say that a factor graph has no cycles if removing any (regular) edge partitions the graph into two disconnected subgraphs. A cycle of length ℓ is a path through the graph that includes ℓ edges and closes back on itself. The *girth* of a graph is the minimum cycle length of the graph.

The definition of normality assumes implicitly that no variable appears in more than two factors. For example, the graph of Fig. 5.3(a) does not satisfy our definition: in fact, the variable x_1 appears as a factor of g_1, g_2, and g_3, and as a result it corresponds to more than one edge. To be able to include in our graphical description also functions that factor as in Fig. 5.3(a), we need to "clone" the variables appearing in more than two factors. By doing this, any factor graph can be transformed into a normal one without any loss of generality or of efficiency [18]. We now explain how this can be done.

The Iverson function

Let P denote a proposition that may be either true or false; we denote by $[P]$ the *Iverson function*

$$[P] \triangleq \begin{cases} 1, & P \text{ is true} \\ 0, & P \text{ is false} \end{cases}$$

Clearly, if we have n propositions $P_1, \ldots, P_n$, we have the factorization

$$[P_1 \text{ and } P_2 \cdots \text{and } P_n] = [P_1][P_2]\cdots[P_n]$$

This function allows the transformation of any graph into a normal one. In fact, define the "repetition" function $r_=$ (with three arguments, as a special case) as

$$r = (x_1, x'_1, x''_1) \triangleq [x_1 = x'_1 = x''_1] \tag{5.14}$$

This transforms the branching point of Fig. 5.3(a) into a node representing a repetition function. Thus, the graph of Fig. 5.3(a) is transformed into the normal graph of Fig. 5.3(b).

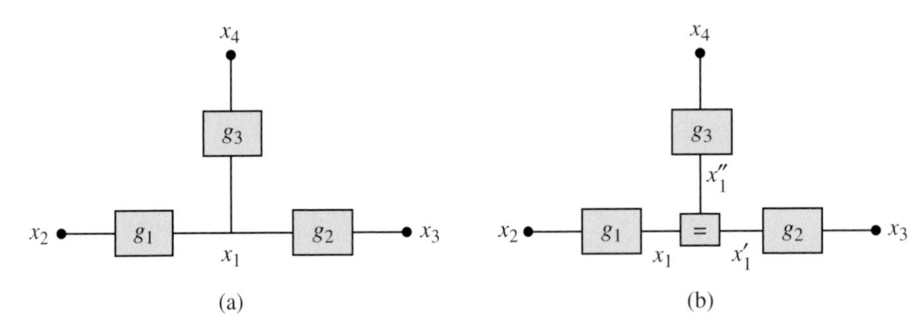

Fig. 5.3. Factor graph of the function $f(x_1, x_2, x_3, x_4) = g_1(x_1, x_2)g_2(x_1, x_3)g_3(x_1, x_4)$. (a) Non-normal form. (b) Normal form.

5.3.2 Examples of factor graphs

Here we describe some examples of factor graphs that play a central role in the study of coded communication systems.

Tanner graphs

A linear binary block code can be represented by its *Tanner graph*. Consider an (N, K) linear binary code $\mathcal{C}$ with words $\mathbf{x}$, and its $(N-K) \times n$ parity-check matrix[1]

$$\mathbf{H} = \begin{bmatrix} \mathbf{h}_1 \\ \vdots \\ \mathbf{h}_{N-K} \end{bmatrix}$$

The condition for the binary N-tuple $\mathbf{x}$ to be a word of $\mathcal{C}$ is that the following $(N-K)$ constraints be satisfied:

$$\mathbf{h}_i \mathbf{x}' = 0, \quad i = 1, \ldots, N-K$$

where $(\cdot)'$ denotes transpose. Using the Iverson function, we can write

$$[\mathbf{x} \in \mathcal{C}] = [\mathbf{h}_1\mathbf{x}' = 0, \ldots, \mathbf{h}_{N-K}\mathbf{x}' = 0] = \prod_{i=1}^{N-K} [\mathbf{h}_i\mathbf{x}' = 0] \tag{5.15}$$

The ith factor in the right-hand side of the equation above can be represented in graphical form as a sum node $\oplus$, connecting branches corresponding to the components of $\mathbf{x}$ that

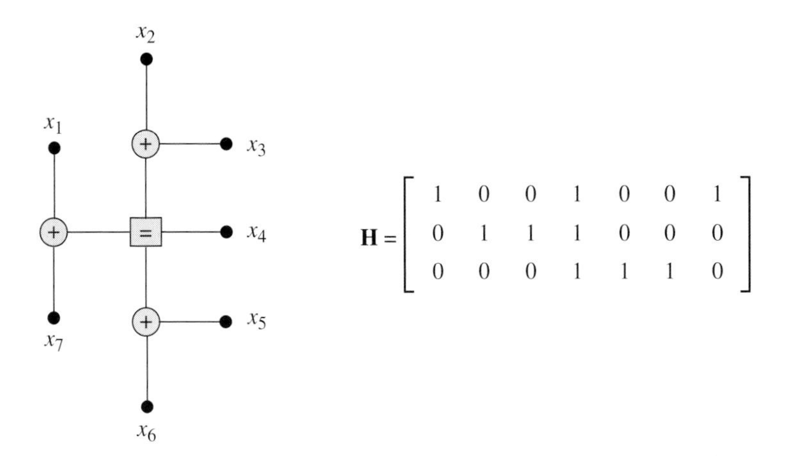

Fig. 5.4. Normal Tanner graph of a linear binary code with parity-check matrix $\mathbf{H}$. The graph has no cycles.

[1] In this chapter we are using the same notation, $\mathbf{H}$, for the MIMO channel matrix and the parity-check matrix of a block code. Both are standard, time-honored notations that we do not want to change here, as we are confident that no confusion will arise.

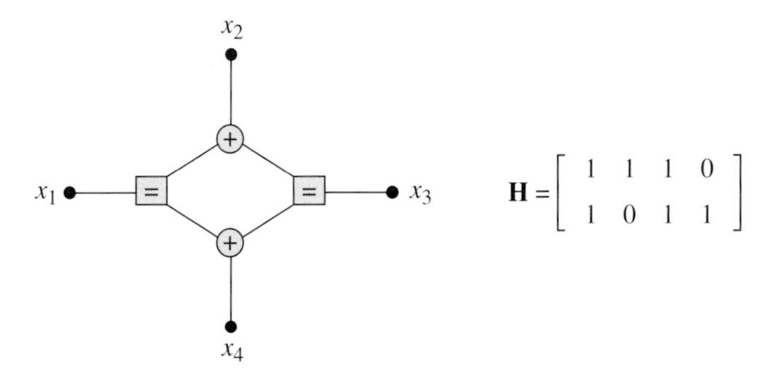

Fig. 5.5. Normal Tanner graph of a linear binary code with parity-check matrix $\mathbf{H}$. The graph has one cycle.

are checked by $\mathbf{h}_i$. Figures 5.4 and 5.5 show examples of Tanner graphs without cycles and with one cycle, respectively.

TWLK (Tanner–Wiberg–Loeliger–Koetter) graphs

Normal graphs can also be used to describe codes originally described by a trellis (e.g., terminated convolutional codes). A trellis can be viewed as a set of triples $(\sigma_{i-1}, x_i, \sigma_i)$ describing which state transitions $\sigma_{i-1} \to \sigma_i$ are driven by the channel symbol x_i at time $i-1$, with $i = 1, \ldots, n$. Let $\mathcal{T}_i$ denote the set of branches in the trellis joining state σ_{i-1} to state σ_i. Then the set of branch labels in $\mathcal{T}_i$ is the domain of a variable x_i, while the set of nodes at time $i-1$ (respectively, i) is the domain of the state variable σ_{i-1} (respectively, σ_i). The initial and final state variables (corresponding to time 0 and time N, respectively) take on a single value. The local function corresponding to the ith trellis section is

$$[(\sigma_{i-1}, x_i, \sigma_i) \in \mathcal{T}_i] \tag{5.16}$$

and the whole trellis corresponds to a product of Iverson functions (Fig. 5.6):

$$[\mathbf{x} \in \mathcal{C}] = \prod_{i=1}^{n} [(\sigma_{i-1}, x_i, \sigma_i) \in \mathcal{T}_i] \tag{5.17}$$

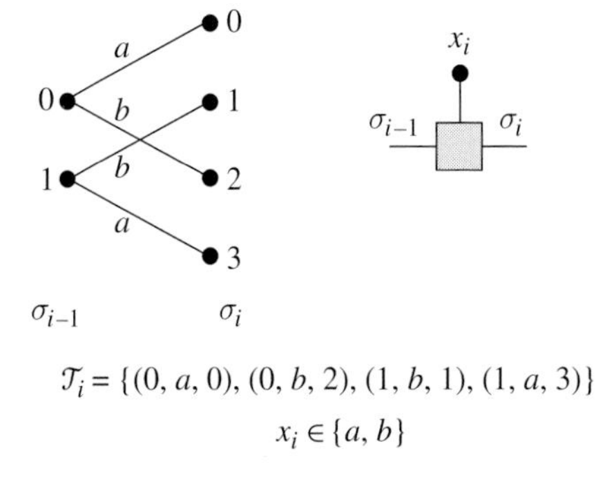

$\mathcal{T}_i = \{(0, a, 0), (0, b, 2), (1, b, 1), (1, a, 3)\}$

$x_i \in \{a, b\}$

Fig. 5.6. A section of a trellis and the node representing it in a normal TWLK graph.

Notice how in this representation the graph edges are associated with states, the filled dots with symbols, and the nodes with constraints. Notice also that the code representation may be made more comprehensive by adding nodes corresponding to the source symbols that drive the transitions between pairs of states. Denoting by u_i these symbols, we have the trellis description

$$[\mathbf{x} \in \mathcal{C}] = \prod_{i=1}^{n} [(\sigma_{i-1}, u_i, x_i, \sigma_i) \in \mathcal{T}_i] \tag{5.18}$$

As a simple example, Fig. 5.7 shows the Tanner graph, the trellis, and the TWLK graph of the binary repetition code with length 4. The repetition nodes in part (b) of the figure illustrate the fact that, in each trellis section, the coded symbol, the starting state, and the ending state coincide.

Factor graph of a dispersive channel

Consider now factor graphs of channels. Our first example refers to a channel affected by additive white Gaussian noise and linear intersymbol interference. Specifically, assume that it responds to the complex N-tuple $\mathbf{x} = (x_1, \ldots, x_N)'$ with the $(N+L)$-tuple $\mathbf{y} = (y_1, \ldots, y_{N+L})'$, where

$$y_k = \sum_{i=0}^{L} h_i x_{k-i} + n_k$$

and $x_i = 0$ for $i < 0$. The samples n_k form a complex Gaussian noise process, and $h_0, \ldots, h_L$ are the channel gains, with L the channel memory. By suitable definition of the $(N+L) \times N$ channel matrix $\mathbf{H}$, we can write

$$\mathbf{y} = \mathbf{H}\mathbf{x} + \mathbf{n}$$

where $\mathbf{n}$ is a complex Gaussian-noise vector, whose components are independent, with zero mean and equal variances $\mathcal{E}|n_i|^2 = N_0$. The input–output relationship of the channel can be described by the conditional probability density function (pdf)

$$f(\mathbf{y} \mid \mathbf{x}) \propto \exp\left(-\|\mathbf{y} - \mathbf{H}\mathbf{x}\|_F^2 / N_0\right) = \prod_{k=1}^{N+L} f(y_k \mid \mathbf{x}) \tag{5.19}$$

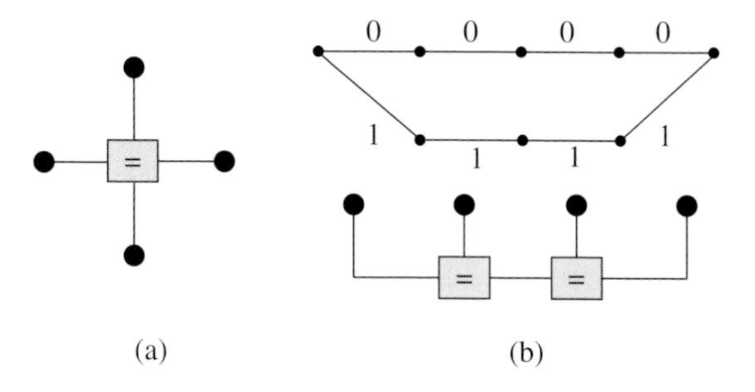

Fig. 5.7. Three representations of the binary (4, 1) repetition code: (a) Normal Tanner graph, and (b) Trellis and TWLK graph.

where

$$f(y_k \mid \mathbf{x}) \triangleq \exp\left(-|y_k - \sum_{i=1}^{N} x_i h_{k-i}|^2 / N_0\right) \tag{5.20}$$

The factor graph corresponding to the factorization (5.19) is shown in Fig. 5.8.

As a special case of the above, a memoryless (i.e., intersymbol-interference-free) channel has $L = 0$, and hence

$$f(\mathbf{y} \mid \mathbf{x}) = \prod_{k=1}^{N} f(y_k \mid x_k) \tag{5.21}$$

The corresponding factor graph is disconnected, as shown in Fig. 5.9.

Factor graph of a MIMO channel

Assume now a MIMO channel with M_T transmit and M_R receive antennas, as in (5.1). The input–output relationship of the channel is described by the conditional pdf

$$f(\mathbf{y} \mid \mathbf{x}) \propto \exp\left(-\|\mathbf{y} - \mathbf{H}\mathbf{x}\|_F^2 / N_0\right) = \prod_{k=1}^{M_R} f(y_k \mid \mathbf{x}) \tag{5.22}$$

where

$$f(y_k \mid \mathbf{x}) \triangleq \exp\left(-|y_k - \mathbf{h}_k \mathbf{x}|^2 / N_0\right) \tag{5.23}$$

and $\mathbf{h}_k$ denotes the kth row of $\mathbf{H}$. The factor graph corresponding to factorization (5.22) is shown in Fig. 5.10. Note the similarity of this factor graph with that of Fig. 5.8: in both cases interference is present (there, intersymbol interference; here, *spatial* interference).[2]

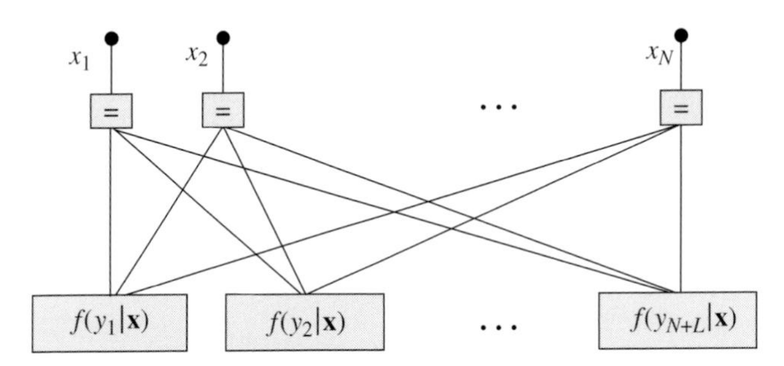

Fig. 5.8. Factor graph of a dispersive channel.

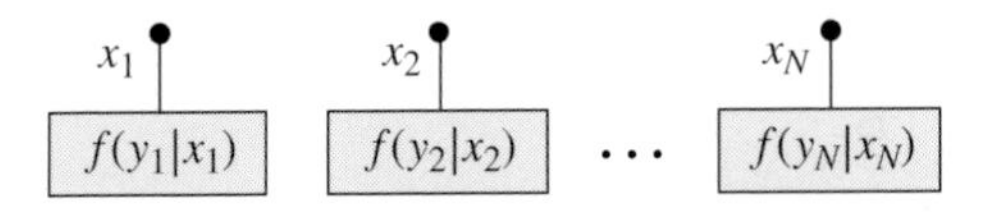

Fig. 5.9. Factor graph of a memoryless channel.

[2] Multi-user detection can also be viewed in this framework, where interference is due to multiple users accessing a common channel (see [13]).

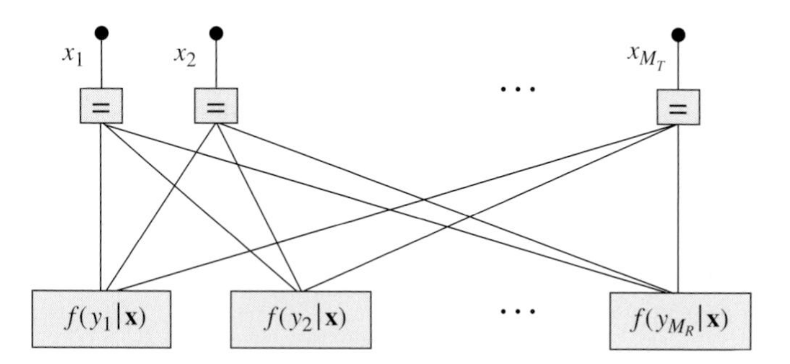

Fig. 5.10. Factor graph of a MIMO channel.

5.3.3 The sum–product algorithm

The sum–product algorithm (SPA) computes efficiently the marginals of a function whose factors are described by a normal factor graph. This works when the graph is cycle-free, and yields, after a finite number of steps, the marginal function corresponding to each variable associated with an edge.

In this algorithm, two *messages* are associated with each edge, one for each direction. Each message, denoted by $\mu(x_i)$, is a function of the variable x_i, and depends on the direction. It is given in the form of a vector, whose components are the values taken on by the message with correspondence to the values of x_i. Messages that are probability distributions of binary variables can be conveniently represented as a single number, the ratio between two probabilities or its logarithm. In fact, as we are interested in MAP decisions, the message (μ_0, μ_1) can be equivalently represented as $(1, \mu_1/\mu_0)$, or as $(0, \log(\mu_1/\mu_0))$, and only the second component of the vector needs processing [18, 25].

Consider the node representing the factor $g(x_1, \ldots, x_n)$ (see Fig. 5.11). The message $\mu_{g\to x_i}(x_i)$ out of this function node along the edge x_i is the function

$$\mu_{g\to x_i}(x_i) = \sum_{\sim x_i} g(x_1, \ldots, x_n) \prod_{\ell\neq i} \mu_{x_\ell\to g}(x_\ell) \tag{5.24}$$

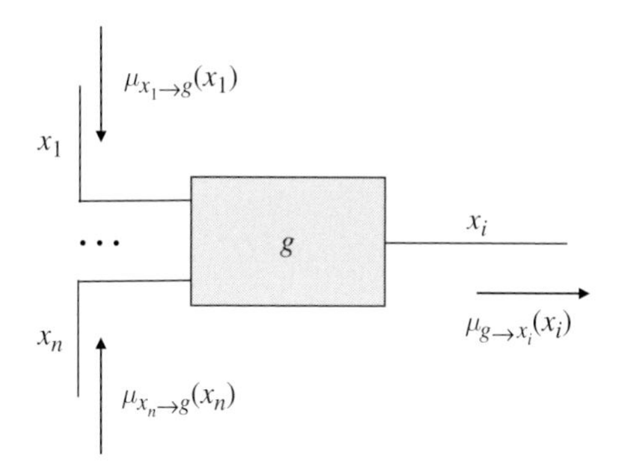

Fig. 5.11. The basic step of the sum–product algorithm.

where $\mu_{x_\ell \to g}(x_\ell)$ is the message incoming on edge x_ℓ. In words, the message $\mu_{g \to x_i}(x_i)$ is the product of g and all messages towards g along all edges except x_i, summed over all variables except x_i. Half-edges, which are connected to a single node, transmit towards it a message with constant value 1.

Two important special cases are as follows.

(1) If g is a function of only one argument x_i, then the product in (5.24) is empty, and we simply have (see Fig. 5.12)

$$\mu_{g \to x_i}(x_i) = g(x_i)$$

(2) If g is the repetition function $f_=$, then we have (see Fig. 5.13)

$$\mu_{f_= \to x_i}(x_i) = \prod_{\ell \neq i} \mu_{x_\ell \to f_=}(x_i) \tag{5.25}$$

A simple example

Consider the function

$$f(x_1, x_2, x_3) = g_1(x_1)g_2(x_1, x_2)g_3(x_2, x_3)$$

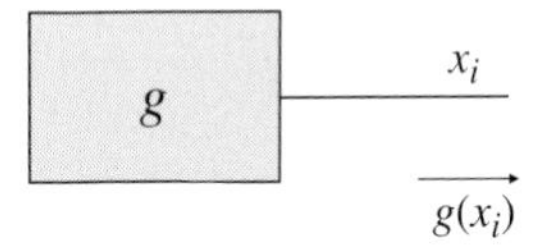

Fig. 5.12. The basic step of the sum–product algorithm when a node is a function of only one argument.

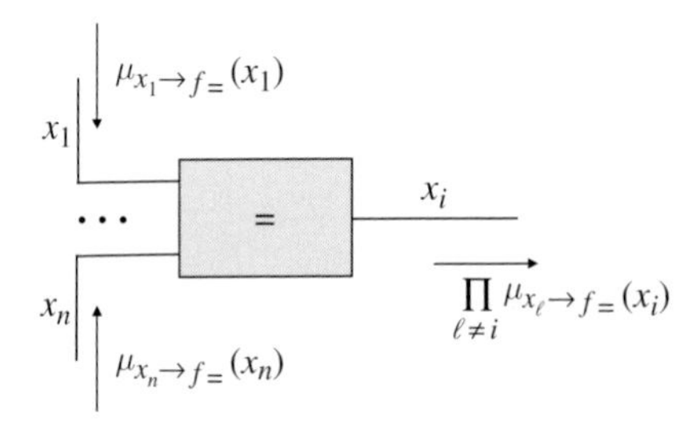

Fig. 5.13. The basic step of the sum–product algorithm when a node represents the repetition function $f_=$.

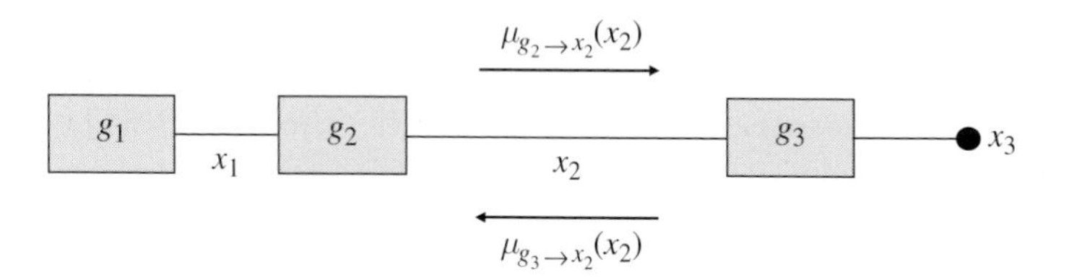

Fig. 5.14. Factor graph of the function $g_1(x_1)g_2(x_1, x_2)g_3(x_2, x_3)$, and messages exchanged by the sum–product algorithm applied to the marginalization with respect to the variable x_2.

whose factor graph is shown in Fig. 5.14. Its marginalization with respect to x_2 can be computed as follows:

$$\begin{aligned} f_2(x_2) &= \sum_{x_1}\sum_{x_3} g_1(x_1)g_2(x_1, x_2)g_3(x_2, x_3) \\ &= \underbrace{\sum_{x_1} g_2(x_1, x_2) \underbrace{g_1(x_1)}_{\mu_{g_1 \to x_1}(x_1)}}_{\mu_{g_2 \to x_2}(x_2)} \cdot \underbrace{\sum_{x_3} g_3(x_2, x_3)}_{\mu_{g_3 \to x_2}(x_2)} \end{aligned}$$

which corresponds to the product of the two messages along edge x_2 exchanged by the SPA.

Scheduling

The messages in the graph must be computed in both directions for each edge. After all of them are computed according to some schedule, the product of the two messages associated with an edge yields the marginal function sought. It should be observed here that the choice of the computational schedule may affect the algorithm efficiency. A possible schedule consists of requiring all nodes to update their outgoing messages whenever their incoming messages are updated. In a graph without cycles, message computation may start from the leaves, and proceed from node to node as the necessary terms in (5.24) become available. In the *flooding* schedule, the messages are transmitted along all edges simultaneously [25].

5.3.4 *Factor graph with cycles: iterative algorithms*

On a graph with cycles, a version of the SPA can still be applied, by implementing the sum–product step (5.24) locally at any node after specifying a set of initial values, a computational schedule, and a stopping rule [25]. However, this iterative ("turbo") algorithm may not converge, or it may converge to an incorrect APP distribution: exact conditions for convergence are a topic of current research. Although a formal proof is still missing, it is commonly accepted that the presence of short cycles should be avoided, as they hinder convergence, but if the girth of the factor graph is very large, the loop-free approximation can be made.[3] To guarantee that the girth is large, so that the graph can

[3] In [31], it is proven that the assumption of a graph without cycles holds asymptotically, as n grows large, for low-density parity-check codes, while for turbo codes it has only a heuristic justification.

be made locally cycle-free within some radius of any vertex, the systems should have a large number of variables, and in addition an interleaver should be introduced in the system: this is done, for example, with low-density parity-check codes and with turbo codes. Under these conditions, in many practical cases the algorithm *does* converge to a probability distribution yielding correct decisions: for this reason it is used in practice to decode powerful codes (see, e.g., [18] and [7, Chapter 9]).

Approximations of the basic iterative algorithm, based on a complexity/performance trade-off, are also possible. We shall describe *infra* some of them for the specific problem of MIMO receivers. Approximations may be aimed at transforming the original factor graph into one without cycles, which is especially useful when the graph girth cannot be increased due to the small number of variables.[4] The elimination of some cycles can also be achieved. For example, messages can be transformed as follows [18, 20, 43].

(1) If all components of a message, except one, are nulled (this operation corresponds to a "hard decision" on a symbol), then the message in the opposite direction need not be computed.
(2) If all components of a message are made equal (this operation corresponds to the "erasure" of a symbol), then only the message in the opposite direction need be computed.

In both cases (1) and (2), the original undirected graph is converted into a partially directed one. Node "clustering" [25] can also eliminate cycles, or the original graph may be transformed into its *dual*, with the dual version of the sum–product algorithm used on it [18].

5.3.5 *Factor graphs and receiver structures*

Let us now return to the problem of MAP detection of a transmitted symbol x_i. We have observed that we should find the maximum over x_i of the conditional pdf $f(x_i \mid \mathbf{y})$, where $\mathbf{y}$ denotes the observation corresponding to the transmission of vector $\mathbf{x}$. The computation of $f(x_i \mid \mathbf{y})$ can be done in two steps: first, by factoring $f(\mathbf{x} \mid \mathbf{y})$, next by marginalizing it with respect to x_i. Observe that

$$f(\mathbf{x} \mid \mathbf{y}) \propto f(\mathbf{x}) f(\mathbf{y} \mid \mathbf{x})$$

The pdf $f(\mathbf{x})$ describes our knowledge of the transmitted-signal statistics, while $f(\mathbf{y} \mid \mathbf{x})$ describes what we know about the channel. Two cases are especially interesting.

[4] Removal of cycles without approximations is always possible, but it may come at the expense of an unacceptable increase in computational complexity.

(1) The transmitted signals are uncoded. If we assume them independent, with known a priori probabilities, then

$$f(\mathbf{x}) = \prod_{i=1}^{N} f(x_i)$$

(2) The transmitted vectors $\mathbf{x}$ are (equally likely) words of code $\mathcal{C}$. Since $f(\mathbf{x}) = 1/|\mathcal{C}|$ if $\mathbf{x}$ is a codeword, while $f(\mathbf{x}) = 0$ otherwise, we write

$$f(\mathbf{x}) \propto [\mathbf{x} \in \mathcal{C}]$$

Special cases were examined in Section 5.3.2.

Decoding over a general channel

Since we have

$$f(\mathbf{x} \mid \mathbf{y}) \propto [\mathbf{x} \in \mathcal{C}] \times f(\mathbf{y} \mid \mathbf{x})$$

the factor graph of the factorization of $f(\mathbf{x} \mid \mathbf{y})$ is obtained by joining the code graph (which describes the function $[\mathbf{x} \in \mathcal{C}]$) to the channel graph (which describes the function $f(\mathbf{y} \mid \mathbf{x})$). The corresponding sum–product algorithm is schematized in Fig. 5.15, where $i(x_i)$ is called the *intrinsic* message, and $e(x_i)$ the *extrinsic* message of the code function (if reference is made to the channel function, then the terms "intrinsic" and "extrinsic" are interchanged).

We have, for $i = 1, \ldots, N$,

$$\begin{cases} i(x_i) = \sum_{\sim x_i} f(\mathbf{y} \mid \mathbf{x}) \prod_{j \neq i} e(x_j) \\ e(x_i) = \sum_{\sim x_i} [\mathbf{x} \in \mathcal{C}] \prod_{j \neq i} i(x_j) \end{cases}$$

In the special case of a memoryless channel, we have explicitly

$$i(x_i) = f(y_i \mid x_i)$$

which shows how $e(x_i)$ depends on the code structure, and on the observation of all components of $\mathbf{y}$ except y_i (this explains why message $e(x_i)$ is called *extrinsic*).

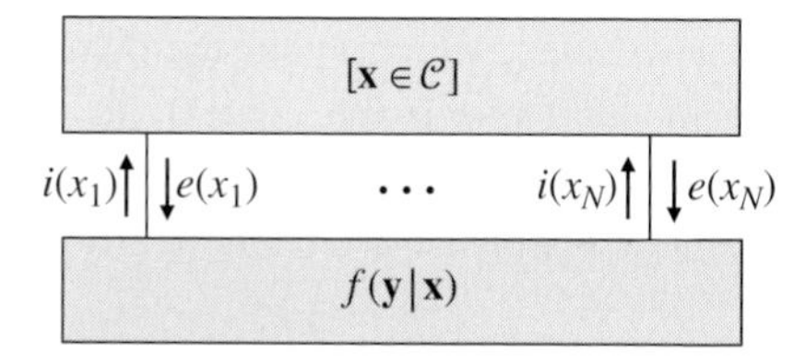

Fig. 5.15. Sum–product algorithm corresponding to decoding over a general channel.

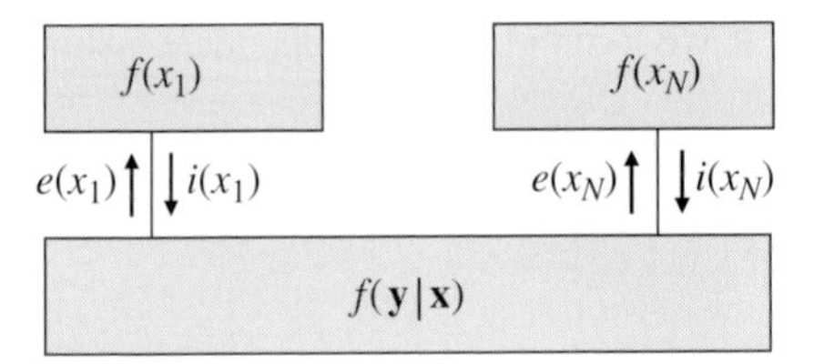

Fig. 5.16. Sum–product algorithm corresponding to equalizing a general channel.

Equalizing a dispersive channel

Without coding, the assumption is that the components of vector $\mathbf{x}$ are independent random variables, so we can write

$$f(\mathbf{x} \mid \mathbf{y}) \propto \prod_{i=1}^{N} f(x_i) \times f(\mathbf{y} \mid \mathbf{x})$$

The corresponding sum–product algorithm is schematized in Fig. 5.16, where

$$\begin{cases} i(x_i) = f(x_i) \\ e(x_i) = \sum_{\sim x_i} f(\mathbf{y} \mid \mathbf{x}) \prod_{j \neq i} i(x_j) \end{cases}$$

5.4 MIMO receivers for uncoded signals

As mentioned before, detection of uncoded signals at the output of a MIMO channel can be viewed as a special case of equalization of a general channel. We have the normal graph shown in Fig. 5.17, which corresponds to the factorization

$$\begin{aligned} f(\mathbf{x} \mid \mathbf{y}) &\propto \prod_{i=1}^{M_T} f(x_i) \times f(\mathbf{y} \mid x_1, \ldots, x_{M_T}) \\ &= \prod_{j=1}^{M_T} f(x_i) \times \prod_{j=1}^{M_R} f(y_j \mid \mathbf{x}) \end{aligned} \tag{5.26}$$

with $f(y_j \mid \mathbf{x})$ given in (5.23).

The graph of Fig. 5.17 has cycles, but in practice iterative algorithms may not be suitable, unless a very large number of antennas produces a very large girth. Thus, a number of noniterative receivers have been advocated, which avoid the complexity of direct marginalization. We categorize them in our factor-graph context. In particular, we classify suboptimum MIMO receivers by considering three types of possible simplifications, whose combination generates a taxonomy of detection algorithms.

(1) The structure of the factor graph is made simpler. This can be obtained by preprocessing the received signal, typically in order to limit the effects of spatial

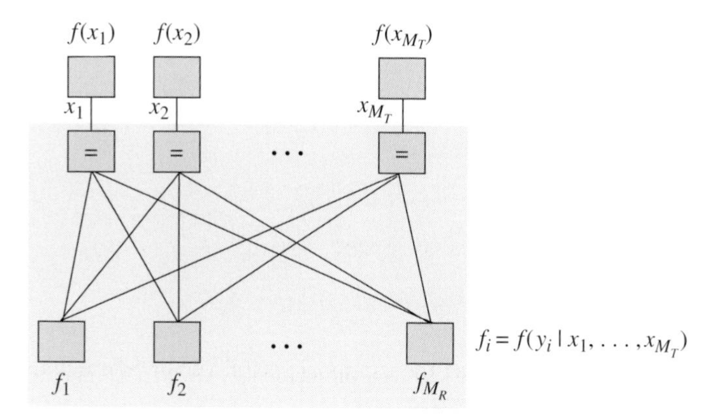

Fig. 5.17. Factor graph corresponding to the APP detection of uncoded signals at the output of a MIMO channel.

interference. We call *interface* a preprocessor transforming the observed vector $\mathbf{y}$ into vector $\widetilde{\mathbf{y}}$. *Linear* interfaces, which we denote by $\mathbf{A}(\mathbf{H})$ to stress their dependence on the channel matrix $\mathbf{H}$, operate by transforming $\mathbf{y}$ into

$$\widetilde{\mathbf{y}} \triangleq \mathbf{A}(\mathbf{H})\mathbf{y} = \mathbf{A}(\mathbf{H})\mathbf{H}\mathbf{x} + \mathbf{A}(\mathbf{H})\mathbf{n} \tag{5.27}$$

as we described briefly in Section 5.1.

(2) The messages exchanged among the graph nodes are approximated by some simplified versions.

(3) The detection algorithm consists of a single sweep involving a finite number of steps.

5.4.1 *Linear interfaces*

We re-examine here the two linear interfaces examined in Section 5.1, and describe the modifications they induce on the factor graph, as well as the resulting algorithms.

Zero-forcing interface

The linear transformation

$$\widetilde{\mathbf{y}} = \mathbf{H}^{\dagger}\mathbf{y}$$

yields

$$\widetilde{\mathbf{y}} = \mathbf{x} + \mathbf{H}^{\dagger}\mathbf{n}$$

which shows that the spatial interference is completely removed from the received signal, and justifies the name *zero-forcing* (ZF) used for this interface. Thus, the factor graph becomes the one shown in Fig. 5.18, which is disconnected, and hence allows symbol-by-symbol decisions. A well-understood drawback of ZF is the resulting *noise enhancement* that turns out from the modification of the noise covariance matrix.

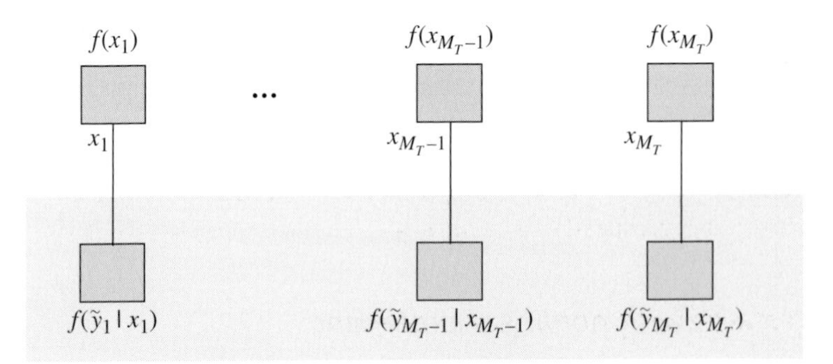

Fig. 5.18. Factor graph corresponding to MIMO receiver factorization with a ZF interface.

Linear MMSE interface

Another preprocessing technique is based on the minimization of a mean-square error (MSE). The linear minimum MSE (LMMSE) interface is obtained from the minimization of the MSE

$$\mathcal{E}[\|\mathbf{Ay}-\mathbf{x}\|_F^2]$$

under the assumption of independent, identically distributed (iid) transmitted symbols. In the general setting of (5.27), LMMSE is characterized by the matrix

$$\mathbf{A}(\mathbf{H}) = \left(\mathbf{H}^H\mathbf{H} + \frac{N_0}{E_s}\mathbf{I}\right)^{-1}\mathbf{H}^H \tag{5.28}$$

where E_s denotes the average energy of a transmitted symbol. Matrix $\mathbf{A}(\mathbf{H})$ exists for all possible pairs M_T, M_R.

Spatial interference is mitigated since the off-diagonal terms in $\mathbf{A}(\mathbf{H})\mathbf{H}$ are smaller than the diagonal terms and can be neglected in a suboptimal receiver. The resulting factor graph simplifies as illustrated by Fig. 5.19, where the neglected interfering links are drawn as dotted lines. As a result, some residual spatial interference remains, but as

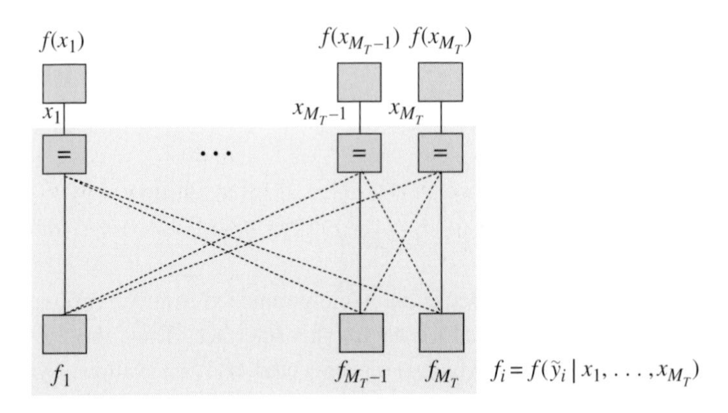

Fig. 5.19. Factor graph corresponding to MIMO receiver factorization with an MMSE interface.

an approximation it can be disregarded. The resulting detection algorithm is based on the approximation

$$f(\tilde{y}_i \mid x_1, x_2, \ldots, x_{M_T}) \approx f(\tilde{y}_i \mid x_i) \tag{5.29}$$

As shown in [11]–[29, p. 152 ff.], ZF and LMSSE interfaces perform well under the condition that M_R is significantly larger than M_T.

5.4.2 *Linear interfaces with nonlinear processing*

A class of suboptimum receivers called *V-BLAST* is based on the following three operations.

(1) *Nulling of spatial interference.* This is obtained by modifying the factor graph through a linear preprocessing operation that transforms y_i into $\tilde{y}_i$ for all i.
(2) *Cancellation of spatial interference.* This is obtained by simplifying the function nodes as follows:

$$f(\tilde{y}_i \mid x_i, x_{i+1}, \ldots, x_{M_T}) \approx f(\tilde{y}_i \mid x_i, \hat{x}_{i+1}, \ldots, \hat{x}_{M_T})$$

sequentially for $i = M_T - 1, M_T - 2, \ldots, 1$, where $\hat{x}$ denotes a decision made on the value of x.
(3) *Ordering.* This consists of ordering the antennas on which the above two steps are performed, as discussed for example in [4, 12]. Here it suffices to observe that, since V-BLAST is prone to error propagation, antenna ordering may affect considerably the receiver performance.

Zero-forcing V-BLAST

Several variants of V-BLAST exist. Zero-forcing V-BLAST can be derived from the QR factorization of the matrix $\mathbf{H}$ according to (5.5), and is characterized by the matrix

$$\mathbf{A}(\mathbf{H}) = \mathbf{Q}^\dagger \tag{5.30}$$

Hence,

$$\mathbf{A}(\mathbf{H})\mathbf{y} = \mathbf{R}\mathbf{x} + \mathbf{Q}^\dagger\mathbf{n}$$

Since $\mathbf{R}$ is an upper triangular matrix, every $\tilde{y}_i$, $i = 1, \ldots, \min\{M_T, M_R\}$, depends only on x_j for $j = i, \ldots, t$. The resulting factor graph simplifies as illustrated by Fig. 5.20.

A detection algorithm on the graph of Fig. 5.20 is based on the approximation (5.29), and works with a single sweep, as follows:

$$\hat{x}_{M_T} = \arg\max_{x \in \mathcal{X}} f(\tilde{y}_{M_T} \mid x_{M_T}) \tag{5.31}$$

$$\hat{x}_i = \arg\max_{x \in \mathcal{X}} f(\tilde{y}_i \mid x_i, \hat{x}_{i+1}, \ldots, \hat{x}_{M_T}),\ i = M_T - 1, \ldots, 1 \tag{5.32}$$

This detection algorithm yields a solution in M_T steps.

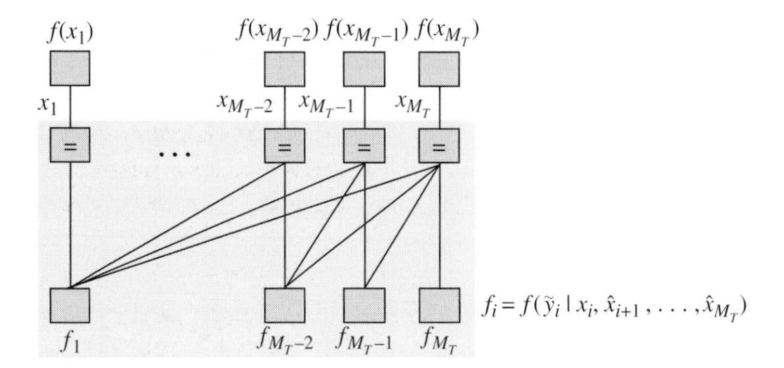

Fig. 5.20. Factor graph corresponding to MIMO receiver factorization with a V-BLAST interface.

LMMSE V-BLAST

LMMSE V-BLAST can be obtained by the minimization of the following MSE (where $\mathbf{G}$ and $\widehat{\mathbf{R}}$ are unknown matrices and $\widehat{\mathbf{R}}$ is strictly upper triangular, i.e., $R_{i,j} = 0$ for $i \leq j$):

$$\begin{aligned}\varepsilon^2(\mathbf{G}, \widehat{\mathbf{R}}) &\triangleq \mathcal{E}[\|\mathbf{Gy} - \widehat{\mathbf{R}}\mathbf{x} - \mathbf{x}\|_F^2] \\ &= \mathcal{E}[\|(\mathbf{GH} - \widehat{\mathbf{R}} - \mathbf{I})\mathbf{x} + \mathbf{Gn}\|_F^2] \\ &= \mathcal{E}[\|\mathbf{GH} - \widehat{\mathbf{R}} - \mathbf{I}\|_F^2 + N_0\|\mathbf{G}\|_F^2] \end{aligned} \tag{5.33}$$

The optimum matrices are obtained (after some algebra, as shown in [12]) as follows. First, calculate the Cholesky factorization

$$\mathbf{H}^\dagger\mathbf{H} + \delta_s\mathbf{I}_t = \mathbf{S}^\dagger\mathbf{S}$$

where $\mathbf{S}$ is an upper triangular matrix. Next, $\mathbf{G}$ and $\widehat{\mathbf{R}}$ are given by

$$\begin{cases}\mathbf{G} = \text{diag}^{-1}(\mathbf{S})(\mathbf{S}^\dagger)^{-1}\mathbf{H}^\dagger \\ \widehat{\mathbf{R}} = \text{diag}^{-1}(\mathbf{S})\mathbf{S} - \mathbf{I}\end{cases} \tag{5.34}$$

In the general setting of (5.27), MMSE V-BLAST is characterized by the matrix

$$\mathbf{A}(\mathbf{H}) = \mathbf{G} \tag{5.35}$$

Hence, the detection algorithm is the same as the one illustrated for ZF V-BLAST with matrix $\mathbf{R} = \mathbf{I} + \widehat{\mathbf{R}}$ obtained from (5.34). As discussed in [4], MMSE V-BLAST performs better than ZF V-BLAST, especially at intermediate signal-to-noise ratios.

5.5 MIMO receivers for coded signals

With coded MIMO, long codewords and the presence of an interleaver allow the use of receivers based on iterative SPA, although noniterative algorithms have also been proposed. The latter can be viewed as extensions of those described above in the context of uncoded MIMO (for details, see [11, 12]).

A space–time codeword with block length N is described by the $M_T \times N$ matrix $\mathbf{X} \triangleq (\mathbf{x}_1, \ldots, \mathbf{x}_N)$. The row index of $\mathbf{X}$ indicates space, while the column index indicates time: to wit, the ith component of the M_T-vector $\mathbf{x}_n$, denoted $x_{i,n}$, is a complex number representing the two-dimensional signal transmitted by the ith antenna at discrete time n, $n = 1, \ldots, N$, $i = 1, \ldots, M_T$. The received signal is the $M_R \times N$ matrix $\mathbf{Y} = (\mathbf{y}_1, \ldots, \mathbf{y}_N)$, with

$$\mathbf{Y} = \mathbf{H}\mathbf{X} + \mathbf{N} \tag{5.36}$$

where $\mathbf{N}$ is a matrix of independent, zero-mean, circularly symmetric complex Gaussian random variables (RVs) each having variance N_0. Thus, the noise affecting the received signal is spatially and temporally white, with $\mathcal{E}[\mathbf{N}\mathbf{N}^\dagger] = NN_0\mathbf{I}_{M_R}$. The channel is described by the $M_R \times M_T$ matrix $\mathbf{H}$. Here we assume, as stated at the onset of this chapter, that $\mathbf{H}$ is independent of both $\mathbf{X}$ and $\mathbf{N}$, it remains constant during the transmission of an entire codeword, and its realization (the channel-state information) is known at the receiver.

If the space–time code $\mathcal{C}$ is used, we have the factorization

$$f(\mathbf{X} \mid \mathbf{Y}) \propto [\mathbf{X} \in \mathcal{C}] \prod_{n=1}^{N} f(\mathbf{y}_n \mid \mathbf{x}_n) \tag{5.37}$$

corresponding to the factor graph of Fig. 5.21 where the interleaver $\boldsymbol{\pi}$ is included so as to maximize the girth of the factor graph (a tilde denotes the variables after interleaving). Notice also that, with reference to the block diagram of Fig. 1.2, the lower blocks of Fig. 5.21 correspond to the symbol demapper.

5.5.1 Iterative sum–product algorithm

Consider again Fig. 5.21, and disregard for notational simplicity the presence of the interleaver (thus, $\tilde{x}_{i,n} = x_{i,n}$). According to the basic step of the sum–product algorithm,

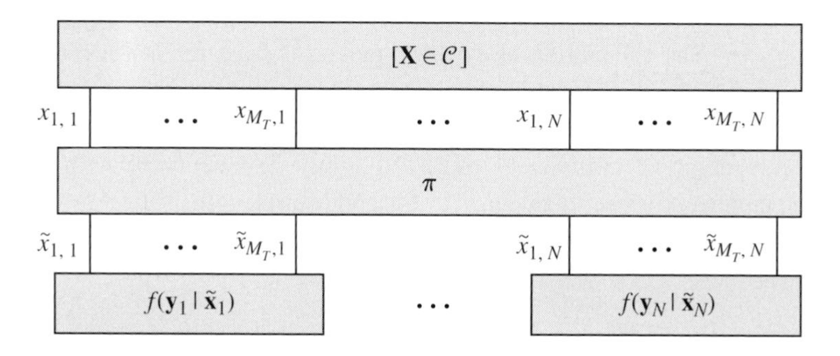

Fig. 5.21. Factor graph corresponding to space–time encoded MIMO with an interleaver.

the upper node, corresponding to the function $[\mathbf{X} \in \mathcal{C}]$, outputs messages that we denote by $\mu_\downarrow(x_{i,n})$ and are given by

$$\mu_\downarrow(x_{i,n}) = \sum_{\sim x_{i,n}} [\mathbf{X} \in \mathcal{C}] \prod_{(j,m) \neq (i,n)} \mu_\uparrow(x_{j,m}) \tag{5.38}$$

where $\mu_\uparrow(x_{i,n})$ are the messages output by the functional nodes $f(\mathbf{y}_i \mid \mathbf{x}_i)$, given by

$$\mu_\uparrow(x_{i,n}) = \sum_{\sim x_{i,n}} f(\mathbf{y}_n|\mathbf{x}_n) \prod_{j \neq i} \mu_\downarrow(x_{j,n}) \tag{5.39}$$

If the above messages are known exactly, the APPs can be computed as

$$f(x_{i,n} \mid \mathbf{Y}) \propto \mu_\downarrow(x_{i,n}) \mu_\uparrow(x_{i,n}) \tag{5.40}$$

To interpret (5.40), let us focus first on the upper functional block of Fig. 5.21, and on its edge corresponding to symbol $x_{i,n}$. The APP of this symbol is proportional to the product of two quantities:

(1) The extrinsic message $e(x_{i,n}) \triangleq \mu_\downarrow(x_{i,n})$
(2) The intrinsic message $i(x_{i,n}) \triangleq \mu_\uparrow(x_{i,n})$

(When we consider the lower functional blocks of Fig. 5.21, $\mu_\downarrow(x_{i,n})$ plays the role of the intrinsic message, and $\mu_\uparrow(x_{i,n})$ that of the extrinsic message.) We have

$$f(x_{i,n} \mid \mathbf{Y}) \propto e(x_{i,n}) i(x_{i,n}) \tag{5.41}$$

In this context, the iterative sum–product ("turbo") algorithm is one that exchanges extrinsic messages after suitable interleaving/deinterleaving. This algorithm computes repeatedly the two-way messages associated with the edges of the graph, until a termination criterion stops the iterative process, after which the APPs are computed and used for MAP decoding. The generation of extrinsic messages from the upper block may be called "soft decoding" of $\mathcal{C}$. For a code described through its TWLK graph, soft decoding can be done through the *BCJR algorithm* or its approximation (see, e.g., [5] or [7, Chapter 8]).[5] The calculation of extrinsic messages from the lower blocks, called "demapping," or "APP equalization," is more complex, as the factor graph corresponding to the function $f(\mathbf{y}_n \mid \widetilde{\mathbf{x}}_n)$ has no special structure enabling efficient calculation of (5.39). Hence, the computational complexity of (5.39) grows exponentially with $M_T N$, the product of the number of transmit antennas by the codeword length. Thus, low-complexity iterative receivers should focus on approximations of this operation.

[5] An interesting special case occurs when $\mathcal{C}$ is itself a turbo code, and hence requires iterative soft decoding: see *infra*.

5.5.2 *Low-complexity approximations*

From the iterative sum–product algorithm described above, we can derive several low-complexity approximations. The procedure to do this can be described as the combination of two steps.

(1) The received signal $\mathbf{Y}$ is linearly preprocessed to transform it into the matrix

$$\widetilde{\mathbf{Y}} = \mathbf{A}(\mathbf{H})\mathbf{Y} = \mathbf{A}(\mathbf{H})\mathbf{H}\mathbf{X} + \mathbf{A}(\mathbf{H})\mathbf{N} \tag{5.42}$$

A typical consequence of preprocessing is a simplification in the structure of the factor graph. As an example, if $\mathbf{A}(\mathbf{H})$ is the left pseudoinverse of $\mathbf{H}$, then the factor graph corresponding to each function $f(\mathbf{y}_i \mid \widetilde{\mathbf{x}}_i)$ in Fig. 5.21 becomes disconnected, as the function factors into the product of M_T terms. This considerably simplifies the calculation of (5.39).

(2) The messages exchanged among the graph nodes are approximated by some simplified versions. We classify these message approximations as *hard* and *soft* interference cancellation (IC).

Message approximation: hard and soft decisions

An approach to simplifying the messages exchanged in the iterative algorithm symbols consists of approximating them with vectors with only one nonzero component. This can be done by replacing the random interfering symbols by their corresponding *hard decisions*, i.e., by replacing the messages $\mu_\downarrow(x_{i,n})$ with the Iverson functions

$$\widetilde{\mu}_\downarrow(x_{i,n}) \triangleq [x_{i,n} = \arg\max_x \mu_\downarrow^{(k)}(x_{i,n} = x)]$$

With this approximation, the summation in (5.39) reduces to a single term and hence a single probability $f(\mathbf{y}_n \mid \mathbf{x}_n)$ is computed.

Another way of simplifying the message functions consists of approximating them by Gaussian distributions with the same mean and variance of the original (discrete) distribution. The corresponding "soft-decision" approximation leads to a simple result when the conditional distribution of the observations, given the transmitted signals, is Gaussian as well. If this is the case,

$$f(\mathbf{y}_n \mid \mathbf{x}_n) = (\pi N_0)^{-M_R} \exp(-\|\mathbf{y}_n - \mathbf{H}\mathbf{x}_n\|_F^2 / N_0) \tag{5.43}$$

The mean and variance of $x_{i,n}$ are given by

$$m_{i,n} \triangleq \sum_{x \in \mathcal{X}} x \mu_\downarrow(x_{i,n} = x)$$

and

$$\sigma_{i,n}^2 \triangleq \left\{ \sum_{x \in \mathcal{X}} |x|^2 \mu_\downarrow(x_{i,n} = x) \right\} - |m_{i,n}|^2$$

respectively. Then, in order to calculate (5.39), we need to average $f(\mathbf{y}_n \mid \mathbf{x}_n)$ with respect to all the random Gaussian approximations of $x_{j,n}$ corresponding to $j \neq i$. Assuming that $\mathbf{x}$ is circularly distributed normal with mean $\mathbf{m}$ and covariance matrix $\boldsymbol{\Sigma}$, which we write $\mathbf{x} \sim \mathcal{N}_c(\mathbf{m}, \boldsymbol{\Sigma})$, we have

$$\mathbf{x} = \det(\pi\boldsymbol{\Sigma})^{-1} \exp(-(\mathbf{x}-\mathbf{m})^{\dagger}\boldsymbol{\Sigma}^{-1}(\mathbf{x}-\mathbf{m})),$$

and we obtain, from $\mathbf{y} = \mathbf{H}\mathbf{x} + \mathbf{n}$,

$$\mathcal{E}_{\mathbf{x}}[f(\mathbf{y} \mid \mathbf{x})] = \det(\pi(\mathbf{H}\boldsymbol{\Sigma}\mathbf{H}^{\dagger} + N_0\mathbf{I}_r))^{-1} \cdot \exp(-(\mathbf{y}-\mathbf{H}\mathbf{m})^{\dagger}(\mathbf{H}\boldsymbol{\Sigma}\mathbf{H}^{\dagger} + N_0\mathbf{I}_r)^{-1}(\mathbf{y}-\mathbf{H}\mathbf{m})) \tag{5.44}$$

Next, by setting

$$(\mathbf{m}_{i,n}(x))_j = \begin{cases} m_{j,n} & j \neq i \\ x & j = i \end{cases} \quad \text{and} \quad (\boldsymbol{\Sigma}_{i,n}(x))_{j,k} = \begin{cases} \sigma^2_{j,n} & j = k, j \neq i \\ 0 & \text{otherwise} \end{cases}$$

we obtain the following approximation to the messages $\mu_{\uparrow}(x_{i,n})$:

$$\widetilde{\mu}_{\uparrow}(x_{i,n} = x) \propto \mathcal{N}_c(\mathbf{m}_{i,n}(x), \mathbf{H}\boldsymbol{\Sigma}_{i,n}(x)\mathbf{H}^{\dagger} + N_0\mathbf{I}_r), \quad x \in x$$

5.5.3 *EXIT-charts*

Since turbo algorithms operate on extrinsic probabilities, their convergence behavior can be studied by examining how these evolve in time. A convenient graphical description of this process is given by EXIT-charts [33], which yield approximate, yet reasonably accurate, results. An EXIT-chart is a graph that illustrates the convergence of the iterative SDA by showing the transformations induced on a single parameter associated with input and output extrinsic messages. Let us focus for simplicity on a binary alphabet $\mathcal{X} = \{\pm 1\}$. The rationale behind EXIT-charts stems from the observation that the logarithmic likelihood ratio (LLR)[6]

$$\Lambda(x) \triangleq \ln \frac{e(x = +1)}{e(x = -1)}$$

is well approximated by a conditionally normal random variable (we write $\Lambda | x \sim \mathcal{N}(\mu, \sigma^2)$) whose pdf $f(\Lambda \mid x)$ satisfies the "consistency condition"

$$|\mu| = \frac{\sigma^2}{2} \tag{5.45}$$

where μ and σ^2 denote the conditional mean and the variance, respectively. Hence, under this condition, a single parameter (e.g., σ^2) completely defines $f(\Lambda \mid x)$.

[6] Here we drop the subscripts i, n to simplify the notation.

To justify the above, assume an AWGN channel so that the observed signal is

$$y = x + z$$

with $z \sim \mathcal{N}(0, \sigma_z^2)$. Since

$$f(y \mid x) = \frac{1}{\sqrt{2\pi}\sigma_z} e^{-(y-x)^2/2\sigma_z^2}$$

the *log-likelihood ratio*

$$\Lambda(y) \triangleq \ln \frac{f(y \mid x = +1)}{f(y \mid x = -1)}$$

takes value

$$\Lambda(y) = \frac{2}{\sigma_z^2}(x + z) \tag{5.46}$$

and hence, given x, Λ is conditionally Gaussian:

$$\Lambda(y) \mid x \sim \mathcal{N}\left(\frac{2}{\sigma_z^2}x, \frac{4}{\sigma_z^2}\right) \tag{5.47}$$

The observation that the conditional mean value of Λ equals the variance multiplied by $x/2$ allows us to write the pdf of the LLR in the form

$$f(\Lambda \mid x) = \frac{1}{\sqrt{2\pi}\sigma} e^{-(\Lambda - x\sigma^2/2)^2/2\sigma^2} \tag{5.48}$$

EXIT-charts describe the evolution of $f(\Lambda \mid x)$ by showing the evolution of one parameter derived from it. A common, convenient choice [38] is the mutual information $I(x; \Lambda)$ between x and Λ, defined as[7]

$$I(x; \Lambda) = \frac{1}{2} \sum_{x \in \{\pm 1\}} \int f(\Lambda \mid x) \log_2 \frac{f(\Lambda \mid x)}{f(\Lambda)} d\Lambda \tag{5.49}$$

with $f(\Lambda) = 0.5[f(\Lambda \mid x = -1) + f(\Lambda \mid x = +1)]$.

If condition (5.45) is satisfied, then $\Lambda | x \sim \mathcal{N}(x\sigma^2/2, \sigma^2)$, and hence $I(x; \Lambda)$ depends only on σ^2. We have, explicitly, $I(x; \Lambda) = J(\sigma^2)$, where

$$J(\sigma^2) \triangleq 1 - \int_{-\infty}^{\infty} \frac{1}{\sqrt{2\pi}\sigma} e^{-[(w - x\sigma^2/2)^2/2\sigma^2]} \log_2(1 + e^{-xw}) dw \tag{5.50}$$

The function $J(\sigma^2)$ is plotted in Fig. 5.22.

[7] The notation here is not the most felicitous one, as it does not distinguish between the random variable x and the values it takes on. We put up with it, as it is commonly used in the literature.

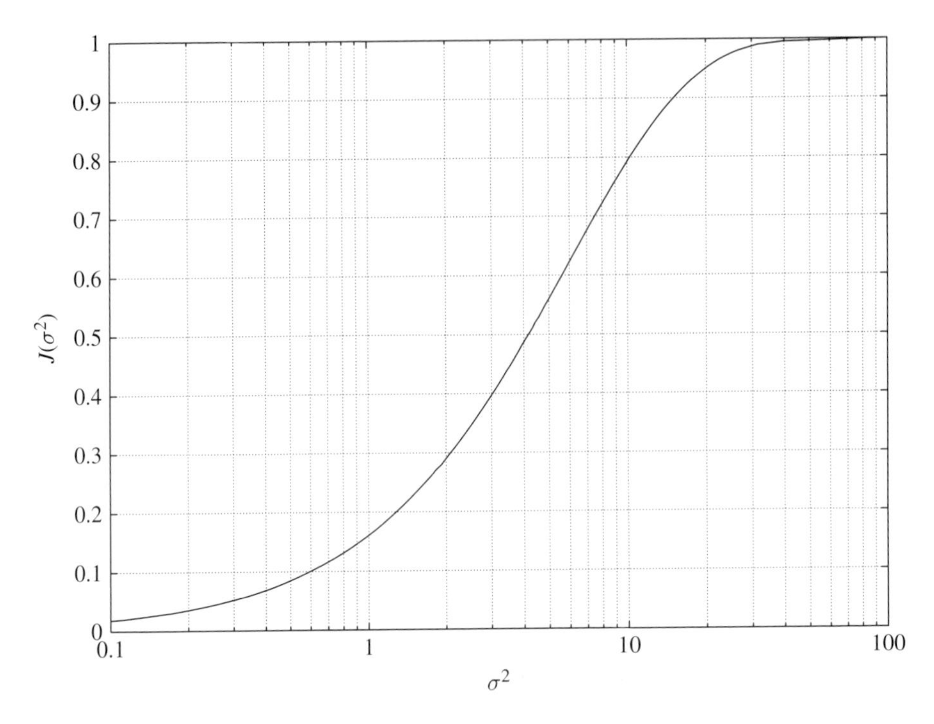

Fig. 5.22. Plot of the function $J(\sigma^2)$ defined in (5.50).

If $f(\Lambda|x)$ is not known, then an approximation of the mutual information (5.49), proposed in [36], is the following:

$$I(x; \Lambda) \approx 1 - \frac{1}{S} \sum_{k=1}^{S} \log_2 (1 + \exp(-x_k \Lambda_k)) \tag{5.51}$$

where Λ_k, x_k, $k = 1, \dots, S$, denote samples of the random variables Λ, x.

Refer again to Fig. 5.15. Since we assume $\mathcal{X} = \{\pm 1\}$, we can denote the messages exchanged as binary random vectors $\boldsymbol{\mu}(x) = (\mu(x = +1), \mu(x = -1))$ representing probability distribution estimates. Since $\mu(x = +1) + \mu(x = -1) = 1$, each one of these messages is equivalently represented by the logarithm of the ratio of its components, i.e., by the LLR

$$\Lambda_i = \ln \frac{\mu(x_i = +1)}{\mu(x_i = -1)}$$

This allows us to write $I(x; \boldsymbol{\mu})$ instead of $I(x; \Lambda)$. Specifically, we have two types of messages:

(1) Input *intrinsic* messages $\boldsymbol{\mu}^{\mathrm{i}}(x) = i(x)$
(2) Output *extrinsic* messages $\boldsymbol{\mu}^{\mathrm{e}}(x) = e(x)$

It follows that we can define the intrinsic and extrinsic mutual informations as $I^{\mathrm{i}} \triangleq I(x; \boldsymbol{\mu}^{\mathrm{i}})$ and $I^{\mathrm{e}} \triangleq I(x; \boldsymbol{\mu}^{\mathrm{e}})$, respectively.

We are now ready to describe the behavior of each one of the functional blocks of Fig. 5.15 by giving its *extrinsic information transfer* (EXIT) function

$$I^{\mathrm{e}} = T(I^{\mathrm{i}}) \tag{5.52}$$

EXIT functions can be obtained by Monte Carlo simulation. The general algorithm used to derive the values of I^{e} from those of I^{i}, and hence the EXIT function T, can be outlined as follows (examples can be found in [23, 33, 37], and will be illustrated below).

(1) Generate a sample input vector $\mathbf{x}$ with K random entries in $\{\pm 1\}$.
(2) Choose a value of I^{i}, and generate the vector of messages $\boldsymbol{\mu}^{\mathrm{i}}(\mathbf{x})$ entering the functional block, under the constraint

$$I(x; \boldsymbol{\mu}^{\mathrm{i}}(x)) = I^{\mathrm{i}}$$

(3) Operate the SPA to obtain the extrinsic messages $e(x)$ at the output of the block.
(4) Estimate I^{e} by using the approximation (5.51).

Notice that the EXIT-chart analysis is approximate, as it is based on the assumption of independent extrinsic probabilities, which holds for an infinite-length interleaver. Thus, some inaccuracies must be expected [26, 33, 37]. Nevertheless, the practical usefulness of EXIT-charts for convergence predictions is unquestioned.

We now specialize to decoders and to demappers the algorithm for the derivation of EXIT functions.

EXIT-charts of decoders

Under the assumption of a stationary memoryless channel, the conditional pdf $f(\mathbf{y} \mid \mathbf{x})$ can be factored into the product $\prod_i f(y_i \mid x_i)$, and we have, from (5.50),

$$I^{\mathrm{i}} = J(\sigma_{\mathrm{i}}^2)$$

Here the intrinsic information comes from channel observations, and σ_{i}^2 is the variance of the additive noise.

A random vector $\mathbf{u} \in \{\pm 1\}^K$ of uncoded symbols, $K \leq N$, is generated, and passed to the encoder of an (N, K) code. This outputs the codeword $\mathbf{x} \in \{\pm 1\}^N$. A Gaussian random noise generator outputs, for each component x of $\mathbf{x}$, the LLR Λ^{i} such that

$$\Lambda^{\mathrm{i}} | x \sim \mathcal{N}\left(x\frac{\sigma_{\mathrm{i}}^2}{2}, \sigma_{\mathrm{i}}^2\right)$$

where $\sigma_{\mathrm{i}}^2 = J^{-1}(I^{\mathrm{i}})$. The decoder outputs the LLRs Λ^{e}. S values of it are used to approximate I^{e} through (5.51), so that no Gaussian assumption is imposed on Λ^{e}.

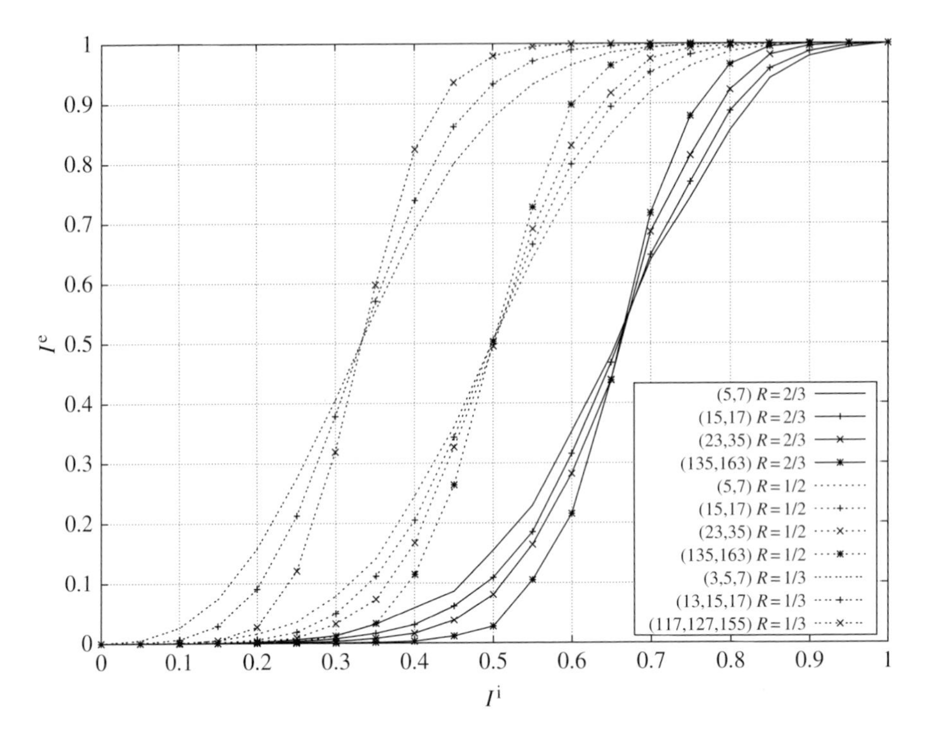

Fig. 5.23. EXIT functions of rate-1/3, rate-1/2, and rate-2/3 RSC codes specified in the legend (rate-2/3 codes are obtained by puncturing corresponding rate-1/2 codes). The curves plot I^e against I^i.

Figure 5.23 shows the EXIT functions referring to rate-1/3, rate-1/2, and rate-2/3 recursive systematic convolutional (RSC) codes with different generators and number of states. Rate-2/3 codes are obtained by puncturing corresponding rate-1/2 codes. The curves plot the mutual information I^e against I^i. A notable common feature of these EXIT-charts is that they can be regarded as smoother versions of a unit step function whose level transition occurs at a value of I^i equal to the code rate R. This can be interpreted by observing that I^i is equivalent to the mutual information exchanged between the transmitted symbol x and the received signal y, and hence equals the capacity. A capacity-achieving code can attain reliable communication if and only if $I^i > R$, and hence its EXIT curve would exhibit a sharp transition of the extrinsic mutual information from 0 (unreliable communication) to 1 (reliable communication) in correspondence of $I^i = R$. Finite-complexity codes generate the smoother behavior exhibited by the EXIT curves of Fig. 5.23.

Figure 5.24 refers to a rate-1/2 parallel turbo-code whose constituent RSC encoders have generators $(5, 7)$. The curves plot the mutual information I^e vs. I^i for different numbers of iterations of the turbo decoding algorithm. Notice also how the transition near R, which is symmetric for convolutional decoders, becomes asymmetric for turbo decoders. These also show a migration from the "unreliable communication" condition slower than convolutional codes of similar rate, but, as the number of iterations increases, a faster acquisition of the "reliable communication" condition.

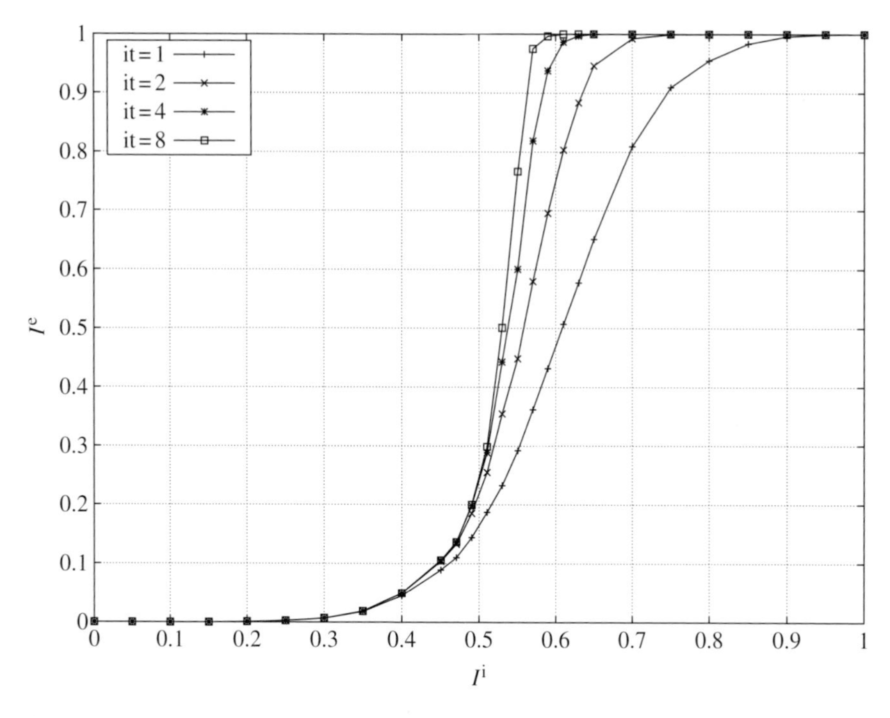

Fig. 5.24. EXIT functions of a rate-1/2 parallel turbo-code whose constituent RSC encoders have generators (5, 7). The curves plot I^e against I^i for different numbers of iterations of the turbo decoding algorithm.

EXIT-charts of demappers

To evaluate the EXIT-chart of the demapper, we define a map (modulator) from the set of binary m-vectors

$$\mathbf{x}_i = (x_{i1}, \ldots, x_{im})^T$$

where $x_{ij} \in \{\pm 1\}$, to a signal set $\mathcal{S}$:

$$\phi_m : \ \{\pm 1\}^m \mapsto \mathcal{S}$$

Then, vector $\mathbf{x} = (\mathbf{x}_1^T, \ldots, \mathbf{x}_t^T)^T$ is first generated and then passed through the modulator to yield the vector

$$\mathbf{s} = \mathbf{m}(\mathbf{x}) \triangleq (\phi_m(\mathbf{x}_1), \ldots, \phi_m(\mathbf{x}_t))^T$$

which is passed through the MIMO channel to obtain the received vector

$$\mathbf{y} = \mathbf{Hm}(\mathbf{x}) + \mathbf{n}$$

Vector $\mathbf{y}$ provides the message $\boldsymbol{\mu}^{\mathrm{i}}$ consisting of the conditional pdf

$$f(\mathbf{y} \mid \mathbf{x}) = (\pi\sigma_{\mathrm{i}}^2)^{-M_R} \exp(-\|\mathbf{y} - \mathbf{Hm}(\mathbf{x})\|_F^2 / \sigma_{\mathrm{i}}^2) \tag{5.53}$$

sampled at all possible values of $\mathbf{x} \in \{\pm 1\}^{mt}$. The evaluation of the extrinsic probability distribution (message $\boldsymbol{\mu}^{\mathrm{e}}$) depends on the specific approximation of the APP demapper considered.

No approximation

When an APP demapper is used without approximations, exact calculation of the extrinsic messages is generally complex. We may apply the approximation (5.51) to the samples $\Lambda^{\mathrm{e}}_{ij}$.

Interference cancelers with linear filtering

Interference cancellation (IC) is based on the generation of soft estimates $\widehat{\mathbf{s}}$ of the transmitted symbol vector $\mathbf{s}$ that are used to eliminate, in an iterative fashion, the spatial interference. For each transmit antenna, $i = 1, \ldots, M_T$, the soft estimates are computed as follows:

$$\hat{s}_i = \sum_{s_i \in \mathcal{S}} s_i f(s_i) \tag{5.54}$$

where, assuming that the bits contributing to the transmission of s are independent, $f(s_i) = f(\mathbf{x}_i) = \prod_{j=1}^{m} f(x_{ij})$ if $s_i = \phi_m(\mathbf{x}_i)$.

Then, the IC block outputs, for each antenna i, the following soft values

$$\begin{aligned} \widehat{\mathbf{y}}_i &= \mathbf{y} - \mathbf{H}\widehat{\mathbf{s}} + \mathbf{h}_i \hat{s}_i \\ &= \mathbf{h}_i s_i + \sum_{j \neq i} \mathbf{h}_j \left(s_j - \hat{s}_j\right) + \mathbf{n} \end{aligned} \tag{5.55}$$

which are subsequently processed by the antenna-specific linear filters as described in the following.

MMSE filter. The MMSE filter operates by minimizing over $\mathbf{f}$ the mean square error $\mathcal{E}[|\mathbf{f}_i^H \widehat{\mathbf{y}}_i - x_i|^2]$. As a result, the filter vector $\mathbf{f}_i$ is obtained as

$$\mathbf{f}_i = \left[\sigma_z^2 \mathbf{I}_r + \mathbf{H}\boldsymbol{\Sigma}_i^2 \mathbf{H}^H\right]^{-1} \mathbf{h}_i \tag{5.56}$$

where $\boldsymbol{\Sigma}_i^2 = \mathrm{diag}(\sigma_1^2, \ldots, \sigma_{i-1}^2, 1, \sigma_{i+1}^2, \ldots, \sigma_t^2)$ and the variances σ_i^2 are given by

$$\begin{aligned} \sigma_i^2 &= \mathcal{E}[|s_i - \hat{s}_i|^2] \\ &= \sum_{s_i \in \mathcal{S}} |s_i|^2 f(s_i) - |\hat{s}_i|^2 \end{aligned} \tag{5.57}$$

Recalling (5.54), the output of the ith filter is given by

$$\widetilde{y}_i = \alpha_i c_i + \beta_i \tag{5.58}$$

where $\mu_i = \mathbf{f}_i^H \mathbf{h}_i$ and where β_i is a complex Gaussian random variable with zero mean and variance

$$\sigma_{\beta_i}^2 = \alpha_i - \alpha_i^2$$

Extrinsic probabilities are finally computed as follows:

$$e(x_{ij}) = \sum_{\mathbf{x}_{i\sim j}} f(\widetilde{y}_i|\mathbf{x}_i) \prod_{j' \neq j} f(x_{ij'}) \tag{5.59}$$

where $\mathbf{x}_{i\sim j}$ denotes the vector $\mathbf{x}_i$ without its jth component. The computational complexity involved in the calculation of $e(x_{ij})$ is linear in M_T and exponential in m, the number of bits per symbol.

Maximum ratio combining filter. The *maximum ratio combining* (MRC) filter is based on the filter vector $\mathbf{f}_i = \mathbf{h}_i$. Again, the filter output can be written as in (5.58) where $\alpha_i = \mathbf{h}_i^H \mathbf{h}_i$ and

$$\sigma_{\beta_i}^2 = \sum_{j \neq i} |\mathbf{h}_i^H \mathbf{h}_j|^2 \sigma_j^2 + \sigma_z^2 \mathbf{h}_i^H \mathbf{h}_i$$

Figure 5.25 shows the EXIT function of the APP, IC+MMSE, and IC+MRC demappers considered here. We assume four transmit and four receive antennas, a complex channel

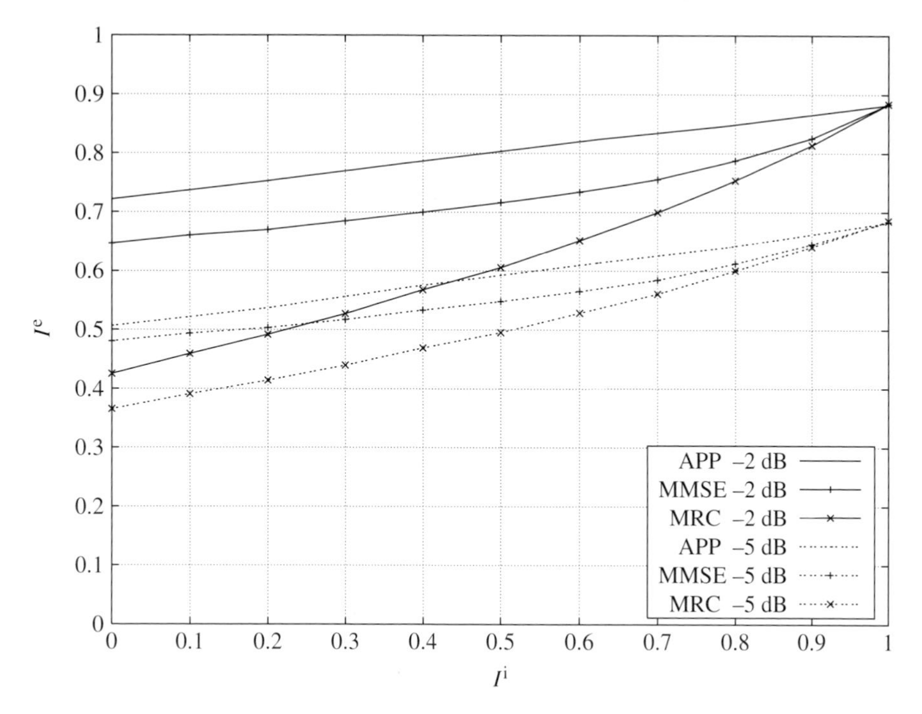

Fig. 5.25. Example of mutual information transfer function for different demappers, static channel, QPSK modulation, and $E_b/N_0 = -2$ dB, -5 dB.

matrix **H** as in [23], a quadrature phase-shift keying (QPSK) signal set, and $1/\sigma_{\mathrm{i}}^2 = E_b/N_0 = -2$ dB (solid lines) or -5 dB (dashed lines). The curves show that the APP demapper outperforms all other processors as it achieves a better value of I^{e} at any given I^{i}. Increasing the signal-to-noise ratio corresponds to an approximate upward shift of the transfer function curves, while increasing the number of antennas increases their slope (see also [4, p. 77 ff.]).

EXIT-chart convergence analysis

Once the EXIT functions of two functional blocks have been obtained, they are drawn on a single chart. Since the output of a block is the input of the other one, the second transfer function is drawn after swapping its axes. The behavior of the iterative decoding algorithm is described by a trajectory, i.e., a sequence of moves, along horizontal and vertical steps, through the pair of EXIT functions.

Figure 5.26 shows qualitatively two examples of convergence behavior. For small SNR, the two EXIT curves intersect, the trajectory is blocked, and no convergence occurs to large values of mutual information (which correspond to small error probabilities). For a higher value of SNR, instead, we have convergence, which is faster when the opening between the two curves is wider.

For a finer analysis of the convergence of the iterative algorithm, estimates of the error probability of a coded system can be superimposed to EXIT-charts to yield insight on the receiver performance. Consider the APP distribution; by assuming its random conditional

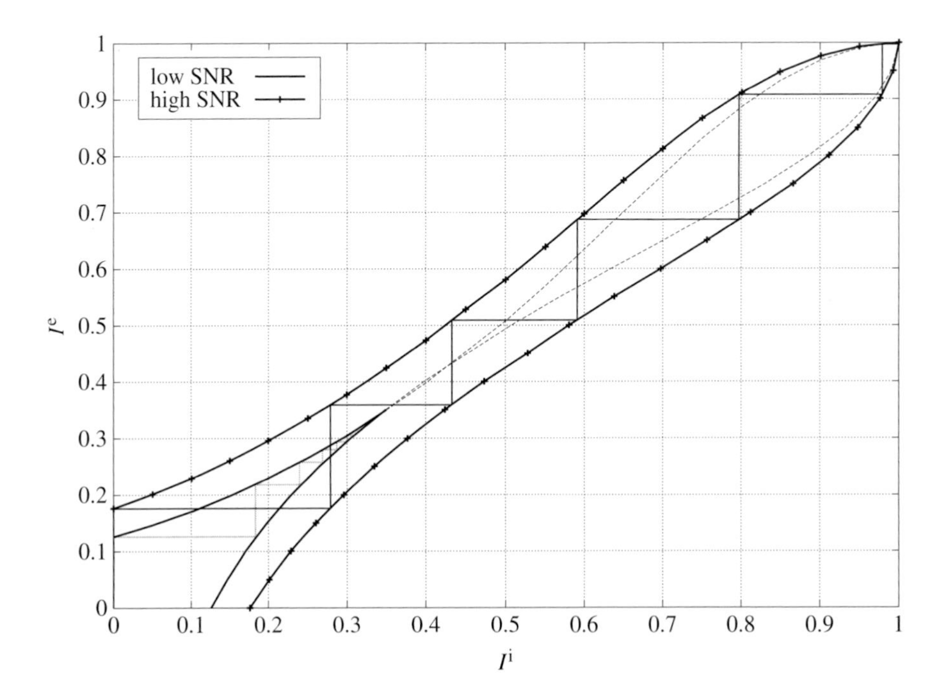

Fig. 5.26. EXIT-chart for an iterative algorithm and two values of SNR.

LLR $\Lambda^{\mathrm{p}}|x$ to be Gaussian, with mean $\sigma_{\mathrm{p}}^2/2$ and variance σ_{p}^2, the bit error rate (BER) $P_b(e)$ can be approximated by

$$P_b(e) \approx Q\left(\frac{\mu_{\mathrm{p}}}{\sigma_{\mathrm{p}}}\right) = Q\left(\frac{\sigma_{\mathrm{p}}}{2}\right) \tag{5.60}$$

where $Q(\cdot) \triangleq (2\pi)^{-1/2} \int_0^\infty \exp(-z^2/2)dz$ is the Gaussian tail function. Since $\Lambda^{\mathrm{p}} = \Lambda^{\mathrm{i}} + \Lambda^{\mathrm{e}}$, the assumption of independent LLRs leads to [33]

$$\sigma_{\mathrm{p}}^2 = \sigma_{\mathrm{i}}^2 + \sigma_{\mathrm{e}}^2$$

which in turn yields

$$P_b(e) \approx Q\left(\frac{\sqrt{J^{-1}(I^{\mathrm{i}}) + J^{-1}(I^{\mathrm{e}})}}{2}\right) \tag{5.61}$$

Figure 5.27 shows the BER plotted as a function of I^{i} and I^{e}.

An example

Consider the combination of a decoder based on a rate-1/2 convolutional code with generators (5, 7) and an MMSE-IC demapper. The MIMO channel has four transmit and

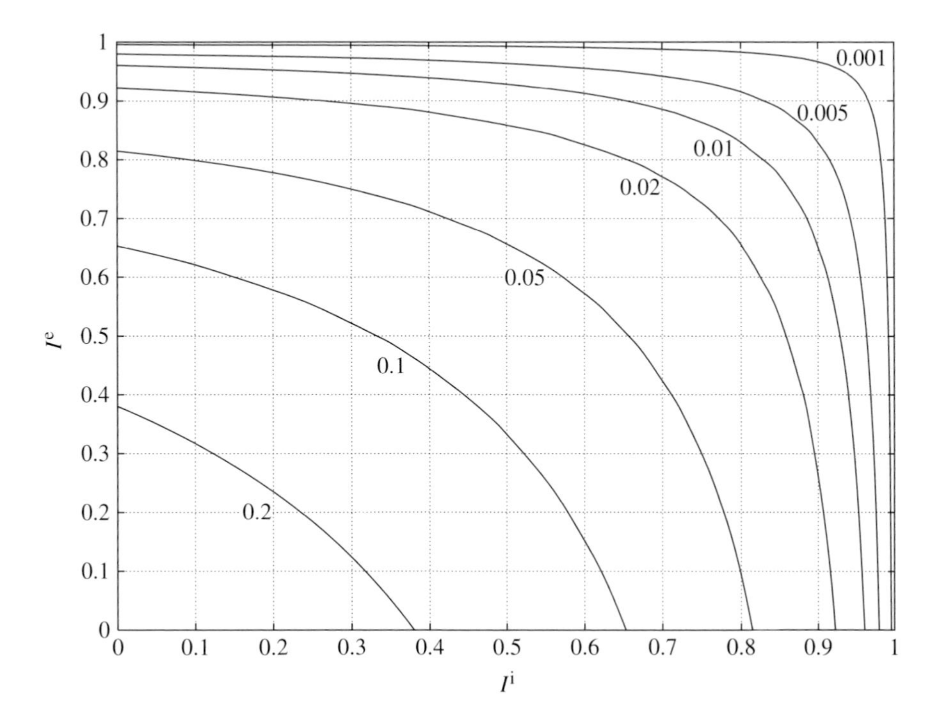

Fig. 5.27. BER chart of an iterative receiver plotted as a function of I^{e} and I^{i}.

four receive antennas. Additionally, QPSK modulation is assumed, and the channel matrix **H** is chosen as in [23], with $E_b/N_0 = -5$ dB. Figure 5.28 illustrates the first few iterations of the turbo algorithm. The figure shows the EXIT functions of the decoder (dashed line) and of the demapper (solid line). They are taken from Figs. 5.23 (after coordinate swapping) and 5.25, respectively. The dotted lines plot the constant-BER curves computed by using (5.61). The arrows indicate the first few iterations of the turbo algorithm: vertical arrows correspond to IC, while horizontal arrows correspond to decoding. The points labeled $k = 0, 1, 2$ correspond to the extrinsic mutual information at the output of the decoder after k iterations. Finally, BER values are reported in the figure (bottom left) obtained by Monte Carlo simulation for comparisons with the values computed by using (5.61) (dotted curves). Figure 5.29 shows the BER for the same system, as obtained by simulation (solid lines) and by EXIT-chart analysis (points), for $k = 0, 1, 2, 8$ iterations.

5.5.4 *Quasi-static channel*

Our previous analysis was concerned with constant channels. In *quasi-static* channel conditions, the channel matrix **H** is random, and changes independently from codeword

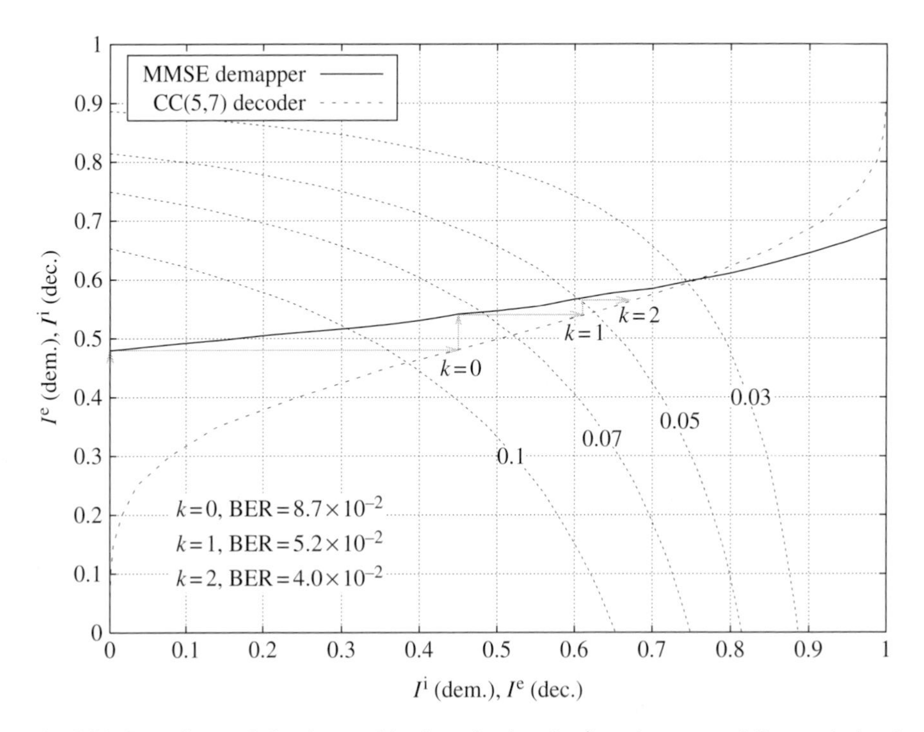

Fig. 5.28. Decoding path for the combination of a decoder (based on a rate-1/2 convolutional code with generators (5,7)) and a demapper (MMSE IC on a MIMO system with four transmit and four receive antennas), with QPSK modulation and $E_b/N_0 = -5$ dB.

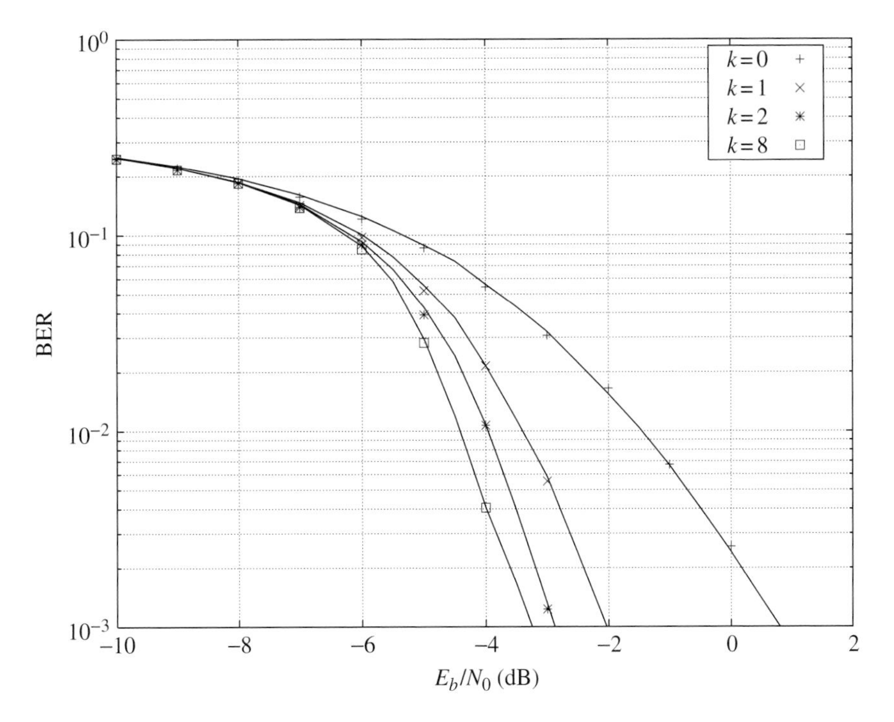

Fig. 5.29. Comparison of BER obtained by simulation and by EXIT-chart analysis, for a system with $M_T = M_R = 4$, QPSK, and a deterministic channel.

to codeword. This implies that the demapper EXIT function changes with $\mathbf{H}$, and should be evaluated for a large number of samples in order to estimate the error performance of the system. The computational burden necessary for convergence analysis might be heavy, but can be substantially alleviated by observing (through computational experience) that the EXIT function of the demapper exhibits in most cases an almost linear behavior and, consequently, only two points are needed to approximate it as a straight line.

This approximation is illustrated by Fig. 5.30, which considers the same system of Fig. 5.28, and plots in addition the straight-line approximation of the EXIT function of the demapper. The convergence points (obtained by the intersection of the demapper and decoder EXIT functions) are denoted by C and C' for the exact and approximate demapper EXIT functions, respectively. Obviously, these points lie on the decoder EXIT function, and represent the asymptotic performance attainable with an infinite number of iterations. It must be noted that the straight-line approximation leads to nonconservative convergence estimates, due to the upward convexity of the exact EXIT function of the demapper. Nevertheless, numerical results show that the approximation is fairly accurate.

A sample set of convergence points is plotted in Fig. 5.31 to show their distribution for the same system parameters. The points have been obtained by using different, randomly

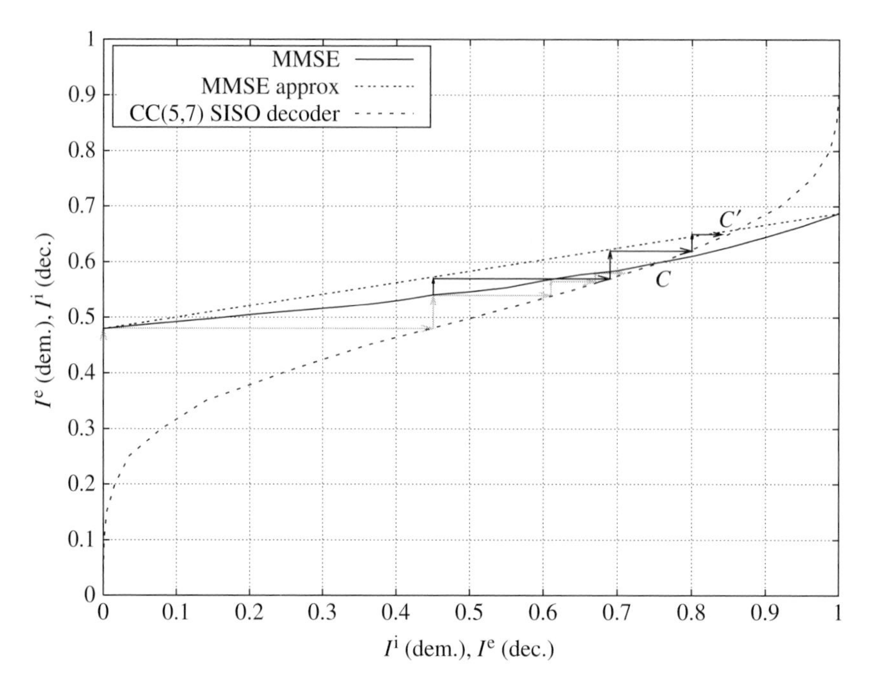

Fig. 5.30. Approximate and exact decoding trajectories for the combination of an MMSE interference canceler with $M_R = M_T = 4$, rate $R = 1/2$ CC(5,7) convolutional code, QPSK modulation, $E_b/N_0 = -5$ dB.

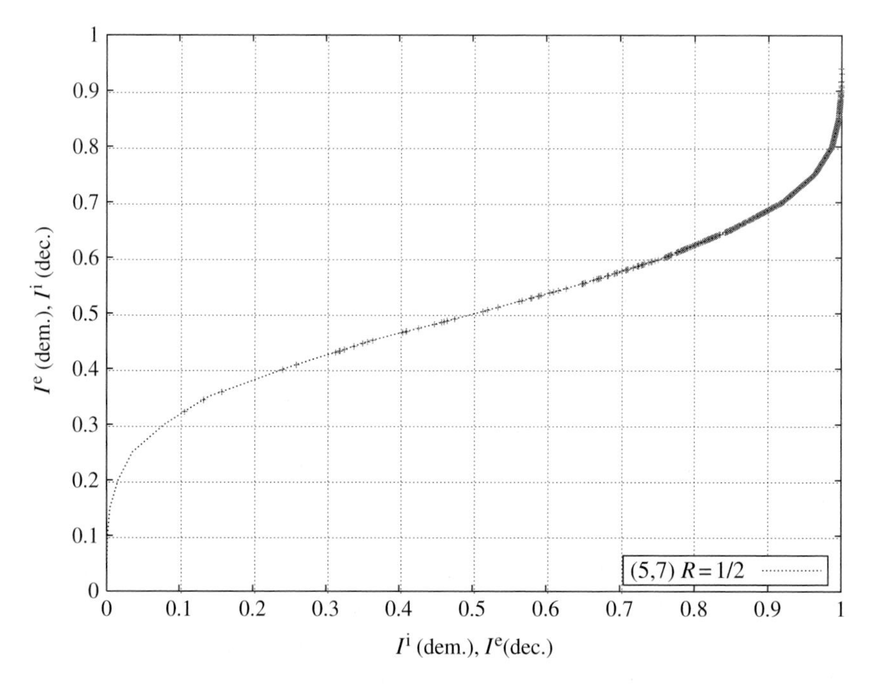

Fig. 5.31. Distribution of the convergence points for a system with $M_R = M_T = 4$, MMSE filter, rate $R = 1/2$ CC(5,7) convolutional code, QPSK modulation, $E_b/N_0 = -5$ dB.

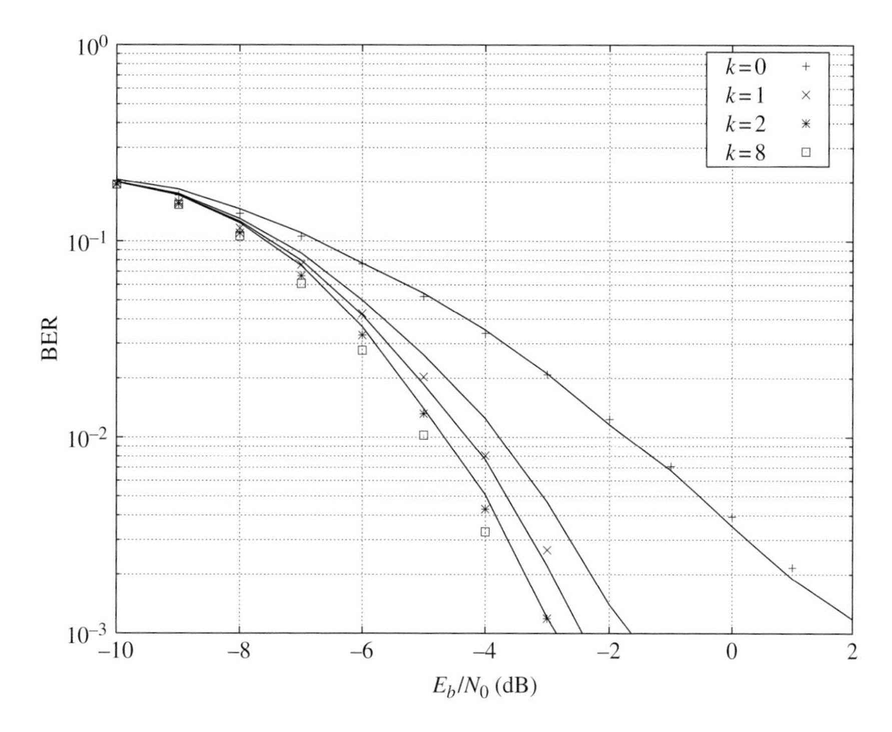

Fig. 5.32. BER obtained by simulation and by "linearized" EXIT-chart analysis for an $M_T = M_R = 4$ MIMO channel with MMSE interference cancellation, QPSK, and quasi-static independent Rayleigh fading.

generated matrices **H** with iid circularly symmetric complex Gaussian random entries with zero mean and unit variance (independent MIMO Rayleigh fading channel). It is seen that the distribution of the points is quite concentrated, thus validating the assumption that their variance is close enough to zero. Finally, Fig. 5.32 compares the BER obtained by simulation (solid lines) and by the "linearized" EXIT-chart analysis (dots) and for $k = 0, 1, 2, 8$ iterations.

5.6 Some iterative receivers

In this section we examine two specific implementations of the iterative receivers, whose general idea was expounded above. Their block diagrams are shown in Figs. 5.33 and 5.34. The former, referred to in the following as *MMSE+IC*, has the MMSE filter located before the IC loop. The latter (*IC+MMSE*) has the MMSE filter located inside the IC loop. The IC+MMSE complexity is higher than for the MMSE+IC, since a bank of MMSE filters has to be calculated at each iteration.

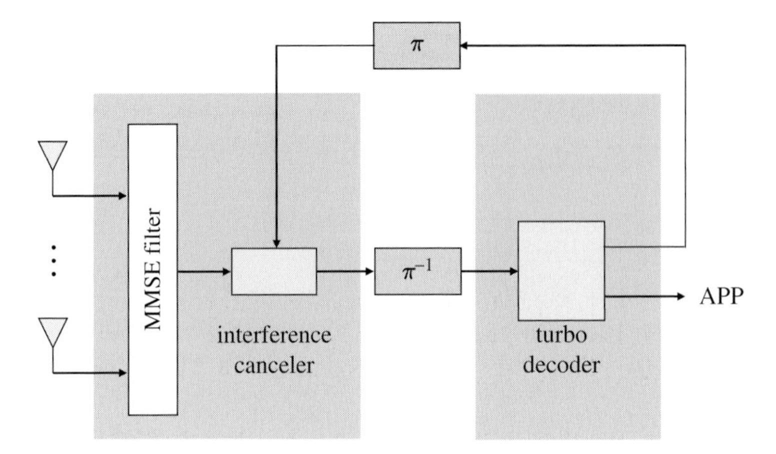

Fig. 5.33. A suboptimum implementation of a turbo receiver: the MMSE+IC receiver.

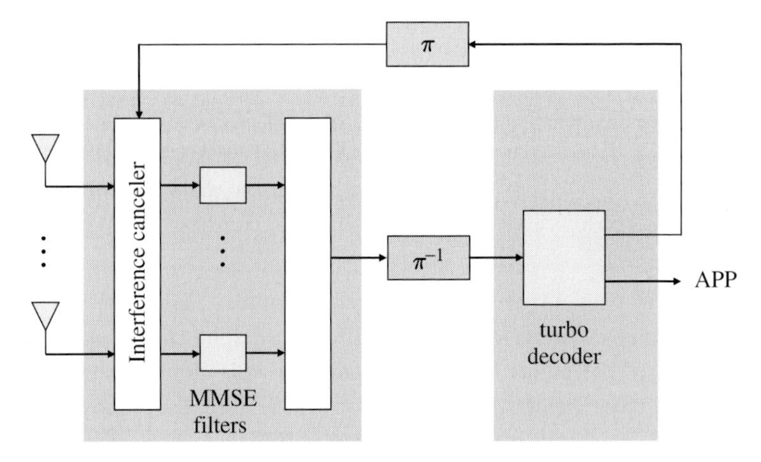

Fig. 5.34. Another suboptimum implementation of a turbo receiver: the IC+MMSE receiver.

5.6.1 *MMSE+IC receiver*

The received signal is first passed through an MMSE filter whose output is

$$\widetilde{\mathbf{Y}} = \mathbf{GY} = \mathbf{X} + \mathbf{LX} + \mathbf{GN} \tag{5.62}$$

where

$$\mathbf{G} \triangleq \mathbf{D}^{-1}\mathbf{A} \tag{5.63}$$

and $\mathbf{A} \triangleq (\mathbf{H}^H\mathbf{H}+\delta_s\mathbf{I})^{-1}\mathbf{H}^H$, $\mathbf{D} \triangleq \text{diag}(\mathbf{AH})$, $\mathbf{L} \triangleq (\mathbf{D}^{-1}\mathbf{AH}-\mathbf{I}_t)$, and $\delta_s = (E_s/N_0)^{-1}$. At iteration k, the turbo decoder provides soft estimates $\widehat{\mathbf{X}}^{(k)}$ of the transmitted signal $\mathbf{X}$. The output of the IC block at iteration k is

$$\widetilde{\mathbf{Y}}^{(k)} = \widetilde{\mathbf{Y}} - \mathbf{L}\widehat{\mathbf{X}}^{(k)} \tag{5.64}$$

5.6.2 IC+MMSE receiver

This receiver of Fig. 5.34 is based on a bank of M_T MMSE filters, one for each transmit antenna, located inside the IC loop and hence to be updated at each iteration step. This is expected to outperform MMSE+IC, since now the filters can also mitigate the residual interference.

Considering the ℓth signaling interval, let the output at iteration k of the interference canceler corresponding to antenna $i = 1, \ldots, M_T$, be given by

$$\widetilde{\mathbf{y}}_i^{(k+1)} = \mathbf{H}(\mathbf{x} - \hat{\mathbf{x}}^{(k)}) + \hat{x}_i^{(k)}\mathbf{h}_i + \mathbf{n} \tag{5.65}$$

where $\hat{\mathbf{x}}^{(k)} = [\hat{x}_1^{(k)}, \ldots, \hat{x}_t^{(k)}]^T$ is the ℓth column of $\widehat{\mathbf{X}}^{(k)}$ and represents the decoder output at iteration $k > 0$ (for $k = 0$ we set $\hat{\mathbf{x}}^{(0)} \triangleq \mathbf{0}$).

As illustrated in [10], the signal output from the ith MMSE filter at iteration k is given by

$$\tilde{y}_i^{(k)} = \mathbf{f}_i^{(k)H}\mathbf{y}_i^{(k)}$$

where the ith normalized MMSE filter vector is

$$\mathbf{f}_i^{(k)} = (\alpha_i^{(k)})^{-1}\left[\sum_{j\neq i}(1-v_j^{(k)})\mathbf{h}_j\mathbf{h}_j^H + \mathbf{h}_i\mathbf{h}_i^H + \delta_s\mathbf{I}_r\right]^{-1}\mathbf{h}_i$$

$v_j^{(k)} \triangleq \mathcal{E}[|\hat{x}_j^{(k)}|^2]/E_s$, and the normalization constant $\alpha_i^{(k)}$ is given by

$$\alpha_i^{(k)} = \mathbf{h}_i^H\left[\sum_{j\neq i}(1-v_j^{(k)})\mathbf{h}_j\mathbf{h}_j^H + \mathbf{h}_i\mathbf{h}_i^H + \delta_s\mathbf{I}_r\right]^{-1}\mathbf{h}_i$$

(see [10] for further details).

5.6.3 Numerical results

Figure 5.35 compares the performance of both receivers for $M_T = M_R = 16$. QPSK is used, along with a rate-1/2 turbo code obtained by parallel concatenation and puncturing of two rate-1/2, four-state equal recursive systematic convolutional codes. At the receiver, eight turbo decoder iterations are performed for each IC iteration. The codeword length is $N = 130$. Detailed calculations show that the overall complexity increase of the IC+MMSE versus the MMSE+IC receiver does not exceed 20%, and, in the case of Fig. 5.35, it amounts to about 5%. The corresponding performance enhancement is more than 1 dB at $k = 4$ iterations of the IC algorithm.

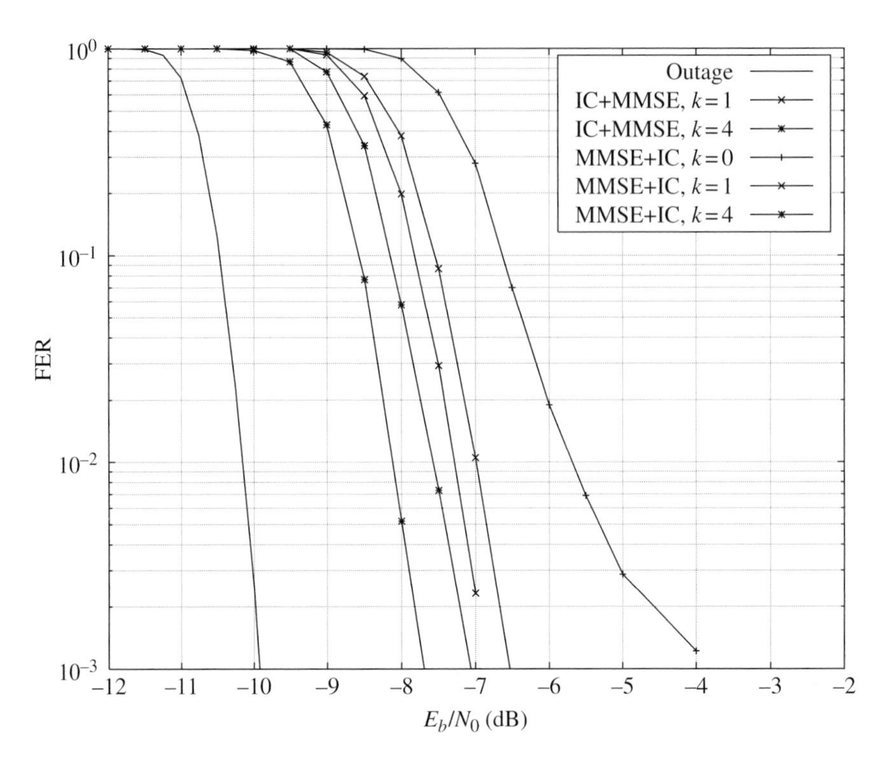

Fig. 5.35. Performance comparison, in terms of the frame-error rate (FER), of MMSE+IC and IC+MMSE receivers on a quasi-static fading channel with $M_T = M_R = 16$. The solid line without markers gives the outage-probability lower bound. Solid lines with markers describe the receiver performances for $k = 0, 1, 4$ interference-cancellation iterations (see [10] for further details).

5.7 Bibliographical notes

Sphere detection was first applied to digital detection problems by Viterbo and Biglieri [41]. An explicit flowchart of SDA in its original form can be found in [42], where it is applied to the detection of rotated lattice constellations for single antenna, independent-Rayleigh-fading channels. Its application to MIMO problems was advocated in [14]. For recent developments, see [2, 6, 15, 21, 30, 34, 39, 40] and the references therein. A VLSI implementation is described in the recent paper [6].

For an introduction to factor graphs and to the sum–product algorithm, see [25, 27] and [7, Chapter 8]. Normal graphs were introduced by Forney [18]. Reference [28] explores the connection between *belief-propagation theory* and turbo algorithms.

BLAST architectures were introduced in [19]. Turbo algorithms for MIMO receivers are advocated in [3, 9, 22, 32].

For recent work on turbo algorithms for frequency-selective MIMO channels, see [1, 16, 35]. *Sequential Monte Carlo* processors for use in iterative receivers are described in [17].

References

[1] T. Abe and T. Matsumoto, "Space–time turbo equalization in frequency-selective MIMO channels," *IEEE Trans. Vehicular Technol.*, vol. 52, no. 3, pp. 469–475, May 2003.

[2] E. Agrell, T. Eriksson, A. Vardy, and K. Zeger, "Closest point search in lattices," *IEEE Trans. Inform. Theory*, vol. 48, no. 8, pp. 2201–2214, Aug. 2002.

[3] S. L. Ariyavisitakul, "Turbo space–time processing to improve wireless channel capacity," *IEEE Trans. Commun.*, vol. 48, no. 8, pp. 1347–1359, Aug. 2000.

[4] S. Bäro, *Iterative Detection for Coded MIMO Systems*. Fortschritt-Berichte VDI, Reihe 10, Nr. 752. Düsseldorf: VDI Verlag, 2005.

[5] L. Bahl, J. Cocke, F. Jelinek, and J. Raviv, "Optimal decoding of linear codes for minimizing symbol error rate," *IEEE Trans. Inform. Theory*, vol. 20, no. 2, pp. 284–287, March 1974.

[6] A. Burg, M. Borgmann, M. Wenk, M. Zellweger, W. Fichtner, and H. Bölcskei, "VLSI implementation of MIMO detection using the sphere decoding algorithm," *IEEE J. Solid-State Circuits*, vol. 40, no. 7, pp. 1566–1577, July 2005.

[7] E. Biglieri, *Coding for Wireless Channels*. New York: Springer, 2005.

[8] E. Biglieri, A. Nordio, and G. Taricco, "Doubly-iterative decoding of space–time turbo codes with a large number of antennas," *IEEE Intl. Conf. Commun. (ICC 2004)*, Paris, France, June 20–24, 2004.

[9] E. Biglieri, A. Nordio, and G. Taricco, "Iterative receivers for coded MIMO signaling," *Wireless Commun. Mob. Comput.*, vol. 4, no. 7, pp. 697–710, Nov. 2004.

[10] E. Biglieri, A. Nordio, and G. Taricco, "MIMO doubly-iterative receivers: pre- vs. post-cancellation filtering," *IEEE Commun. Letters*, vol. 9, no. 2, pp. 106–108, Feb. 2005.

[11] E. Biglieri, G. Taricco, and A. Tulino, "Performance of space–time codes for a large number of antennas," *IEEE Trans. Inform. Theory*, vol. 48, no. 7, pp. 1794–1803, July 2002.

[12] E. Biglieri, G. Taricco, and A. Tulino, "Decoding space–time codes with BLAST architectures," *IEEE Trans. Signal Processing*, vol. 50, no. 10, pp. 2547–2552, Oct. 2002.

[13] J. Boutros and G. Caire, "Iterative multiuser joint detection: unified framework and asymptotic analysis," *IEEE Trans. Inform. Theory*, vol. 48, no. 7, pp. 1772–1793, July 2002.

[14] M. O. Damen, A. Chkeif, and J.-C. Belfiore, "Lattice codes decoder for space–time codes," *IEEE Commun. Letters*, vol. 4, pp. 161–163, May 2000.

[15] M. O. Damen, H. El Gamal, and G. Caire, "On maximum-likelihood detection and the search for the closest lattice point," *IEEE Trans. Inform. Theory*, vol. 49, no. 10, pp. 2389–2402, Oct. 2003.

[16] B. Dong and X. Wang, "Sampling-based soft equalization for frequency-selective MIMO channels," *IEEE Trans. Commun.*, vol. 53, no. 2, pp. 278–288, Feb. 2005.

[17] B. Dong, X. Wang, and A. Doucet, "A new class of soft MIMO demodulation algorithms," *IEEE Trans. Signal Processing*, vol. 51, no. 11, pp. 2752–2763, Nov. 2003.

[18] G. D. Forney, Jr., "Codes on graphs: normal realizations," *IEEE Trans. Inform. Theory*, vol. 47, no. 2, pp. 520–548, Feb. 2001.

[19] G. J. Foschini, "Layered space–time architecture for wireless communication in a fading environment when using multi-element antennas," *Bell Labs. Tech. J.*, vol. 1, no. 2, pp. 41–59, Autumn 1996.

[20] B. J. Frey and F. R. Kschischang, "Early detection and trellis splicing: reduced-complexity iterative decoding," *IEEE J. Select. Areas Commun.*, vol. 16, no. 2, pp. 153–159, Feb. 1998.

[21] B. Hassibi and H. Vikalo, "On sphere decoding algorithm. I. Expected complexity," *IEEE Trans. Signal Processing*, vol. 53, no. 8, pp. 2806–2818, August 2005.

[22] S. Haykin, M. Sellathurai, Y. de Jong, and T. Willink, "Turbo-MIMO for wireless communications," *IEEE Commun. Magazine*, vol. 42, no. 10, pp. 48–53, Oct. 2004.

[23] C. Hermosilla and L. Szczeciński, "EXIT charts for turbo receivers in MIMO systems," *Proc. 7th Intl. Symp. Signal Processing and its Applications (ISSPA 2003)*, pp. 209–212, July 1–4, 2003.

[24] R. A. Horn and C. R. Johnson, *Matrix Analysis*. Cambridge: Cambridge University Press, 1991.

[25] F. R. Kschischang, B. J. Frey, and H.-A. Loeliger, "Factor graphs and the sum–product algorithm," *IEEE Trans. Inform. Theory*, vol. 47, no. 2, pp. 498–519, Feb. 2001.

[26] S.-J. Lee, A. C. Singer, and N. R. Shanbhag, "Analysis of linear turbo equalizer via EXIT chart," *Proc. IEEE Global Telecomm. Conf. (GLOBECOM 2003)*, vol. 4, pp. 2237–2242, Dec. 2003.

[27] H.-A. Loeliger, "An introduction to factor graphs," *IEEE Signal Processing Magazine*, vol. 21, no. 1, pp. 28–41, Jan. 2004.

[28] R. J. McEliece, D. J. C. MacKay, and J.-F. Cheng, "Turbo decoding as an instance of Pearl's 'belief propagation' algorithm," *IEEE J. Select. Areas Commun.*, vol. 16, no. 2, pp. 140–152, Feb. 1998.

[29] A. Paulraj, R. Nabar, and D. Gore, *Introduction to Space–Time Wireless Communications*. Cambridge: Cambridge University Press, 2003.

[30] G. Rekaya and J.-C. Belfiore, "Complexity of ML lattice decoders for the decoding of linear full-rate space–time codes," *IEEE Trans. Wireless Commun.*, to be published.

[31] T. J. Richardson and R. L. Urbanke, "The capacity of low-density parity-check codes under message-passing decoding," *IEEE Trans. Inform. Theory*, vol. 47, no. 2, pp. 599–618, Feb. 2001.

[32] M. Sellathurai and S. Haykin, "TURBO-BLAST for wireless communications: theory and experiments," *IEEE Trans. Signal Processing*, vol. 50, no. 10, pp. 2538–2546, Oct. 2002.

[33] S. ten Brink, "Convergence behavior of iteratively decoded parallel concatenated codes," *IEEE Trans. Commun.*, vol. 49, no. 10, pp. 1727–1737, October 2001.

[34] L. M. G. M. Tolhuizen, "Soft-decision sphere decoding for systems with more transmit antennas than receive antennas," *12th Annual Symp. of the IEEE/CVT*, Enschede, The Netherlands, Nov. 3, 2005.

[35] A. Tonello, "MIMO MAP equalization and turbo decoding in interleaved space–time coded systems," *IEEE Trans. Commun.*, vol. 51, no. 2, pp. 155–160, Feb. 2003.

[36] M. Tüchler and J. Hagenauer, "EXIT charts of irregular codes," in *2002 Conf. on Information Sciences and Systems*, Princeton, NJ, March 2002.

[37] M. Tüchler, R. Koetter, and A. Singer, "Turbo-equalization: principles and new results," *IEEE Trans. Commun.*, vol. 50, no. 5, pp. 754–767, May 2002.

[38] M. Tüchler, S. ten Brink, and J. Hagenauer, "Measures for tracing convergence of iterative decoding algorithms," *Proc. 4th IEEE/ITG Conf. on Source and Channel Coding*, Berlin, Germany, pp. 53–60, Jan. 2002.

[39] H. Vikalo and B. Hassibi, "Maximum-likelihood sequence detection of multiple antenna systems over dispersive channels via sphere decoding," *EURASIP J. Appl. Signal Processing*, no. 5, pp. 525–531, May 2002.

[40] H. Vikalo, B. Hassibi, and T. Kailath, "Iterative decoding for MIMO channels via modified sphere decoding," *IEEE Trans. Wireless Commun.*, vol. 3, no. 6, pp. 2299–2311, Nov. 2004.

[41] E. Viterbo and E. Biglieri, "A universal lattice decoder," in *14-ème Colloque GRETSI*, Juan-les-Pins, France, Sept. 1993.

[42] E. Viterbo and J. Boutros, "A universal lattice code decoder for fading channels," *IEEE Trans. Inform. Theory*, vol. 45, no. 5, pp. 1639–1642, July 1999.

[43] A. P. Worthen and W. E. Stark, "Unified design of iterative receivers using factor graphs," *IEEE Trans. Inform. Theory*, vol. 47, no. 2, pp. 843–849, Feb. 2001.

6 Multi-user receiver design

6.1 Introduction

The preceding chapter considered the design of receivers for MIMO systems operating as single-user systems. Increasingly however, as noted in Chapters 2 and 4, wireless communication networks operate as shared-access systems in which multiple transmitters share the same radio resources. This is due largely to the ability of shared-access systems to support flexible admission protocols, to take advantage of statistical multiplexing, and to support transmission in unlicensed spectrum. In this chapter we will extend the treatment of Chapter 5 to consider receiver structures for multi-user, and specifically, multiple-access MIMO systems. We will also generalize the channel model considered to include more general situations than the flat-fading channels considered in Chapter 5. To treat these problems, we will first describe a general model for multi-user MIMO signaling, and then discuss the structure of optimal receivers for this signal model. This model will generally include several sources of interference arising in MIMO wireless systems, including multiple-access interference caused by the sharing of radio resources noted above, inter-symbol interference caused by dispersive channels, and inter-antenna interference caused by the use of multiple transmit antennas. Algorithms for the mitigation of all of these types of interference can be derived in this common framework, leading to a general receiver structure for multi-user MIMO communications over frequency-selective channels. As we shall see, these basic algorithms will echo similar algorithms that have been described in Chapters 3 and 5. Since optimal receivers in this situation are often prohibitively complex, the bulk of the chapter will focus on useful lower complexity sub-optimal iterative and adaptive receiver structures that can achieve excellent performance in mitigating interference in such systems. This discussion is organized as follows.

Section 6.2 will introduce a simple, yet useful, model for the signals received by the receiver in a MIMO system. This model is rich enough to capture the important behavior of most wireless communication channels, while being simple enough to allow for the straightforward motivation and understanding of the basic receiver elements arising in practical situations. This section also derives a canonical multi-user MIMO receiver structure, discusses several specific receivers that can be explained within this structure, and provides a digital receiver implementation that will be useful in the discussion of adaptive systems later in the chapter.

As noted above, complexity is a major issue in multi-user receiver design and implementation, and the remainder of this chapter addresses the problem of complexity reduction in multi-user MIMO systems. This complexity takes two forms: computational, or implementational, complexity; and informational complexity.

The first type of complexity refers to the amount of resources needed to implement a given receiver algorithm. Optimal MIMO multi-user receiver algorithms are typically prohibitively complex in this sense, and thus a major issue in this area is complexity reduction. Sections 6.3 and 6.4 address the principal method for complexity reduction in practical multi-user receivers, namely the use of iterative algorithms in which tentative decisions are made and updated iteratively. There are a number of basic iterative techniques, involving different trade-offs between complexity and performance, and depending on the type of system under consideration, and these are described in Section 6.3. In Section 6.4, we tackle the additional complexity that arises in receiving space–time coded transmissions, such as those described in Chapter 4, in multi-user MIMO systems. Here, iterative algorithms similar to those discussed in Chapter 5 provide the answer to finding algorithms that can exploit the space–time coded structure with only moderate increases in complexity.

The second type of complexity refers to the amount of knowledge that a given receiver needs to have about the structure of received signals in order to effect signal reception. Although, as we will see shortly, optimal MIMO multi-user reception requires knowledge of the waveforms being transmitted by all users sharing the channel and the structure of the physical channel intervening between transmitters and the receiver, this type of knowledge is rarely available in practical wireless multi-user systems. Thus, it is necessary to consider adaptive receiver algorithms that can operate without such knowledge, or with only limited such knowledge. Such algorithms are the topic of Section 6.5, in which the structure of adaptive MIMO multi-user receivers is reviewed.

The chapter will conclude in Sections 6.6 and 6.7 with a summary and pointers to additional reading of interest in this general area.

6.2 Multiple-access MIMO systems

As noted above, this section will provide a general treatment of the multi-user MIMO receiver design problem. Here we will focus on modeling and on the structure of optimal receivers. In doing so, we will expose the principal issues underlying the reception of signals in multi-user MIMO systems, and also will set the stage for more practical algorithms developed in succeeding sections.

6.2.1 Signal and channel models

In order to discuss multi-user MIMO receiver structures, it is useful to first specify a general model for the signal received by a MIMO receiver in a multi-user environment (see Fig. 6.1). In doing so, we will build on the signaling model developed in Chapter 1,

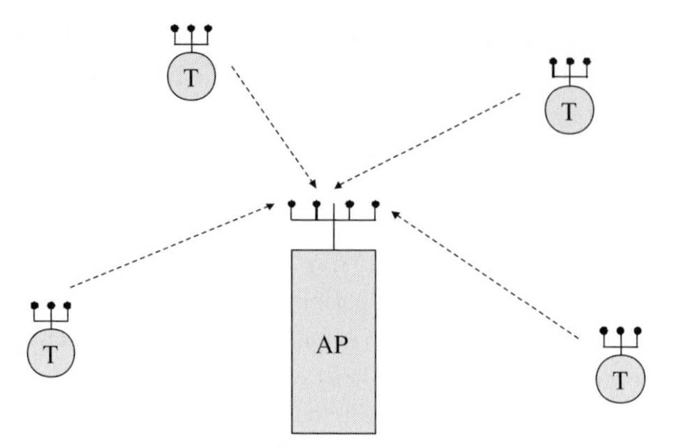

Figure 6.1. A multi-user MIMO system.

and in particular our model is an abstraction of the physical channel described there that is especially useful for the purposes of this chapter. Specifically, a useful received signal model for a multi-user MIMO system having K active users, M_T transmit antennas and M_R receive antennas, and transmitting over a frame of B symbol periods, can be written as follows:

$$r_p(t) = \sum_{k=1}^{K} \sum_{m=1}^{M_T} \sum_{i=0}^{B-1} b_{k,m}[i] g_{k,m,p}(t - iT_s) + n_p(t), \qquad p = 1, \ldots, M_R, \tag{6.1}$$

where the various quantities are as follows:

- $r_p(\cdot)$ = the signal received at the output of the pth receive antenna,
- $b_{k,m}[i]$ = the symbol transmitted by user k from its mth antenna in the ith symbol interval,
- $g_{k,m,p}(\cdot)$ = the waveform on which symbols from the mth antenna of user k arrive at the output of the pth receive antenna,
- T_s = the symbol period, and
- $n_p(\cdot)$ = ambient noise at the pth receive antenna.

Each of the waveforms $g_{k,m,p}(\cdot)$ can be modeled as

$$g_{k,m,p}(t) = \int_{-\infty}^{\infty} s_{k,m}(u) f_{k,m,p}(t - u) \, du, \tag{6.2}$$

where

- $s_{k,m}(\cdot)$ = the signaling waveform used by user k on its mth antenna and
- $f_{k,m,p}(\cdot)$ = the impulse response of the channel between the mth transmit antenna of user k and the pth receive antenna output.

Thus, we are assuming linear modulation and a linear channel model, both of which are reasonable assumptions for wireless systems. Note that, since $g_{k,m,p}(\cdot)$ does not depend on the symbol index i in this model, we are implicitly assuming here that the channel is stable (and time-invariant) over the transmission frame (which is BT_s seconds long) and that the transmitters use the same signaling waveforms in each symbol-period. The first of these assumptions is valid for the coherence times and signaling parameters arising in most systems of interest, while the latter is often violated, particularly in cellular systems. However, with the exception of the adaptive methods of Section 6.5, this time variation is not difficult to incorporate into any of the results described in this chapter, and is omitted here for the sake of notational simplicity (see, e.g., [46]).

In order to minimize the number of parameters in this model, we will assume that the signaling waveforms are normalized to have unit total energy, i.e.,

$$\int_{-\infty}^{\infty} \left[s_{k,m}(t)\right]^2 dt = 1, \qquad k = 1, \ldots, K, \quad m = 1, \ldots, M_T. \tag{6.3}$$

In reality, the actual transmitted waveforms will carry differing and non-unit energies, reflecting the transmitted powers of the various users' terminals. However, from the vantage point of *receiver* design, the critical scale parameter is the received power of a user, which will depend on the user's transmitter power and the gain of the intervening channel. Thus, it is convenient to lump all scaling of the signals into the channel impulse response $f_{k,m,p}(\cdot)$, and to simply assume normalized waveforms (6.3) at the transmitter. Again, from the receiver's point of view, it is impossible anyway to separate the effects of channel gain and transmit power on the received power. Also for convenience, we will assume that the transmitted waveforms have a duration of only a single symbol interval; i.e.,

$$s_{k,m}(t) = 0, \, t \notin [0, T_s]. \tag{6.4}$$

As with the normalization constraint (6.3), this assumption does not remove any generality since received waveforms that extend beyond a single symbol interval can be modeled via dispersion in the channel response.

A typical and useful model for the channel response is as a discrete multi-path model:

$$f_{k,m,p}(t) = \sum_{\ell=1}^{L} h_{k,m,p,\ell}\,\delta(t - \tau_{k,m,p,\ell}), \tag{6.5}$$

where $\delta(\cdot)$ denotes the Dirac delta function, and where $h_{k,m,p,\ell}$ and $\tau_{k,m,p,\ell} \geq 0$ denote the channel gain and propagation delay, respectively, of the ℓth path of the channel between the mth transmit antenna of user k and the output of the pth receive antenna.[1] In this case, the waveforms $g_{k,m,p}(\cdot)$ are of the form

$$g_{k,m,p}(t) = \sum_{\ell=1}^{L} h_{k,m,p,\ell}\, s_{k,m}(t - \tau_{k,m,p,\ell}). \tag{6.6}$$

[1] For simplicity, we lump the effects of the radio channel itself and the antenna response into the same term $h_{k,m,p,\ell}$. Often these two terms can be separated (see, e.g., [46]). However, no generality is lost in lumping these effects together for the purposes of analysis and exposition.

That is, in this model, the waveform received at a given receive antenna p from a given transmit antenna m of a particular user k is the superposition of L scaled and delayed copies of the waveform $s_{k,m}(\cdot)$ transmitted from that antenna. Except where noted otherwise, we will assume this particular model for the channel response in the following.

The signaling waveforms $s_{k,m}(\cdot)$ can take many forms. Although these waveforms can be thought of as being generic in our discussion, a quintessential example is the case in which the transmitted signals are in direct-sequence code-division multiple-access (DS/CDMA) format. This is a very widely used signaling format in wireless systems (used notably in both major 3G cellular standards), and is the example used in the simulations discussed in succeeding sections of this chapter. In the notation of this section, this format can be described as follows.

DS/CDMA signaling

In the DS/CDMA format, the signaling waveforms used by all transmitters are in the form of spread-spectrum signals; i.e., the waveforms $\{s_{k,m}(\cdot)\}$ of (6.1) are of the form

$$s_{k,m}(t) = \frac{1}{\sqrt{N}} \sum_{j=0}^{N-1} c_{k,m}^{(j)} \psi(t-(j-1)T_c), \qquad 0 \le t \le T_s, \tag{6.7}$$

where N is the spreading gain of the system, $c_{k,m}^{(0)}, c_{k,m}^{(1)}, \ldots, c_{k,m}^{(N-1)}$ is the spreading code (or signature sequence) associated with the mth transmit antenna of user k, $T_c = T_s/N$ is the chip interval, and $\psi(\cdot)$ is a chip waveform having unit-energy and approximate duration T_c. (For a general discussion of spread-spectrum signaling, see, e.g., [48].) In studying this format, the chip waveform $\psi(\cdot)$ is often modeled as a unit-energy pulse of duration T_c; i.e.,

$$\psi(t) = \begin{cases} \frac{1}{\sqrt{T_c}}, & t \in [0, T_c] \\ 0, & \text{otherwise}. \end{cases} \tag{6.8}$$

Again, most of the results of this chapter apply to general signaling waveforms, and it is not necessary to particularize to this specific format except where noted. It should also be mentioned that these signaling waveforms, the symbols, the noise, and the channel responses may be taken to be complex (rather than real as is tacitly assumed here). We will not need this generality here until Section 6.5, and so we will defer discussions of needed modifications (which are minor) until then. A complex version of the above model can be found in [46], which allows for two-dimensional signaling constellations, such as QPSK and QAM, to fit within this model.

As an additional assumption, we assume that the ambient noise processes $n_p(\cdot)$, $p = 1, \ldots, M_R$, are mutually independent white Gaussian processes with common spectral height σ^2. We also assume that the transmitted symbols take values in a finite alphabet $\mathcal{A}$ containing $|\mathcal{A}|$ elements. Beginning in Section 6.3, we will specialize this to the binary antipodal case $\mathcal{A} = \{-1, +1\}$. This is primarily for convenience, as most of the results in this chapter hold for more general signaling alphabets.

Finally, we note that M_T, B and L in the above model could vary from user to user, while L could also vary from antenna pair to antenna pair. However, again for simplicity, we will assume them to be constants, as the extensions of the discussions in the chapter to these non-constant cases are quite straightforward.

6.2.2 *Canonical receiver structure*

A basic MIMO multi-user receiver structure can be usefully decomposed into two parts: a front-end (or hardware) part and a decision algorithm (or software) part. In practice, these pieces may not be completely distinct, as much of the front-end may be implemented in software; but for the purposes of exposition, it is a useful decomposition.

A canonical front-end for such a system can be derived based on the theory of statistical inference. In particular, it is of interest to examine the so-called *likelihood function* of the observations (6.1) given the collection of transmitted symbols: $\{b_{k,m}[i]\}_{k=1,\ldots,K;\, m=1,\ldots,M_T;\, i=0,\ldots,B-1}$. Owing to the assumption of white, Gaussian noise, the logarithm of this likelihood function is given (up to a scalar multiple) by the Cameron–Martin formula [29] to be

$$\sum_{k=1}^{K}\sum_{m=1}^{M_T}\sum_{i=0}^{B-1} b_{k,m}[i]z_{k,m}[i] - \frac{1}{2}\sum_{k,k'=1}^{K}\sum_{m,m'=1}^{M_T}\sum_{i,i'=0}^{B-1} b_{k,m}[i]b_{k',m'}[i']C(k,m,i;k',m',i'), \tag{6.9}$$

where, for $k = 1, \ldots, K$, $m = 1, \ldots, M_T$, and $i = 0, \ldots, B-1$,

$$z_{k,m}[i] = \sum_{\ell=1}^{L}\sum_{p=1}^{P} h_{k,m,p,\ell}\int_{-\infty}^{\infty} r_p(t)s_{k,m}(t-\tau_{k,m,p,\ell}-iT_s)\,dt, \tag{6.10}$$

and for $k, k' = 1, \ldots, K$, $m, m' = 1, \ldots, M_T$, and $i, i' = 0, \ldots, B-1$,

$$\begin{aligned} C(k,m,i;k',m',i') = &\sum_{p=1}^{P}\sum_{\ell,\ell'=1}^{L} h_{k,m,p,\ell}h_{k',m',p,\ell'} \\ &\times \int_{-\infty}^{\infty} s_{k,m}(t-\tau_{k,m,p,\ell}-iT_s)s_{k',m'}(t-\tau_{k',m',p,\ell'}-i'T_s)\,dt. \end{aligned} \tag{6.11}$$

Although the expression (6.9) may seem somewhat complicated, the key thing to note about it is that the antenna outputs, $r_1(t), r_2(t), \ldots, r_P(t)$, enter into the likelihood function only through the collection of "observables" $\{z_{k,m}[i]\}_{k=1,\ldots,K;\, m=1,\ldots,M_T;\, i=0,\ldots,B-1}$. This means that this collection of variables is a *sufficient statistic* [29] for making inferences about the corresponding set of transmitted symbols $\{b_{k,m}[i]\}_{k=1,\ldots,K;\, m=1,\ldots,M_T;\, i=0,\ldots,B-1}$, which implies in turn that all attention can be restricted to this set of observables when designing and building systems or algorithms for demodulating and detecting the transmitted symbols.

Before turning to some types of algorithms that we might use for this purpose, it is worthwhile to examine the structure of this set of observables a bit more closely. In particular, it can be seen that (6.10) consists of three basic operations:

1. integration to obtain: $x_{k,m,p,\ell}[i] = \int_{-\infty}^{\infty} r_p(t) s_{k,m}(t - \tau_{k,m,p,\ell} - iT_s)\, dt$;
2. correlation to obtain: $y_{k,m,\ell}[i] = \sum_{p=1}^{P} h_{k,m,p,\ell}\, x_{k,m,p,\ell}[i]$; and
3. summation to obtain: $z_{k,m}[i] = \sum_{\ell=1}^{L} y_{k,m,\ell}[i]$.

The first operation is a *matched filtering* operation, so that we see that each received antenna output is filtered with a filter that is matched to the waveform received on each path from each transmit antenna in each symbol interval of each user. Thus, there are $K \times M_R \times B \times L \times M_T$ matched filter outputs, which we can think of as being produced by a bank of linear filters, each of which is sampled at the end of each signaling interval; i.e., samples are taken at times iT_s for $i = 0, \ldots, B-1$.

The second operation, in which the matched filter outputs $\{x_{k,m,p,\ell}[i]\}$ are correlated across the receive antenna array with the channel/antenna gains $\{h_{k,m,p,\ell}\}$, can be viewed as a form of *beamforming*, through which the spatial dimension afforded by the receive array is exploited. Since the terms $h_{k,m,p,\ell}$ also incorporate channel gains, this is not strictly a simple beamforming operation in general, but it has a similar effect of coherently collapsing the spatial dimension of the array. Note that, after beamforming, there are $K \times B \times L \times M_T$ observables.

Finally, the third operation, in which the beamformer outputs $\{y_{k,m,\ell}[i]\}$ are added, is a *multi-path combiner*, or Rake operation through which the spatial dimension introduced by the multi-path channel is exploited. Typically a Rake receiver also includes a correlation with the channel multi-path coefficients. This is being done here as part of the beamforming operation. So, the combination of the second and third operations is equivalent to beamforming followed by Rake combining, and this combination might be decomposed in other ways in practice. After this third operation, there are $K \times M_T \times B$ observables, one for each symbol in the frame of each user.

These three operations constitute the (hardware) front-end of a canonical multi-user receiver, as illustrated in Fig. 6.2. This front-end is sometimes known as a *space–time matched filter*. Note that, although this structure may seem complicated, it is essentially composed of standard communication-system components: matched filters, beamformers, and Rake receivers.

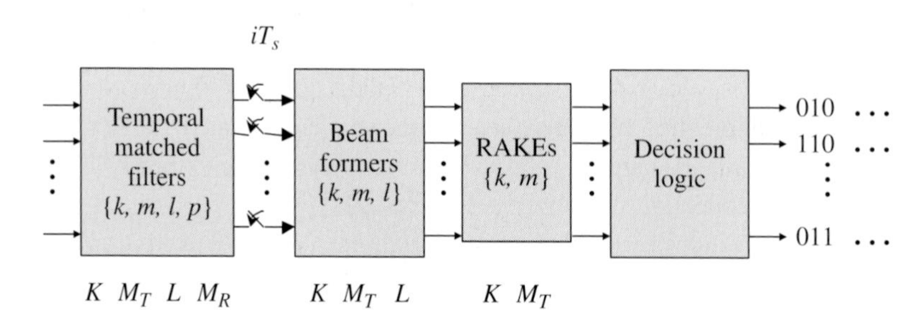

Figure 6.2. A canonical MIMO multi-user receiver structure.

It is noteworthy that this formalism and general front-end structure encompasses three standard interference-mitigation problems in communications. To discuss this point, it is useful to define the parameter

$$\Delta = \left\lceil \frac{\max_{k,m,p,\ell}\left\{\tau_{k,m,p,\ell}\right\}}{T_s} \right\rceil, \tag{6.12}$$

where $\lceil x \rceil$ denotes the smallest integer not less than x. Δ is the maximum delay spread of the wireless channels (6.5) in units of symbol intervals, and is thus the maximum extent to which symbols of a given user interfere with one another. Returning to the general receiver structure, the case in which $K = M_T = 1$ and $\Delta > 1$ is the channel equalization problem studied notably in the 1970s; the case $M_T = \Delta = 1$ and $K > 1$ is the traditional multi-user detection problem, studied notably in the 1980s; and finally the case in which $K = \Delta = 1$ and $M_T > 1$ is the standard MIMO communications problem, exemplified by the BLAST architecture studied notably in the 1990s. Combinations of these problems and refinements on them have been mainstays of research and development in digital communications throughout the past few decades and continuing to the present day. The applicability of the results in this chapter to these various problems, both individually and jointly, is worth keeping in mind in the subsequent discussions. Thus, the receiver architectures described herein can be applied other than in the multi-user MIMO communications setting, and many of them generalize solutions to the more particular cases noted above.

6.2.3 *Basic MUD algorithms*

As illustrated in Fig. 6.2, the KM_TB outputs of the canonical multi-user front-end are operated upon by a decision algorithm whose purpose is to infer the values of the KM_TB transmitted symbols $\{b_{k,m}[i]\}$. This decision algorithm can take many forms, ranging through the full toolbox of statistical signal processing: optimal algorithms based on maximum-likelihood or maximum *a posteriori* probability criteria, linear algorithms, iterative algorithms, and adaptive algorithms. Each of these techniques will be discussed briefly in the following paragraphs, and counterparts to these algorithms are discussed in Sections 6.3 and 6.5. However, before discussing these types of algorithms, it is useful to first examine the relationship between the observables $\{z_{k,m}[i]\}$ and the corresponding symbols $\{b_{k,m}[i]\}$ to be inferred. To do so, it is convenient to collect the symbols into a KM_TB-long column vector $\mathbf{b}$ by sorting the symbols $\{b_{k,m}[i]\}$ first by symbol number, then by user number, and finally by antenna number. That is,

$$\mathbf{b} = \begin{pmatrix} \mathbf{b}[0] \\ \mathbf{b}[1] \\ \vdots \\ \mathbf{b}[N-1] \end{pmatrix} \tag{6.13}$$

where

$$\mathbf{b}[i] = \begin{pmatrix} \mathbf{b}_1[i] \\ \mathbf{b}_2[i] \\ \vdots \\ \mathbf{b}_K[i] \end{pmatrix} \tag{6.14}$$

with

$$\mathbf{b}_k[i] = \begin{pmatrix} \mathbf{b}_{k,1}[i] \\ \mathbf{b}_{k,2}[i] \\ \vdots \\ \mathbf{b}_{k,M_T}[i] \end{pmatrix}. \tag{6.15}$$

Similarly, we can denote by $\mathbf{z}$ the set of observables $\{z_{k,m}[i]\}$ collected into a KM_TB-long column vector indexed conformally with $\mathbf{b}$. We can also define a $KM_TB \times KM_TB$ cross-correlation matrix $\mathbf{R}$ whose (n, n')th element is given by the cross-correlation $C(k, m, i; k', m', i')$ from (6.11) where the indices are determined by matching with the corresponding elements of $\mathbf{b}$ (or, equivalently, $\mathbf{z}$); i.e., $b_n = b_{k,m}[i]$ and $b_{n'} = b_{k',m'}[i']$ with $n = [iK + (k-1)]M_T + m$ and $n' = [i'K + (k'-1)]M_T + m'$.

With these definitions, the observables and transmitted symbols can be related to one another through the relationship

$$\mathbf{z} = \mathbf{R}\mathbf{b} + \mathbf{n}, \tag{6.16}$$

where $\mathbf{n}$ denotes a KM_TB-long noise vector having the $\mathcal{N}(\mathbf{0}, \sigma^2\mathbf{R})$ distribution. (Here, $\mathbf{0}$ denotes a KM_TB-long vector having all components equal to zero.)

As a simple example, we can consider the flat-fading, synchronous case, in which all signals arrive at the receive array with the same symbol timing. This corresponds to the discrete multi-path model of (**??**) with $L = 1$ and $\tau_{k,m,p,\ell} \equiv 0$.

$$f_{k,m,p}(t) = h_{k,m,p,1}\delta(t). \tag{6.17}$$

In this case, the matrix $\mathbf{R}$ is a block-diagonal matrix having B identical blocks along its diagonal, each of dimension $KM_T \times KM_T$. These square sub-matrices contain the cross-correlations between the signals received from the different antennas of the different users. So, for example, in this case, the first block is given by

$$\mathbf{R}_{n,n'} = \int_{\infty}^{\infty} s_{k,m}(t)s_{k',m'}(t)\,dt \times \sum_{p=1}^{P} h_{k,m,p,1}h_{k',m',p,1}, \quad n, n' = 1, 2, \ldots, KM_T, \tag{6.18}$$

where the indices n and n' correspond, respectively, to antenna m of user k and antenna m' of user k', both in the zeroth symbol interval. This block is then repeated B times along the diagonal of $\mathbf{R}$. This example is further illuminated by considering the single-receive-antenna case ($M_R = 1$), in which this first diagonal block simplifies to

$$\mathbf{R}_{n,n'} = \int_{\infty}^{\infty} s_{k,m}(t)s_{k',m'}(t)\,dt\,A_{k,m}A_{k',m'}, \qquad n, n' = 1, 2, \ldots, KM_T, \tag{6.19}$$

with $A_{k,m} = h_{k,m,1,1}$ for $k, m = 1, \ldots, K, M_T$, $n = (k-1)M_T + m$, and $n' = (k'-1)M_T + m'$. This block is thus of the form

$$\mathbf{A}\overline{\mathbf{R}}\mathbf{A} \tag{6.20}$$

where $\mathbf{A}$ is a diagonal matrix having the received amplitudes $A_{1,1}, \ldots, A_{1,M_T}, A_{2,1}, \ldots, A_{2,M_T}, A_{K,1}, \ldots, A_{K,M_T}$ on its diagonal, and where $\overline{\mathbf{R}}$ is the *normalized* cross-correlation matrix of the signaling multiplex:

$$\overline{\mathbf{R}}_{n,n'} = \int_{\infty}^{\infty} s_{k,m}(t)s_{k',m'}(t)dt, \qquad n, n' = 1, 2, \ldots, KM_T. \tag{6.21}$$

For example, in the DS/CDMA case of (6.7) and (6.8), this normalized cross-matrix is given by

$$\overline{\mathbf{R}}_{n,n'} = \frac{1}{N}\sum_{j=0}^{N-1} c_{k,m}^{(j)} c_{k',m'}^{(j)}, \qquad n, n' = 1, 2, \ldots, KM_T; \tag{6.22}$$

that is, the normalized cross-correlation matrix is determined by the cross-correlations of the spreading sequences used by the system. The specific structure of this matrix depends on how the spreading sequences are allocated to the various users' antennas. In some systems, all antennas of the same user use the same spreading code, while in others, different spreading codes are used for all antennas. As an example, if the spreading codes are so-called m-sequences (see, e.g., [48]), then $\overline{\mathbf{R}}_{n,n'} = 1$ for antennas using identical spreading codes and $\overline{\mathbf{R}}_{n,n'} = -1/N$, for antennas (and users) using different spreading codes.

In the general case in which the channel is not flat or the users are not synchronous, the block diagonal form of this example becomes a block Toeplitz form, as will be discussed in Section 6.3. From (6.16) we see that the basic relationship between $\mathbf{z}$ and $\mathbf{b}$ is that of a noisy linear model, and so the basic problem to be solved by the decision algorithm in Fig. 6.2 is that of fitting such a model. At first glance, this appears to be a rather straightforward problem, as the fitting of linear models is a classical problem in statistics. However, the difficulty in this problem arises because the vector $\mathbf{b}$ to be chosen in this fit has discrete-valued elements (e.g., ± 1), and this significantly increases the complexity of fitting this model (6.16).

In general, the most powerful techniques for data detection are maximum-likelihood (ML) and maximum *a posteriori* probability (MAP) detection. ML detection makes inferences about the transmitted symbols in (6.1) by choosing those symbol values that maximize the log-likelihood function of (6.9). To get a sense of this task, it is useful to use the compact notation of (6.16) to re-write the log-likelihood function (6.9) as

$$\mathbf{b}^T\mathbf{z} - \frac{1}{2}\mathbf{b}^T\mathbf{R}\mathbf{b}. \tag{6.23}$$

So, the ML symbol decisions solve the optimization problem:

$$\max_{\mathbf{b}\in\mathcal{B}}\left[\mathbf{b}^T\mathbf{z}-\frac{1}{2}\mathbf{b}^T\mathbf{R}\mathbf{b}\right], \tag{6.24}$$

where $\mathcal{B}=\mathcal{A}^{KM_TB}$. The optimization problem (6.24) is an integer quadratic program, which is known to be an NP-complete computational problem. Since the size of the search set $\mathcal{B}$ is potentially enormous at $|\mathcal{A}|^{KM_TB}$, solving this problem appears to be impossible.[2] However, for most practical wireless channels, the matrix $\mathbf{R}$ has many zero elements which reduces the complexity of this problem significantly. In particular, assuming that the signaling waveforms $\{s_{k,m}(\cdot)\}$ are limited in duration to a single symbol interval, and given the finite multi-path channel model (6.5), the matrix $\mathbf{R}$ is a banded matrix, meaning that all of its elements are zero except on a certain number of diagonals; i.e., $\mathbf{R}_{n,n'}=0$ if $|n-n'|>KM_T\Delta$, where again Δ is the maximum delay spread of the wireless channels (6.5) in units of symbol intervals (6.12). This bandedness allows for a complexity reduction from the order of $|\mathcal{A}|^{KM_TB}$ needed to exhaustively search for the ML solution, to the order of $|\mathcal{A}|^{KM_T\Delta}$ (per symbol) to search via dynamic programming (see, e.g., [30]). Although in most wireless channels the maximum delay spread Δ is much less than the frame length B, even this reduced complexity is prohibitive for most applications as the exponent $KM_T\Delta$ could still be fairly large in a typical situation with dozens of users, a few antennas per user, and a few symbols of delay spread. The ML detector is sometimes referred to as the jointly optimal (JO) detector.

MAP detection is applicable to situations in which the receiver knows a prior probability distribution governing the values that the transmitted symbols may assume. In this situation, it is possible to consider the posterior probability distribution of a given symbol, conditioned on the observations, and to infer that value for each symbol that has maximum *a posteriori* probability (APP). That is, a given symbol, say b_n is detected as $\hat{b}_n$ according to the following criterion:

$$\hat{b}_n=\arg\left\{\max_{a\in\mathcal{A}}P(b_n=a|\mathbf{z})\right\}. \tag{6.25}$$

Using Bayes' formula, we can write the APP as

$$P(b_n=a|\mathbf{z})=\frac{\sum_{\mathbf{b}\in\mathcal{B}_{n,a}}\ell(\mathbf{z}|\mathbf{b})w(\mathbf{b})}{\sum_{\mathbf{b}\in\mathcal{B}}\ell(\mathbf{z}|\mathbf{b})w(\mathbf{b})}, \tag{6.26}$$

were $\mathcal{B}_{n,a}$ denotes the subset of $\mathcal{B}$ in which the nth coordinate is fixed at a, $w(\mathbf{b})$ is the prior probability of $\mathbf{b}$, and $\ell(\mathbf{z}|\mathbf{b})$ denotes the likelihood function of $\mathbf{z}$ given $\mathbf{b}$:

$$\ell(\mathbf{z}|\mathbf{b})=e^{\left(\mathbf{b}^T\mathbf{z}-\frac{1}{2}\mathbf{b}^T\mathbf{R}\mathbf{b}\right)/\sigma^2}. \tag{6.27}$$

Commonly, it is assumed that the symbol vector $\mathbf{b}$ is uniformly distributed in its range $\mathcal{B}$; i.e., that

$$w(\mathbf{b})\equiv|\mathcal{A}|^{-KM_TB}. \tag{6.28}$$

[2] Typically, K might be dozens, M_T several, and B hundreds.

This assumption is equivalent to assuming that all the symbols are independent and identically distributed (i.i.d.) from time to time, from user to user, and from antenna to antenna, and that each symbol is chosen equiprobably among the elements of $\mathcal{A}$. This assumption is not always valid, as we will discuss below. However, when it is valid, the prior distribution drops out of the computation of the APP, and the MAP criterion becomes

$$\hat{b}_n = \arg\left\{\max_{a\in\mathcal{A}} \sum_{\mathbf{b}\in\mathcal{B}_{n,a}} \ell(\mathbf{z}|\mathbf{b})\right\}. \tag{6.29}$$

(Note that the denominator in the APP (6.26) is irrelevant to the maximization since it does not depend on the value of any individual symbol.) The MAP detector is sometimes termed the individually optimal (IO) detector since it chooses each symbol decision according to a single-symbol criterion.

Like the ML detector, the computation of symbol decisions using (6.29) is generally prohibitively complex. In particular, we note that computation of the APP for each individual symbol value involves a summation over $|\mathcal{A}|^{KM_TB-1}$ values of the symbol vector. Also like ML detection, however, this complexity can be reduced via dynamic programming to the order of $|\mathcal{A}|^{KM_T\Delta}$ operations per symbol when the channel has delay spread of Δ symbol intervals [30, 39].

As we see from the above discussion, the basic complexity of ML (JO) or MAP (IO) data detection is quite complex, on the order of $|\mathcal{A}|^{KM_T\Delta}$ operations per detected symbol. So, the complexity grows with the number of users, the number of antennas, and the channel length. It is noteworthy that this issue is present even in the single-user ($K = 1$) case, the single-antenna case ($M_T = 1$), or in the flat-fading case ($\Delta = 1$). Only if all of these conditions is missing do we get a simple detector structure, which reduces in either the ML or MAP case to a simple quantization:

$$\hat{b}_n = Q(z_n), \tag{6.30}$$

where the quantizer $Q : \mathbb{R} \rightarrow \mathcal{A}$. For example, in the case of binary antipodal symbols ($\mathcal{A} = \{-1, +1\}$), we take Q to be the signum function:

$$Q(z) = \text{sgn}(z) = \begin{cases} -1 & z < 0 \\ +1 & z \geq 0. \end{cases} \tag{6.31}$$

Since data detection must be performed on a relatively limited computing platform (i.e., a communications receiver) at essentially the rate of data transmission (i.e., tens to thousands of kilobits per second), it is of interest to consider alternatives to the optimal detectors described above. One family of such detectors are the *linear* multi-user detectors, which seek to balance the simplicity of the simple detector (6.30) with the power of IO or JO detection. In linear detection, this is accomplished by first multiplying the sufficient statistic $\mathbf{z}$ by a suitably chosen square matrix, and then quantizing the result:

$$\hat{b}_n = Q(v_n), \tag{6.32}$$

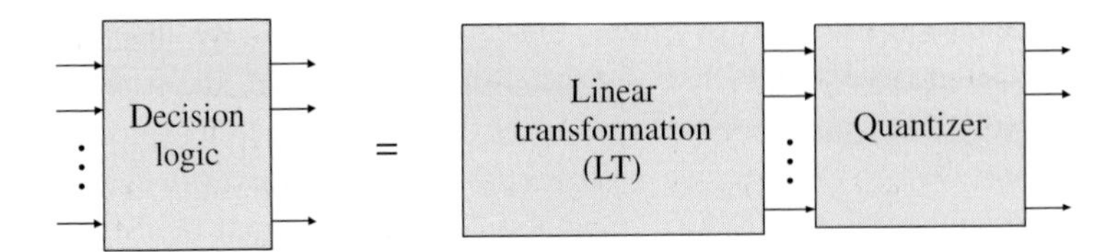

Figure 6.3. Linear multi-user algorithm.

where

$$\mathbf{v} = \mathbf{M}\mathbf{z} \tag{6.33}$$

and where $\mathbf{M}$ is a $KM_TB \times KM_TB$ matrix. This type of detector is illustrated in Fig. 6.3. Various types of detectors can be implemented through different choices of the matrix $\mathbf{M}$. Three key ones can be described as follows.

Space–time matched filter/rake receiver

The simplest example of a linear detector arises from choosing $\mathbf{M}$ to be the $KM_TB \times KM_TB$ identity matrix $\mathbf{I}$, in which case the linear detector reduces to the simple detector of (6.30). This detector is a classical space–time matched filter receiver which is optimal in an additive white Gaussian noise (AWGN) channel. A flaw of this receiver is that it addresses only the ambient noise, while ignoring the cross-correlations between the signals affecting different symbols; i.e., it ignores the off-diagonal elements of $\mathbf{R}$.

Decorrelating (zero-forcing) receiver

Noting from (6.16) that the mapping from transmitted symbols $\mathbf{b}$ to the observables $\mathbf{z}$ is in the form of a (square) linear transformation plus noise, a natural detection strategy is a zero-forcing detector that eliminates the interference embodied in the cross-correlation matrix $\mathbf{R}$. Assuming that $\mathbf{R}$ is non-singular, this can be implemented as a linear detector with $\mathbf{M} = \mathbf{R}^{-1}$. The resulting detector is known as the *decorrelator*. The decorrelator thus quantizes the variables $\mathbf{v} = \mathbf{R}^{-1}\mathbf{z}$, which are given by

$$\mathbf{v} = \mathbf{b} + \mathbf{R}^{-1}\mathbf{n}. \tag{6.34}$$

Note that, as expected, these transformed observables are free of (inter-user, inter-antenna, and inter-symbol) interference. However, this receiver is the opposite extreme of the matched filter receiver, in that it is tantamount to ignoring the ambient noise to suppress the interference. Using standard properties of the multivariate Gaussian distribution, the noise terms in (6.34) are distributed according to

$$\mathbf{R}^{-1}\mathbf{n} \sim \mathcal{N}\left(\mathbf{0}, \sigma^2\mathbf{R}^{-1}\right). \tag{6.35}$$

Depending on the structure of $\mathbf{R}$ the inverse $\mathbf{R}^{-1}$ can have very large diagonal values, leading to noise enhancement and consequently a high error rate. (This problem is well-known in the context of equalization [33].) The assumption that $\mathbf{R}$ be invertible is not overly restrictive in general, as $\mathbf{R}$ is at least non-negative definite. However, there are non-trivial situations in which $\mathbf{R}$ can be singular, in which case the decorrelator is not a viable structure.

MMSE receiver

While the matched filter addresses the ambient noise and the decorrelator addresses the interference, the minimum-mean-square-error (MMSE) multi-user detector effects a compromise between these two impairments by selecting the transformation $\mathbf{M}$ such as the vector $\mathbf{v} = \mathbf{Mz}$ is an MMSE estimate of the symbol vector $\mathbf{b}$. For this criterion to make sense, it is necessary to provide a prior model for $\mathbf{b}$. On making the common assumption that the elements of $\mathbf{b}$ are of zero-mean and mutually uncorrelated, the MMSE detector corresponds to the following choice of the matrix $\mathbf{M}$:

$$\mathbf{M} = \left(\mathbf{R} + \sigma^2 \mathbf{I}\right)^{-1} \tag{6.36}$$

where, as before, $\mathbf{I}$ denotes the $KM_TB \times KM_TB$ identity matrix. Note that, it is clear from this form that the MMSE detector represents a compromise between the matched filter ($\mathbf{M} = \mathbf{I}$) and the decorrelator ($\mathbf{M} = \mathbf{R}^{-1}$), in which the action of each is tempered with the action of the other. The relative mix of these two is controlled by the noise level (or more properly by the signal-to-noise ratio (SNR), as the signal strength is incorporated into $\mathbf{R}$). When the interference is dominant (i.e., for high SNR), the MMSE detector mimics the decorrelator, while when the ambient noise is dominant (i.e., for low SNR) it mimics the matched filter. More generally, it balances between these two.

In general, the complexity of linear multi-user detectors is that of matrix inversion, which is on the order of $(KM_TB)^3$. As with the ML and MAP solutions, this complexity can be reduced by exploiting bandedness in the case of short delay spread. In some cases, this complexity may also be amortized over many frames. However, for most wireless systems, such amortization is not possible as the signaling waveforms, the user population, or the channel parameters may change from frame to frame. Thus, although the order of complexity here has been reduced from exponential to polynomial, complexity is still a concern for practical systems. Moreover, in both linear and nonlinear cases, constraints on the transmitted symbols imposed by space–time coding or temporal channel coding can add to this complexity substantially [30].

For these reasons, a number of other techniques for multi-user reception have been developed, with the objective of reducing computational complexity while maintaining good performance in the presence of multiple-access interference. The principal technique for doing this is to make use of iterative algorithms to fit the linear model (6.16). This can be done either linearly with a final quantization (i.e., iterative linear detection), or nonlinearly with inter-iteration quantization (sometimes known as interference cancellation). Section 6.3 will address this issue in some detail for multi-user MIMO systems. When further complexity is introduced by channel coding, iterative algorithms such as those described in Chapter 5 (in this case "turbo" style algorithms) again allow for excellent performance with moderate complexity. This topic is addressed in Section 6.4.

As noted above, another form of complexity is *informational* complexity, which arises from the need to know the received waveforms $\{g_{k,m,p}(\cdot)\}$ in the model (6.1) for the received signal. There are two potential problems with this requirement. One is that the channels intervening the transmitters and receiver are typically dynamic and behave in an apparently random fashion. So, the channel parameters (assuming the channel can be parameterized) are not readily known to the receiver. Another problem is that the signaling waveforms of all users may also not be known to the receiver, because, for example, the receiver may only be intended to receive a subset of the users. In either case, it is thus necessary for the receiver to be able to adapt itself to those properties of the signaling environment that it does not know. Receiver structures for this purpose are described in Section 6.5. In preparation for this latter treatment, we turn briefly, in the following subsection, to a discrete-time model for the received signals considered above that is more suitable for developing and discussing such adaptive receiver algorithms.

6.2.4 *Digital receiver implementation*

For receiver implementation, and particularly for the adaptive algorithms to be discussed in Section 6.5, it is useful to consider a digital representation of the signals and observables that we have described in the preceding paragraphs. This type of representation is typically obtained by projecting the received signals (6.1) onto a finite set of functions arising from a model in which there are finitely many degrees of freedom in the signals of interest. (Most practical signaling methods have this property.) In this subsection, we will particularize the above structures for this situation, and in particular will consider the common case in which the signaling waveforms are in the DS/CDMA format, described above and in Chapter 1. This model will then be used exclusively in Section 6.5. It should be noted, however, that similar techniques can be applied in any system allowing for a finite-degree-of-freedom model. A notable alternative example to DS/CDMA is the case of orthogonal frequency-division multiple-access (OFDMA) systems, in which the incoming signal can be decomposed along orthogonal sub-carrier signals using the discrete Fourier transform (DFT).

Recall that, in the DS/CDMA format, the signaling waveforms used by all transmitters are in the form (6.7). Here, we consider this format in the particular case where the chip waveform is the unit pulse of (6.8). For this type of system, a natural set of observables can be obtained by projecting the received signals of (6.1) onto time shifts of the chip waveform $\psi(\cdot)$:

$$r_p[j] = \int_{\infty}^{\infty} r_p(t)\psi(t - jT_s)\,dt = \int_{jT_c}^{(j+1)T_c} r_p(t)\,dt, \qquad j = 0, 1, \ldots . \tag{6.37}$$

If the system delays are all integer multiples of a chip interval (this is termed the "chip-synchronous" case), then no information is lost in this operation, as the outputs of the matched filter bank of Fig. 6.2 and hence the sufficient statistic $\mathbf{z}$ can be extracted from these observables. In the chip-asynchronous case, inferential information may be lost in performing this operation. However, this loss is often minimal and the signal-processing advantages of reducing the observations to a discrete-time sequence outweigh this. (An alternative for the chip-asynchronous case is to integrate over shorter time intervals and thus effectively to over-sample the signal; however, we will not consider this level of detail here. For further discussion, see [46].)

As noted above, in the chip-synchronous case, the sufficient statistic $\mathbf{z}$ can be written as a function of the observables $\{r_p[j]\}$ and thus the ML and MAP detectors are functions of these observables, as are the linear detectors described in the preceding subsection. In the latter case it is sometimes convenient to combine all of the linear processing of the receiver front end and the decision algorithm of Fig. 6.2 into a single linear transformation, in which case symbol detection is of the form

$$\hat{b}_{k,m}[i] = Q\left(\sum_{p=1}^{M_R} \sum_{j} w_{k,m,p}^{(j)}[i] r_p[j] \right), \tag{6.38}$$

where the coefficients $\left\{ w_{k,m,p}^{(j)}[i] \right\}$ are chosen appropriately. This structure is one that can be adapted using standard adaptive algorithms to adjust the weighting coefficients. Although there are a number of issues surrounding such an adaptation, such as the decomposition of spatial and temporal combining, this structure is the essence of many adaptive algorithms for multiantenna, multi-user receiver design. An extensive treatment of this problem can be found in [46], and we will consider particularly the MIMO case in Section 6.5.

6.3 Iterative space–time multi-user detection

Advanced signal processing such as multi-user detection, typically improves system performance at the cost of computational complexity. As noted in Section 6.1, the optimal maximum-likelihood multi-user detector has prohibitive computational requirements for most current applications, and consequently a variety of linear and nonlinear multi-user detectors have been proposed to ease this computational burden while maintaining satisfactory performance [38, 46]. However, in many situations where the combined system has large dimensions (e.g., large array size, large delay spread, large user population, and combinations of these conditions), direct implementation of these sub-optimal techniques still proves to be very complex. In this section, we discuss iterative techniques for efficient space–time multi-user detection in MIMO systems [7, 8, 45]. Iterative methods are among the most practical techniques for multi-user detection. For example, an implementation for 3G cellular systems is described in [19].

6.3.1 System model

As noted in Section 6.1, we can restrict attention to the following system model (i.e., Eq. (6.16)):

$$\mathbf{z} = \mathbf{R}\mathbf{b} + \mathbf{n}, \tag{6.39}$$

where $\mathbf{R}$ is the cross-correlation matrix, $\mathbf{b}$ is the symbol vector, and $\mathbf{n}$ is the background noise at the input to the decision algorithm of Fig. 6.2. An optimal ML space–time multi-user detector will maximize the log-likelihood function of (6.23), and the computational complexity of this maximization is a major concern, particularly when the system dimension is large. In the following, we will use a multi-path CDMA channel for illustration purpose, but the techniques discussed can be readily applied to other equivalent MIMO scenarios as well. In principle, the computational complexity of ML detection grows exponentially with the size of $\mathbf{R}$, which for a multi-path MIMO multi-user channel is proportional to the number of users K, the number of transmit antennas M_T, and the data frame length B. As the data frame length is typically much larger than the multi-path delay spread Δ, $\mathbf{R}$ exhibits a block Toeplitz structure exemplified as

$$\mathbf{R} \equiv \begin{bmatrix} \underline{R}^{[0]} & \underline{R}^{[1]} & \cdots & \underline{R}^{[\Delta]} & & & \\ \underline{R}^{[-1]} & \underline{R}^{[0]} & \underline{R}^{[1]} & \cdots & \underline{R}^{[\Delta]} & & \\ & \underline{R}^{[-\Delta]} & \cdots & \underline{R}^{[0]} & \cdots & \underline{R}^{[\Delta]} & \\ & & \underline{R}^{[-\Delta]} & \cdots & \underline{R}^{[-1]} & \underline{R}^{[0]} & \underline{R}^{[1]} \\ & & & \underline{R}^{[-\Delta]} & \cdots & \underline{R}^{[-1]} & \underline{R}^{[0]} \end{bmatrix}. \tag{6.40}$$

As noted in Section 6.1, dynamic programming can be used to reduce the computational complexity of ML detection to $O(|\mathcal{A}|^{KM_T\Delta})$ per transmitted symbol. This computational requirement is still prohibitive except for very small values of $|\mathcal{A}|$, M_T, Δ, and K.

6.3.2 Iterative linear space–time multi-user detection

In this section, we consider the application of iterative processing to the implementation of various linear space–time multi-user detectors in algebraic form. After an introduction to the general form of linear space–time multi-user detection (ST MUD), we go on to discuss two general approaches to solving large systems of linear equations iteratively. Subsequent sections will treat nonlinear iterative methods.

As noted in Section 6.1, linear multi-user detectors in the framework of (6.39) are of the form

$$\hat{\mathbf{b}} = \text{sgn}(\text{Re}\{\mathbf{M}\mathbf{z}\}), \tag{6.41}$$

where $\mathbf{M}$ is a linear detection matrix. For the linear decorrelating (zero-forcing) detector, this matrix is given by

$$\mathbf{M}_d = \mathbf{R}^{-1}, \tag{6.42}$$

while for the linear minimum-mean-square-error (MMSE) detector, we have

$$\mathbf{M}_m = (\mathbf{R} + \sigma^2 \mathbf{I})^{-1}. \tag{6.43}$$

Direct inversion of the matrices in (6.42) and (6.43) (after exploiting the block Toeplitz structure) is of complexity $O((KM_T)^2 B\Delta)$ per user per symbol.

The linear multi-user detection estimates of (6.41) can be seen as the solution of a linear equation

$$\mathbf{C}\mathbf{v} = \mathbf{z}, \tag{6.44}$$

with $\mathbf{C} = \mathbf{R}$ for the decorrelating detector and $\mathbf{C} = \mathbf{R} + \sigma^2 \mathbf{I}$ for the MMSE detector. Jacobi and Gauss–Seidel iteration are two common low-complexity iterative schemes for solving linear equations such as (6.44) [14]. If we decompose the matrix $\mathbf{C}$ as $\mathbf{C} = \mathbf{C}_L + \mathbf{D} + \mathbf{C}_U$, where $\mathbf{C}_L$ denotes the lower triangular part, $\mathbf{D}$ denotes the diagonal part, and $\mathbf{C}_U$ denotes the upper triangular part, then Jacobi iteration can be written as

$$\mathbf{v}_m = -\mathbf{D}^{-1}(\mathbf{C}_L + \mathbf{C}_U)\mathbf{v}_{m-1} + \mathbf{D}^{-1}\mathbf{z}, \tag{6.45}$$

and Gauss–Seidel iteration is represented as

$$\mathbf{v}_m = -(\mathbf{D} + \mathbf{C}_L)^{-1}\mathbf{C}_U \mathbf{v}_{m-1} + (\mathbf{D} + \mathbf{C}_L)^{-1}\mathbf{z}. \tag{6.46}$$

From (6.45), Jacobi iteration can be seen to be a form of linear parallel interference cancellation, the convergence of which is not guaranteed in general. One of the sufficient conditions for the convergence of Jacobi iteration is that $\mathbf{D} - (\mathbf{C}_L + \mathbf{C}_U)$ be positive definite. In contrast, Gauss–Seidel iteration, which (6.46) reveals to be a form of linear serial interference cancellation, converges to the solution of the linear equation from any initial value, under the mild conditions that $\mathbf{C}$ be symmetric and positive definite, which is always true for the MMSE detector.

Another general approach to solving the linear equation (6.44) involves the use of gradient methods, among which are steepest descent and conjugate gradient iteration [14]. Note that solving (6.44) is equivalent to minimizing the cost function

$$\Phi(\mathbf{v}) = \frac{1}{2}\mathbf{v}^{\mathsf{H}}\mathbf{C}\mathbf{v} - \mathbf{v}^{\mathsf{H}}\mathbf{z}. \tag{6.47}$$

The idea of gradient methods is to successively minimize this cost function along a set of directions $\{\mathbf{p}_m\}$ via

$$\mathbf{v}_m = \mathbf{v}_{m-1} + \alpha_m \mathbf{p}_m, \tag{6.48}$$

with

$$\alpha_m = \mathbf{p}_m^{\mathsf{H}}\mathbf{q}_{m-1} / \mathbf{p}_m^{\mathsf{H}}\mathbf{C}\mathbf{p}_m, \tag{6.49}$$

and

$$\mathbf{q}_m = -\nabla\Phi(\mathbf{v})|_{\mathbf{v}=\mathbf{v}_m} = \mathbf{z} - \mathbf{C}\mathbf{v}_m. \tag{6.50}$$

Different choices of the set $\{\mathbf{p}_m\}$ give different algorithms. If we choose the search direction $\mathbf{p}_m$ to be the negative gradient of the cost function $\mathbf{q}_{m-1}$ directly, this algorithm is the steepest descent method, global convergence of which is guaranteed. The convergence rate may be prohibitively slow, however, due to the linear dependence of the search directions, resulting in redundant minimization. If instead we choose the search direction to be C-conjugate as follows

$$\mathbf{p}_m = \arg\min_{\mathbf{p}\in\Lambda_{m-1}^{\perp}} \|\mathbf{p} - \mathbf{q}_{m-1}\|, \tag{6.51}$$

where $\Lambda_m = \text{span}\{\mathbf{C}\mathbf{p}_1, \ldots, \mathbf{C}\mathbf{p}_m\}$, then we have the conjugate gradient method, whose convergence is guaranteed and performs well when $\mathbf{C}$ is close to identity either in the sense of being a low-rank perturbation or in the sense of a norm. The computational complexity of Gauss–Seidel and conjugate gradient iteration are similar, which is on the order of $O(KM_T\Delta\overline{m})$ per user per symbol, where $\overline{m}$ is the number of iterations. The numbers of iterations required by the Gauss–Seidel and conjugate gradient methods to achieve a stable solution to the associated large system equations have been found to be of the same order in simulations.

6.3.3 *Iterative nonlinear space–time multi-user detection*

Nonlinear multi-user detectors are often based on bootstrapping techniques, which are iterative in nature. In this section, we will consider the iterative implementation of decision-feedback multi-user detection in the space–time domain. We also discuss briefly the implementation of multistage interference canceling ST MUD, which serves as a reference point for introducing a new expectation-maximization-(EM-) based iterative ST MUD, to be discussed in the next subsection. For simplicity, we now restrict the signaling alphabet to the binary antipodal set: $\mathcal{A} = \{-1, +1\}$.

Cholesky iterative decorrelating decision-feedback ST MUD

Decorrelating decision-feedback multi-user detection (DDF MUD) exploits the Cholesky decomposition $\mathbf{R} = \mathbf{F}^{\mathsf{H}}\mathbf{F}$, where $\mathbf{F}$ is a lower triangular matrix, to determine the feedforward and feedback matrix for detection via the algorithm

$$\hat{\mathbf{b}} = \text{sgn}(\mathbf{F}^{-\mathsf{H}}\mathbf{z} - (\mathbf{F} - \text{diag}(\mathbf{F}))\hat{\mathbf{b}}). \tag{6.52}$$

The discussion here applies readily to the implementation of MMSE decision-feedback multi-user detection as well.

Suppose we are interested in detecting symbol b_n. The purpose of the feedforward matrix $\mathbf{F}^{-\mathsf{H}}$ is to whiten the noise and decorrelate against the "future users"

$\{s_{n+1}, \ldots, s_{KM_TB}\}$; while the purpose of the feedback matrix $(\mathbf{F} - \operatorname{diag}(\mathbf{F}))$ is to cancel the interference from "previous users" $\{s_1, \ldots, s_{n-1}\}$. Note that the performance of DDF MUD is not uniform. While the first "user" is demodulated by its decorrelating detector, the last detected "user" will essentially achieve its single-user lower bound providing the previous decisions are correct. There is another form of Cholesky decomposition, in which the feedforward matrix $\mathbf{F}$ is upper triangular. If we were to use this form instead in (6.52), then the multi-user detection would operate in the reverse order, as would the performance. The idea of *Cholesky iterative DDF ST MUD* is to employ these two forms of Cholesky decomposition alternatively as follows. For lower triangular Cholesky decomposition $\mathbf{F}_1$, first feedforward filtering is applied as

$$\bar{\mathbf{z}}_1 = \mathbf{F}_1^{-\mathsf{H}} \mathbf{z}, \tag{6.53}$$

where it is readily shown that $\bar{z}_{1,i} = \mathbf{F}_{1,ii} b_i + \sum_{j=1}^{i-1} \mathbf{F}_{1,ij} b_j + \bar{n}_{1,i}$, $i = 1, \ldots, KM_TB$, with $\bar{n}_{1,i}$, $i = 1, \ldots, KM_TB$, being independent and identically distributed (i.i.d.) Gaussian noise components with zero-mean and variance σ^2. We can see that the influence of the "future users" is eliminated and the noise component is whitened. Then we use the feedback filtering to cancel the interference from "previous users" as

$$\mathbf{u}_1 = \bar{\mathbf{z}}_1 - (\mathbf{F}_1 - \operatorname{diag}(\mathbf{F}_1))\hat{\mathbf{b}}, \tag{6.54}$$

where it is easily seen that $u_{1,i} = \bar{z}_{1,i} - \sum_{j=1}^{i-1} \mathbf{F}_{1,ij} \hat{b}_j \approx \mathbf{F}_{1,ii} b_i + \bar{n}_{1,i}$, $i = 1, \ldots, KM_TB$. Similarly, for upper triangular Cholesky decomposition $\mathbf{F}_2$, we have

$$\bar{\mathbf{z}}_2 = \mathbf{F}_2^{-\mathsf{H}} \mathbf{z}, \tag{6.55}$$

where $\bar{z}_{2,i} = \mathbf{F}_{2,ii} b_i + \sum_{j=i+1}^{KM_TB} \mathbf{F}_{2,ij} b_j + \bar{n}_{2,i}$, $i = KM_TB, \ldots, 1$, and

$$\mathbf{u}_2 = \bar{\mathbf{z}}_2 - (\mathbf{F}_2 - \operatorname{diag}(\mathbf{F}_2))\hat{\mathbf{b}}, \tag{6.56}$$

where $u_{2,i} = \bar{z}_{2,i} - \sum_{j=i+1}^{KM_TB} \mathbf{F}_{2,ij} \hat{b}_j \approx \mathbf{F}_{2,ii} b_i + \bar{n}_{2,i}$, $i = KM_TB, \ldots, 1$. After the above operations are (alternately) executed, the following log-likelihood ratio is calculated:

$$L_i = 2\,\mathrm{Re}(\mathbf{F}_{1/2,ii}^* u_{1/2,i})/\sigma^2, \tag{6.57}$$

where $\mathbf{F}_{1/2}$ and $\mathbf{u}_{1/2}$ are used to give a shorthand representation for both alternatives. Then the log-likelihood ratio is compared with the last stored value, which is replaced by the new value if the new one is more reliable, i.e.,

$$L_i^{stored} = \begin{cases} L_i^{stored} & \text{if } |L_i^{stored}| > |L_i^{new}|, \\ L_i^{new} & \text{otherwise.} \end{cases} \tag{6.58}$$

Finally we make soft decisions $\hat{b}_i = \tanh(L_i/2)$ at an intermediate iteration, which has been shown to offer better performance than making hard intermediate decisions, and

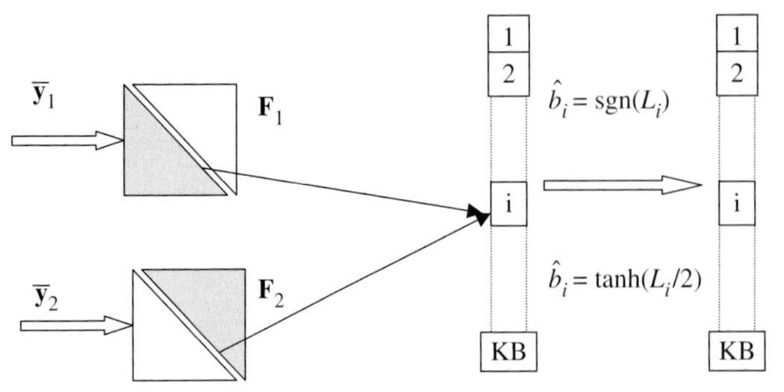

Figure 6.4. Cholesky iterative decorrelating decision-feedback ST MUD.

make hard decisions $\hat{b}_i = \text{sgn}(L_i)$ at the last iteration. Several iterations are usually enough for the system to achieve an improved steady state without significant oscillation. The structure of Cholesky iterative decorrelating decision-feedback ST MUD is illustrated in Fig. 6.4 (assuming $M_T = 1$).

The Cholesky factorization of the block Toeplitz matrix $\mathbf{H}$ (see (6.40)) can be performed recursively. For $\Delta = 1$ we have

$$\mathbf{F} = \begin{bmatrix} \underline{F}_1(0) & 0 & \cdots & 0 & 0 \\ \underline{F}_2(1) & \underline{F}_2(0) & \cdots & 0 & 0 \\ 0 & & & & \\ \vdots & & & & \\ 0 & 0 & \cdots & \underline{F}_M(1) & \underline{F}_M(0) \end{bmatrix}, \tag{6.59}$$

where the element matrices are obtained recursively as follows:

$$\underline{V}_B = \underline{R}^{[0]}, \tag{6.60}$$

and, for $i = B, B-1, \ldots, 1$, we perform Cholesky decomposition of the reduced-rank matrix $\underline{V}_i$ to obtain $\underline{F}_i(0)$

$$\underline{V}_i = \underline{F}_i^{\mathsf{H}}(0)\underline{F}_i(0), \tag{6.61}$$

while $\underline{F}_i(1)$ is obtained as

$$\underline{F}_i(1) = (\underline{F}_i^{\mathsf{H}}(0))^{-1}\underline{R}^{[-1]}. \tag{6.62}$$

Finally we have

$$\underline{V}_{i-1} = \underline{R}^{[0]} - \underline{R}^{[1]}\underline{V}_i^{-1}\underline{R}^{[-1]} \tag{6.63}$$

for use in the next iteration. The extension of this algorithm to $\Delta > 1$ is straightforward and is omitted here.

Multistage interference canceling ST MUD

Multistage interference cancellation (IC) is similar to Jacobi iteration except that hard decisions are made at the end of each stage in place of the linear terms that are fed back in (6.45). Thus we have

$$\hat{\mathbf{b}}_m = \mathrm{sgn}(\mathbf{z} - (\mathbf{C}_L + \mathbf{C}_U)\hat{\mathbf{b}}_{m-1}) = \mathrm{sgn}(\mathbf{z} - (\mathbf{H} - \mathbf{D})\hat{\mathbf{b}}_{m-1}). \tag{6.64}$$

The underlying rationale for this method is that the estimator–subtracter structure exploits the discrete-alphabet property of the transmitted data streams. This nonlinear hard-decision operation typically results in more accurate estimates in high SNR situations. Although the optimal decisions are a fixed point of the nonlinear transformation (6.64), there are problems with the multistage IC such as a possible lack of convergence and oscillatory behavior. In the following section we consider some improvements on space–time multistage IC MUD. Except for the Cholesky factorization, the computational complexity for Cholesky iterative DDF ST MUD is the same as multistage IC ST MUD, which is essentially the same as that of linear interference cancellation, i.e., $O(KM_T\Delta\overline{m})$ per user per symbol.

6.3.4 EM-based iterative space–time multi-user detection

In this section, expectation-maximization-based multi-user detection is introduced to avoid the convergence and stability problem of the multistage IC MUD.

The EM algorithm [10] provides an iterative solution of maximum–likelihood estimation problems such as

$$\hat{\boldsymbol{\theta}}(\mathbf{Z}) = \arg\max_{\boldsymbol{\theta}\in\Lambda} \log f(\mathbf{Z}; \boldsymbol{\theta}), \tag{6.65}$$

where $\boldsymbol{\theta} \in \Lambda$ are the parameters to be estimated, and $f(\cdot)$ is the parameterized probability density function of the observable $\mathbf{Z}$. The idea of the EM algorithm is to consider a judiciously chosen set of "missing data" $\mathbf{W}$ to form the complete data $\mathbf{X} = \{\mathbf{Z}, \mathbf{W}\}$ as an aid to parameter estimation, and then to iteratively maximize the following new objective function:

$$Q(\boldsymbol{\theta}; \overline{\boldsymbol{\theta}}) = E\left[\log f(\mathbf{Z}, \mathbf{W};\ \boldsymbol{\theta})|\mathbf{Z} = \mathbf{z};\ \overline{\boldsymbol{\theta}}\right], \tag{6.66}$$

where it is worth emphasizing again that $\boldsymbol{\theta}$ are the parameters in the likelihood function, which are to be estimated, while $\overline{\boldsymbol{\theta}}$ represent *a priori* estimates of the parameters from the previous iterations. Together with the observations, these previous estimates are used to calculate the expected value of the log-likelihood function with respect to the complete data $\mathbf{X} = \{\mathbf{Z}, \mathbf{W}\}$. To be specific, given an initial estimate $\boldsymbol{\theta}^0$, the EM algorithm alternates between the following two steps:

1. E-step, where the complete-data sufficient statistic $Q(\boldsymbol{\theta}; \boldsymbol{\theta}^i)$ is computed;
2. M-step, where the estimates are refined by $\boldsymbol{\theta}^{i+1} = \arg\max_{\boldsymbol{\theta}\in\Lambda} Q(\boldsymbol{\theta}; \boldsymbol{\theta}^i)$.

It has been shown that EM estimates monotonically increase the likelihood, and converge stably to an ML solution under certain conditions [10].

An issue in using the EM algorithm is the trade-off between the ease of implementation and the convergence rate. One would like to add more "missing data" to make the complete data space more informative so that the implementation of the EM algorithm is simpler than the original setting (6.65). However, the convergence rate of the algorithm is inversely proportional to the Fisher information contained in the complete data space. This trade-off is essentially due to the simultaneous updating nature of the M-step in the original EM algorithm [11]. Consequently, the space-alternating generalized EM (SAGE) algorithm has been proposed in [11] to improve the convergence rate for multidimensional parameter estimation. The idea is to divide the parameters into several groups (subspaces), with only one group being updated at each iteration. Thus, we can associate multiple less-informative "missing data" sets to improve the convergence rate while maintaining the overall tractability of optimization problems. For each iteration, a subset of parameters $\boldsymbol{\theta}_{S_i}$ and the corresponding missing data $\mathbf{W}^{S_i}$ are chosen, which is called the definition step. Then similarly to the EM algorithm, in the E-step we calculate

$$Q^{S_i}(\boldsymbol{\theta}_{S_i}; \boldsymbol{\theta}^i) = E\left[\log f(\mathbf{Z}, \mathbf{W}^{S_i}; \boldsymbol{\theta}_{S_i}, \boldsymbol{\theta}^i_{\tilde{S}_i} | \mathbf{Z} = \mathbf{z}; \boldsymbol{\theta}^i)\right], \tag{6.67}$$

where $\boldsymbol{\theta}_{\tilde{S}_i}$ denotes the complement of $\boldsymbol{\theta}_{S_i}$ in the whole parameter set; in the M-step, the chosen parameters are updated while the others remain unchanged as

$$\begin{cases} \boldsymbol{\theta}^{i+1}_{S_i} = \arg\max\limits_{\boldsymbol{\theta}_{S_i}\in\Lambda_{S_i}} Q^{S_i}(\boldsymbol{\theta}_{S_i}; \boldsymbol{\theta}^i), \\ \boldsymbol{\theta}^{i+1}_{\tilde{S}_i} = \boldsymbol{\theta}^i_{\tilde{S}_i}, \end{cases} \tag{6.68}$$

where Λ_{S_i} denotes the restriction of the entire parameter space to those dimensions indexed by S_i. Like the traditional EM estimates, the SAGE estimates also monotonically increase the likelihood and converge stably to an ML solution under appropriate conditions [11].

The EM algorithm is applied to space–time multi-user detection as follows. Suppose we would like to detect a bit b_n, $n \in \{1, 2, \ldots, KM_TB\}$, which can be viewed as the parameter of interest, while the interfering users' bits $\mathbf{b}_{\tilde{k}} = \{b_j\}_{j\neq n}$ are treated as the missing data. The complete-data sufficient statistic is given by ($\mathbf{R}_{nm}$ is the element of matrix $\mathbf{R}$ at the nth row and mth column)

$$Q(b_n; b_n^i) = \frac{1}{2\sigma^2}\left(-\mathbf{R}_{nn}b_n^2 + 2b_n\left(z_n - \sum_{m\neq n}\mathbf{R}_{nm}\tilde{b}_m\right)\right), \tag{6.69}$$

with

$$\widetilde{b}_m = E\left[b_m | \mathbf{Z} = \mathbf{z}; b_n = b_n^i\right] = \tanh\left(\frac{\mathbf{R}_{mm}}{\sigma^2}(z_m - \mathbf{R}_{mn} b_n^i)\right), \tag{6.70}$$

which forms the E-step of the EM algorithm. The M-step is given by

$$b_n^{i+1} = \arg\max_{b_n \in \Lambda} Q(b_n; b_n^i) = \begin{cases} \operatorname{sgn}(z_n - \sum_{m \neq n} \mathbf{R}_{nm} \widetilde{b}_m), & \Lambda = \{\pm 1\}, \\ \dfrac{1}{\mathbf{R}_{nn}}(z_n - \sum_{m \neq n} \mathbf{R}_{nm} \widetilde{b}_m), & \Lambda = \Re, \end{cases} \tag{6.71}$$

where $\Lambda = \Re$ (the set of real numbers) means a soft decision is needed, e.g., at an intermediate stage. Note that in the E-step (6.70), interference from users $j \neq n$ is not taken into account, since these are treated as "missing data." This shortcoming is overcome by the application of the SAGE algorithm, where the symbol vector of all users $\mathbf{b} = \{b_j\}_{j=1}^{KM_T B}$ is treated as the parameter to be estimated and no missing data are needed. The algorithm is described as follows: for $i = 0, 1, \ldots$.

1. Definition step: $S_i = 1 + (i \mod KM_T B)$
2. M-step:

$$\begin{cases} b_n^{i+1} = \operatorname{sgn}(z_n - \sum_{m \neq n} \mathbf{R}_{nm} b_m^i), & n \in S_i, \\ b_m^{i+1} = b_m^i, \quad m \notin S_i. \end{cases} \tag{6.72}$$

Note that there is no E-step since there are no missing data, and interference from all other users is recreated from previous estimates and subtracted. The resulting receiver is similar to the multistage interference canceling multi-user receiver (see (6.64)), except that the symbol estimates are made sequentially rather than in parallel. However, with this simple concept of sequential interference cancellation, the resulting multi-user receiver is convergent, guaranteed by the SAGE algorithm. The multistage interference canceling multi-user receiver discussed in Section 6.3.3, on the other hand, does not always converge. The computational complexity of this SAGE iterative ST MUD is also $O(K\Delta\overline{m})$ per user per symbol.

6.3.5 *Simulation results*

In this section, the performance of the above described space–time multi-user detectors is examined through simulations on a CDMA example. We assume a $K = 8$-user CDMA system with spreading gain $N = 16$. Each user, equipped with one single antenna, travels through $L = 3$ paths before it reaches a base-station (or access point), equipped with a uniform linear array with $M_R = 3$ elements and half-wavelength spacing. The maximum delay spread is set to be $\Delta = 1$. The complex gains and delays of the multi-path and the directions of arrival are randomly generated and kept fixed for the whole data frame. This corresponds to a slow fading situation. The spreading codes of all users are randomly generated and kept fixed for all the simulations.

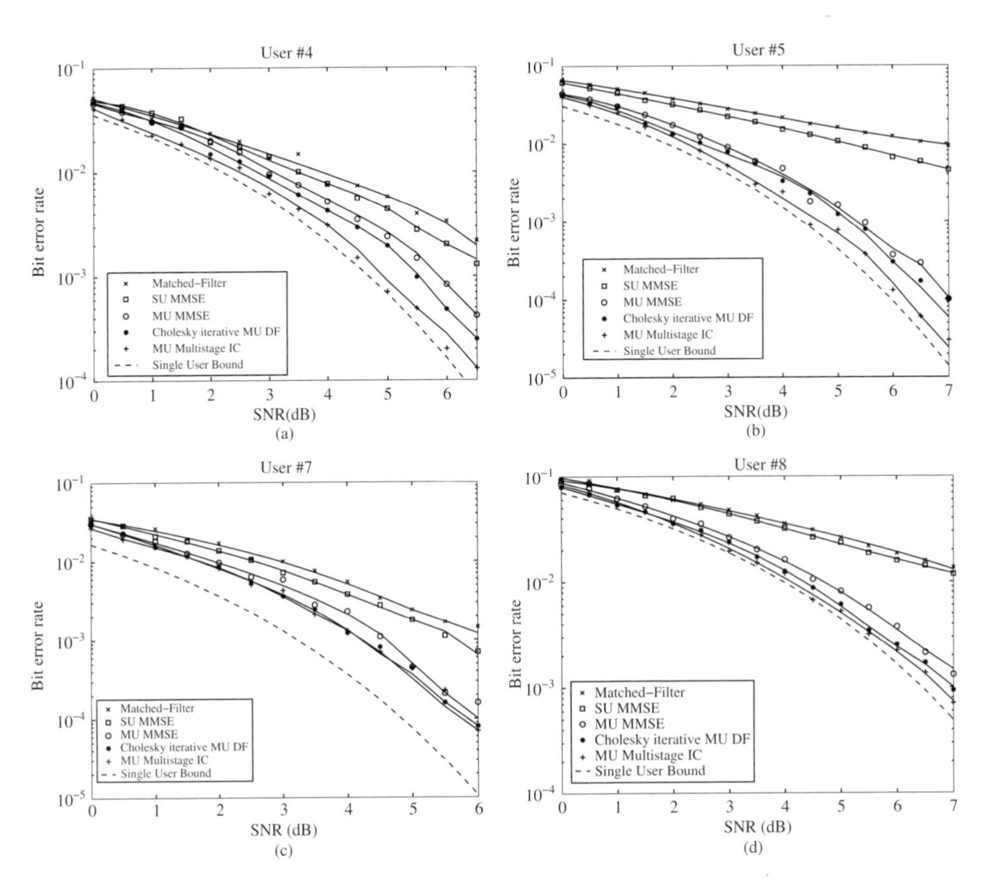

Figure 6.5. Performance comparison of BER versus SNR for five space–time multi-user receivers.

First we compare the performance of various space–time multi-user receivers and some single-user space–time receivers in Fig. 6.5. Five receivers are considered: the single-user matched filter (matched filter), the single-user MMSE receiver (SU MMSE), the multi-user MMSE receiver (MU MMSE) implemented using the Gauss–Seidel or conjugate gradient iteration method (the performance is the same for both), the Cholesky iterative decorrelating decision-feedback multi-user receiver (Cholesky iterative MU DF), and the multistage interference canceling multi-user receiver (MU multistage IC). An interested reader can refer to [45] for derivations of the single-user-based receivers. The performance is evaluated after the iterative algorithms converge. Owing to the poor convergence behavior of the multistage IC MUD, we measure its performance after three stages. The single-user lower bound is also depicted for reference. We can see that the multi-user approach greatly outperforms the single-user-based methods; nonlinear MUD offers further gain over the linear MUD; and the multistage IC seems to approach the optimal performance (not always though, as is seen in Fig. 6.7), when it has good convergence behavior. Note that due to the introduction of spatial (receive antenna) and spectral (RAKE combining) diversity, the SNR for the same BER is substantially lower than that required by normal receivers without these methods.

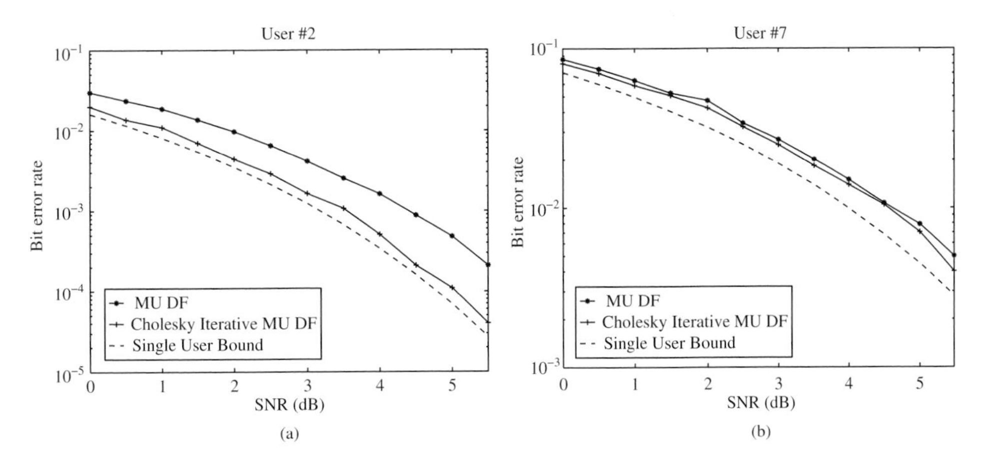

Figure 6.6. Performance comparison of decision-feedback ST MUD and Cholesky iterative ST MUD.

Figure 6.6 shows the performance of Cholesky iterative decorrelating decision-feedback ST MUD for two users, which is also typical for other users. We find that the Cholesky iterative method offers uniform gain over its non-iterative counterpart. This gain may be substantial for some users and negligible for others due to the individual characteristics of signals and channels.

Finally, we show the advantage of the EM-based (SAGE) iterative method over the multistage IC method with regard to the convergence of the algorithms. From Fig. 6.7 we find that, while the multistage interference canceling ST MUD converges slowly and exhibits oscillatory behavior, the SAGE ST MUD converges quickly and outperforms the multistage IC method. The oscillation of the performance of the multistage IC corresponds to performance degradation as no statistically best iteration number can be chosen.

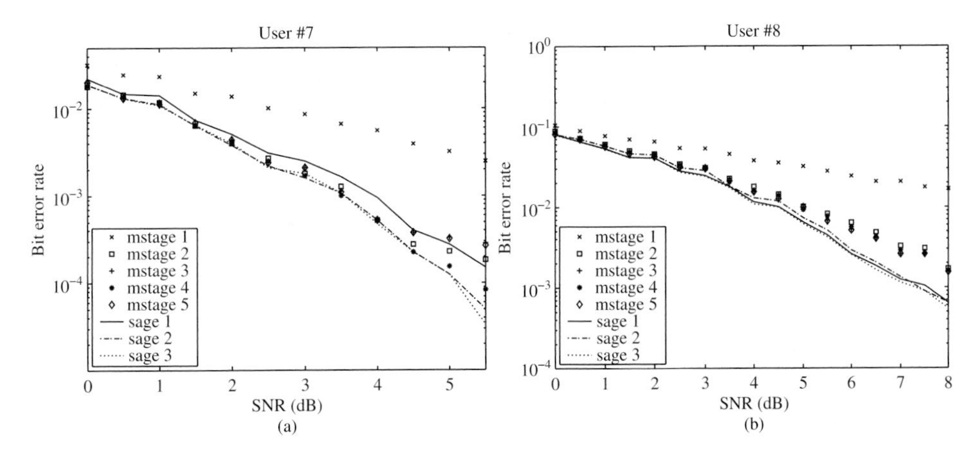

Figure 6.7. Performance comparison of convergence behavior of multistage interference canceling ST MUD and EM-based iterative ST MUD.

6.3.6 Summary

In this section, we have considered several iterative space–time multi-user detection schemes. It is shown that iterative implementation of these linear and nonlinear multi-user receivers approaches the optimum performance with reasonable complexity. Among these iterative implementations the SAGE space–time multi-user receiver outperforms the others while requiring similar complexity. While we focus on single-cell communications, all of the techniques discussed here can be extended to the multicell scenario [6], where the requirement for efficient algorithms only becomes more stringent.

6.4 Multi-user detection in space–time coded systems

With the invention of powerful space–time coding techniques in the late 1990s as described in Chapter 4, there has been a growing interest in adapting these to multiple-access communication systems. Although early space–time code construction was concerned with single-user channels [1, 36, 37] (see Chapter 4), subsequently it has been shown that most of the performance criteria developed can still be used effectively in multi-user channels [21]. Space–time block codes have been applied to multiple-access communication systems in [9, 24, 25]. The receivers that explicitly take into account the structure of the space–time block codes have been shown to perform well in this context [23, 27].

Here we consider multi-user detection in space–time coded multiple-access systems. As we will see, the joint maximum-likelihood decoder for such systems has prohibitively large computational complexity, motivating a search for low-complexity, sub-optimal detector structures. We investigate several partitioned space–time multi-user detectors that separate the multi-user detection and space–time decoding into two stages. Both linear and nonlinear schemes are considered for the first stage of the partitioned receiver and the performance versus complexity trade-offs are discussed.

Inspired by the development of turbo codes [3, 4] that were discussed in Chapter 5, various iterative detection and decoding schemes for multiple-access channels have been proposed in recent years. These proposals show that in general iterative receivers can offer significant performance improvements over their non-iterative counterparts. A good example is [44], in which a soft interference canceling turbo receiver was proposed for convolutionally coded CDMA. The performance results obtained via simulations showed that it is possible to achieve near single-user performance with only a few iterations in an asynchronous, multi-path CDMA channel. In this section, among others, we will show a generalization of this idea to a space–time coded CDMA system as in [20, 21, 26].

The development of turbo multi-user receivers for space–time coded systems here closely follows that of [21]. In particular, we assume a multiple-access system based on DS-CDMA signaling as opposed to space-division multiple-access as in [26]. There are two main implementations of CDMA-based multiple transmit antenna systems. One involves assigning a single spreading code to each user so that the signals transmitted from all its antennas are spread by the same code. We will assume a design of this type. In an alternative implementation, each user is assigned multiple spreading codes so that the signals transmitted from different antennas are spread by different codes [9, 17, 18, 28].

Low-complexity multi-user receiver structures for space–time coded systems have been described in [26, 47]. For example, a multistage receiver suitable for a system employing both turbo and space–time block coding was proposed in [47]. Turbo receiver structures for multiple-access systems with both space–time block and trellis coded systems have been presented in [20, 21, 26]. In general these turbo receivers operate by partitioning the detection and decoding into two separate stages. In the first stage, a multi-user detection technique is employed and a set of soft outputs is generated for each user. The next stage of the receiver is equipped with a bank of decoders (either channel, space–time or both) that decode the individual user channel or space–time codes (or both). These decoders then generate an updated set of soft information about the code symbols which can then be fed back to the first stage to be used as *a priori* information at the next iteration. The process continues by repeating the same steps.

In Section 6.4.1 we present a simplified signal model for a space–time coded, synchronous multi-user system, while in Section 6.4.2 we derive the jointly optimal ML detector/decoder. In Section 6.4.3, we consider low-complexity receiver structures for space–time coded multi-user systems by separating the multi-user detection stage from the space–time decoder stage. We consider both linear and nonlinear multi-user detection stages. In particular, in this section, we consider partitioned space–time multi-user receivers based on the linear decorrelator and on the linear MMSE estimator, as well as two partitioned receiver structures based on nonlinear interference canceling multi-user detection stages. Section 6.4.4 details a soft-input soft-output (SISO) *maximum a posteriori* (MAP) decoder [2] that can be used as the second stage of the interference canceling receivers (for more details on MAP decoding refer to Chapter 5).

6.4.1 Signal model

Consider a system of K independent users, each employing an independent space–time code with M_T transmitter antennas. The binary information sequence $\{d_k[n]\}_{n=0}^{\infty}$ of user k, for $k = 1, \ldots, K$, is first encoded by a space–time encoder, and then the encoded data are divided into M_T streams by passing them through a serial-to-parallel converter. (For simplicity we assume that all the users employ the same number of transmitter antennas, although generalizing to different numbers of transmitter antennas is straightforward.) The code bits in each parallel stream are block interleaved, BPSK symbol-mapped, modulated by an appropriate signature waveform, $s_k(t)$, and are transmitted simultaneously from the M_T transmitter antennas. It should be emphasized that throughout this section we assume that user k employs the same signaling waveform $s_k(t)$ in all its M_T transmitter antennas (i.e., $s_{k,m}(t) = s_k(t)$ for $m = 1, \ldots, M_T$).

The kth user's transmitted signal at time t can thus be written as[3]

$$x_k(t) = \frac{A_k}{\sqrt{M_T}} \sum_{i=0}^{B-1} \sum_{m=1}^{M_T} b_{k,m}[i] s_k(t - iT), \tag{6.73}$$

where $\{b_{k,m}[i] \in \{+1, -1\}\}_{i=0}^{B-1}$ is the symbol-mapped space–time encoder output of the kth user on transmitter antenna m at time i, and B is the number of channel symbols per user in a data frame which is assumed to be the same as the length of a space–time codeword. We assume that the signature waveform of each user is supported only on the interval $0 \leq t \leq T$, and is normalized so that $\int_0^T s_k^2(t)dt = 1$, for $k = 1, \ldots, K$. Thus, A_k^2 represents the transmitted energy per bit of user k, independent of the number of transmitter antennas. Note that the model of (6.73) is otherwise general with regard to the signaling format, and so the following results can be applied to any signaling scheme. However, we are interested here in non-orthogonal signaling schemes such as code-division multiple-access (CDMA).

Assuming that the fading is sufficiently slow to be constant over a received data frame, the corresponding signal received at a single receive antenna can be written as

$$r(t) = \sum_{i=0}^{B-1} \sum_{k=1}^{K} \frac{A_k}{\sqrt{M_T}} \sum_{n_T=1}^{M_T} h_{k,m} b_{k,m}[i] s_k(t - iT) + n(t) \tag{6.74}$$

where $n(t)$ is complex white Gaussian noise with zero-mean and variance $N_0/2$ per dimension. The complex fading coefficient, $h_{k,m}$, between the kth user's mth transmitter antenna and the receiver, is assumed to be a zero-mean unit variance complex Gaussian random variable with independent real and imaginary parts. Equivalently, $h_{k,m}$ has uniform phase and Rayleigh amplitude; i.e., the so-called Rayleigh fading model. These fading coefficients are assumed to be mutually independent with respect to both k and m. In what follows, we assume that all parameters of the model (6.74) are known to the receiver. Only the transmitted symbols are unknown.

6.4.2 *Joint ML multi-user detection and decoding for space–time coded multi-user systems*

We start by considering the joint maximum-likelihood detection and decoding of the symbols in the model of Section 6.4.1. To do so, we first establish some notation.

[3] Elsewhere in this chapter, we have assumed that the transmitted signals are normalized, and have absorbed the transmitter amplitude into the channel response. In this section, we will decompose the channel response to explicitly show the transmitted amplitude as a separate term, similarly to (6.20).

As before, we denote the kth user's transmitted symbol vector (on M_T antennas) at time i by the vector $\mathbf{b}_k[i] = [b_{k,1}[i] \ldots b_{k,M_T}[i]]^\mathsf{T}$. Define, the $BK \times M_T K$ joint codeword matrix $\mathbf{D}$, of all users, as

$$\mathbf{D} = \begin{bmatrix} \mathbf{D}_1 & \mathbf{0}_{B\times M_T} & \cdots & \mathbf{0}_{B\times M_T} \\ \mathbf{0}_{B\times M_T} & \mathbf{D}_2 & \cdots & \mathbf{0}_{B\times M_T} \\ \vdots & \vdots & \ddots & \vdots \\ \mathbf{0}_{B\times M_T} & \mathbf{0}_{B\times M_T} & \cdots & \mathbf{D}_K \end{bmatrix} \tag{6.75}$$

where we have also introduced the notation, for $k = 1, \ldots, K$,

$$\mathbf{D}_k = \begin{bmatrix} \mathbf{b}_k^\mathsf{T}[0] \\ \vdots \\ \mathbf{b}_k^\mathsf{T}[B-1] \end{bmatrix}. \tag{6.76}$$

Note that $\mathbf{D}_k \in \{+1, -1\}^{B\times M_T}$, for $k = 1, \ldots, K$. We will call the joint codeword, $\mathbf{D}$, of all users, the *super codeword*. The space–time coded output from all the users at time i is the $K \times KM_T$ matrix denoted as $\mathbf{D}[i]$, where

$$\mathbf{D}[i] = \begin{bmatrix} \mathbf{b}_1^\mathsf{T}[i] & \mathbf{0}_{1\times M_T} & \cdots & \mathbf{0}_{1\times M_T} \\ \mathbf{0}_{1\times M_T} & \mathbf{b}_2^\mathsf{T}[i] & \cdots & \mathbf{0}_{1\times M_T} \\ \vdots & \vdots & \ddots & \vdots \\ \mathbf{0}_{1\times M_T} & \mathbf{0}_{1\times M_T} & \cdots & \mathbf{b}_K^\mathsf{T}[i] \end{bmatrix}. \tag{6.77}$$

The fading coefficients of the kth user can be collected into a vector $\mathbf{h}_k = \left[h_{k,1}, \ldots, h_{k,M_T}\right]^\mathsf{T} \in \mathbb{C}^{M_T\times 1}$, and we can combine all these fading coefficient vectors into one vector $\mathbf{h} = \left[\mathbf{h}_1^\mathsf{T} \ldots \mathbf{h}_K^\mathsf{T}\right]^\mathsf{T} \in \mathbb{C}^{KM_T\times 1}$. With this notation, the output, $\mathbf{z}[i] = [z_1[i] \ldots z_K[i]]^\mathsf{T}$, of a bank of K matched filters (each matched to a user signature waveform $s_k(t)$) at the ith symbol interval can be written as

$$\mathbf{z}[i] = \overline{\mathbf{R}}\mathbf{A}\mathbf{D}[i]\mathbf{h} + \boldsymbol{\eta}[i] \tag{6.78}$$

where the diagonal matrix $\mathbf{A}$ is defined as $\mathbf{A} = \mathrm{diag}(\frac{A_1}{\sqrt{M_T}}, \ldots, \frac{A_K}{\sqrt{M_T}})$, $\overline{\mathbf{R}}$ is the (normalized) cross-correlation matrix of the users' signature waveforms and $\boldsymbol{\eta}(i) \sim \mathcal{N}(\mathbf{0}, N_0\overline{\mathbf{R}})$.

Let us denote the B-vector of the kth matched filter outputs corresponding to the complete received codeword as $\mathbf{z}_k = [z_k(0) \ldots z_k(B-1)]^\mathsf{T}$ and the BK-vector of outputs of all the matched filters corresponding to a complete codeword as $\mathbf{z} = [\mathbf{z}_1 \ldots \mathbf{z}_K]^\mathsf{T}$. Then we can write

$$\mathbf{z} = (\overline{\mathbf{R}}\mathbf{A} \otimes \mathbf{I}_B)\mathbf{D}\mathbf{h} + \boldsymbol{\eta} \tag{6.79}$$

where $\boldsymbol{\eta} \sim \mathcal{N}(\mathbf{0}, N_0\overline{\mathbf{R}} \otimes \mathbf{I}_B)$, $\mathbf{I}_B$ denotes the $B \times B$ identity matrix and $\otimes$ denotes the Kronecker product. The joint ML multi-user decision rule for the space–time coded CDMA system is then given by

$$\begin{aligned} \hat{\mathbf{D}} &= \arg\max_{\mathbf{D}} p(\mathbf{z}|\mathbf{D}, \mathbf{h}) \\ &= \arg\max_{\mathbf{D}} [2\,\mathrm{Re}\left\{\mathbf{h}^\mathsf{H}\mathbf{D}^\mathsf{T}(\mathbf{A} \otimes \mathbf{I}_B)\mathbf{z}\right\} - \mathbf{h}^\mathsf{H}\mathbf{D}^\mathsf{T}(\mathbf{A} \otimes \mathbf{I}_B)(\overline{\mathbf{R}} \otimes \mathbf{I}_B)(\mathbf{A} \otimes \mathbf{I}_B)\mathbf{D}\mathbf{h}] \end{aligned}$$

where the maximization is over all the valid super codewords and we have used the fact that for general matrices $\mathbf{A}$, $\mathbf{B}$, $\mathbf{C}$, and $\mathbf{D}$ we have, $(\mathbf{A} \otimes \mathbf{B})(\mathbf{C} \otimes \mathbf{D}) = (\mathbf{AC} \otimes \mathbf{BD})$ provided the dimensions of the matrices $\mathbf{A}$, $\mathbf{B}$, $\mathbf{C}$, and $\mathbf{D}$ are such that the various matrix products are well-defined [22]. Note that this joint ML detector and decoder searches over a super trellis made up by combining all the users' space–time code trellises.

The asymptotic performance of a space–time code can be quantified by the so-called diversity gain. The diversity gain determines the asymptotic slope of the probability of error curve on a log scale. As discussed in Chapter 4, in order to maximize the diversity gain for a Rayleigh fading channel one should design the space–time code such that the minimum rank of the codeword difference matrix for any two codewords is as large as possible [15, 36]. When this minimum rank over all pairs of distinct codewords is the largest possible value M_T, then the space–time code is said to achieve full-diversity.

In [21], it was shown that the space–time codes designed to achieve full-diversity in single-user channels will also be able to achieve full-diversity asymptotically in the CDMA multi-user channel, at least when the SNR is sufficiently large. That is, if the minimum rank of all the valid error codewords $\mathbf{E}_k = \mathbf{D}_k - \hat{\mathbf{D}}_k$ is r_k (where $r_k \leq M_T$), then the asymptotic diversity advantage of the kth user's space–time code in the multi-user channel is equal to r_k. In particular, if the kth user's space–time code were to achieve the full-diversity M_T in a single-user environment, then it will also achieve the full-diversity M_T in the multi-user channel, at least asymptotically in SNR, as long as the signature cross-correlation matrix is non-singular.

Figure 6.8 shows the performance results for the joint maximum-likelihood detector in a space–time coded, synchronous, multiple-access system with two equal-power users having a cross-correlation of 0.4. We set the number of receiver antennas to one, ignoring the possibility of exploiting receiver diversity since our primary concern here is to investigate the transmitter diversity schemes. In Fig. 6.8 we have shown the joint ML receiver performance results for two systems: one with two transmit antennas and another with four transmit antennas. We make use of full-diversity BPSK space–time trellis codes with constraint length $\nu = 5$, given in [16], for both systems. Specifically, we employ space–time codes based on the underlying rate-$1/2$ convolutional code with octal generators $(46, 72)$, and the underlying rate-$1/4$ convolutional code with octal generators $(52, 56, 66, 76)$, both given in [16], for the two- and four-antenna systems, respectively. We use the frame error rate (FER) as the measure of performance. Also shown in this figure is the performance of an equivalent system but without space–time coding. Figure 6.8 reveals the significant gains that can be achieved with space–time coding in multi-user systems. Moreover, it shows that the joint ML receiver performance is very close to that of the single-user bound as predicted above.

It is easily seen that the above ML path search can be implemented as a maximum-likelihood path search over a super trellis formed by combining all the users' space–time code trellises using the Viterbi algorithm. This is similar to the optimal decoder for convolutionally coded CDMA channels derived in [12]. Assuming (for simplicity) that all the users employ space–time codes based on underlying convolutional codes that have a constraint length ν, this super trellis will have a total of $K(\nu - 1)$ states, resulting in a total complexity per user of about $\mathcal{O}(2^{K\nu}/K)$, which is exponential in $K\nu$. Note also

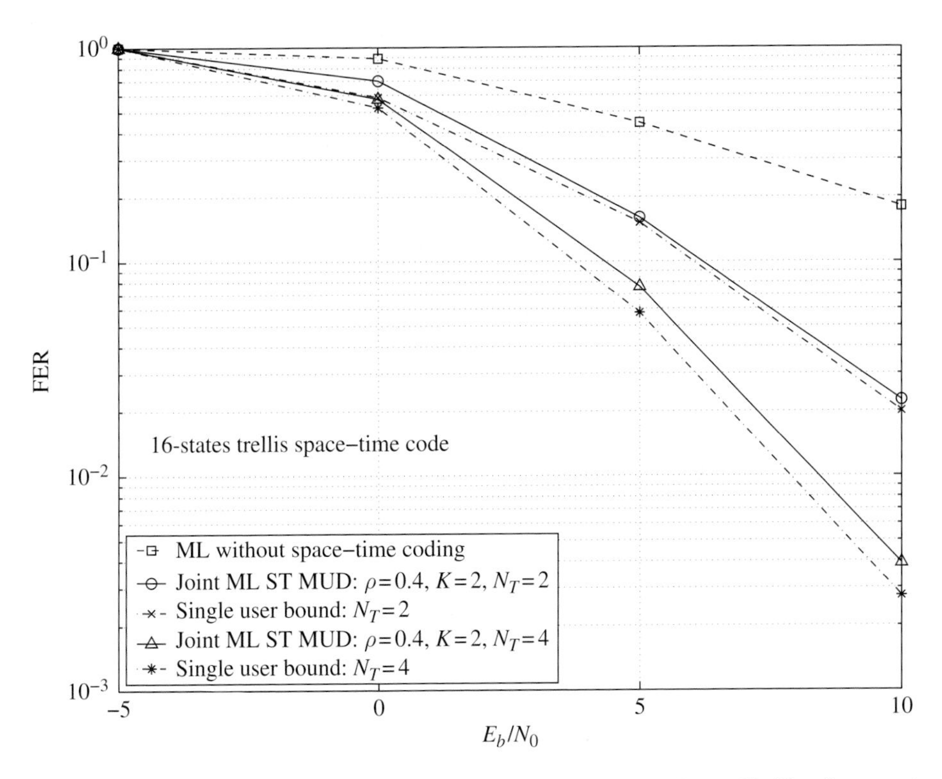

Figure 6.8. FER performance versus E_b/N_0 (in dB) of the joint maximum-likelihood space–time multi-user detector: $K = 2$ and $\rho = 0.4$.

that, in order to achieve full-diversity gain M_T in an M_T transmitter antenna system we must have $\nu \geq M_T$ [16, 37]. Hence, it is clear that even for a small number of users this could easily become a prohibitively large computational burden at the receiver. This motivates us to look for sub-optimal, low-complexity receiver structures for space–time coded multi-user systems.

In order to reduce the computational complexity of joint multi-user detection and space–time decoding while still achieving competitive performance against the joint ML decision rule, one can use partitioned receiver structures. Specifically, the multi-user detection and the space–time decoding can be separated into two stages, as is done in [13] for the case of (single-antenna) convolutionally coded CDMA channels. At the first stage of the partitioned receiver, multi-user detection is performed. The outputs from the multi-user detection stage are then passed onto a bank of single-user space–time decoders corresponding to the K users in the system. Thus, each user's space–time decoder operates independently from the others. Of course, it is possible to employ either an ML or a maximum *a posteriori* probability decoder as the single-user space–time decoder at the second stage of the receiver. Also, it is possible to use any reasonable multi-user detection strategy at the first stage of the receiver. In the following we consider both linear and nonlinear multi-user detectors as the first stage of the partitioned space–time multi-user receiver, and compare the performance of these receivers against the best possible performance.

6.4.3 *Partitioned low-complexity receivers for space–time coded multi-user systems*

We consider linear multi-user detection based partitioned receivers, followed by the nonlinear multi-user detection approaches. For linear multi-user detectors, we investigate both decorrelator and linear MMSE detectors [38]. For nonlinear approaches we consider both a simple iterative receiver based on interference cancellation and the turbo principle, and an improved iterative receiver based on instantaneous MMSE filtering after the interference cancellation step.

Decorrelator-based partitioned space–time multi-user receiver

The decorrelator output at the ith symbol time is given by [38],

$$\hat{\mathbf{z}}[i] = \overline{\mathbf{R}}^{-1}\mathbf{z}[i] = \mathbf{A}\mathbf{D}[i]\mathbf{h} + \hat{\boldsymbol{\eta}}, \tag{6.80}$$

where $\hat{\boldsymbol{\eta}} \sim \mathcal{N}(\mathbf{0}, N_0\overline{\mathbf{R}}^{-1})$. The first stage of the receiver computes soft outputs corresponding to each user's transmitted symbol vectors at time i. The soft outputs are the *a posteriori* probabilities (APPs) of each user's transmitted symbol vectors, defined as below for $l = 1, \ldots, 2^{M_T}$, $k = 1, \ldots, K$ and $i = 0, \ldots, B-1$ (note that 2^{M_T} is the number of possible transmitted symbol vectors):

$$p_{k,l}[i] = \mathrm{P}\left[\mathbf{b}_k[i] = \mathbf{s}_l | \hat{\mathbf{z}}[i], \mathbf{h}\right] \qquad \text{for } \mathbf{s}_l \in \{+1, -1\}^{M_T \times 1}.$$

From (6.80), we can write this *a posteriori* probability as

$$p_{k,l}[i] = C_1 \exp\left(-\frac{1}{N_0(\overline{\mathbf{R}}^{-1})_{kk}}\left|\hat{z}_k[i] - \frac{A_k}{\sqrt{M_T}}\mathbf{s}_l^{\mathsf{T}}\mathbf{h}_k\right|^2\right),$$

where $(\overline{\mathbf{R}}^{-1})_{kk}$ is the (k, k)th element of the matrix $\overline{\mathbf{R}}^{-1}$, $\hat{z}_k[i]$ is the kth component of the vector $\hat{\mathbf{z}}[i]$ and C_1 is a normalizing constant.

The second stage of the partitioned receiver employs a bank of single-user space–time Viterbi decoders that use these *a posteriori* probabilities as inputs. The kth user's decoder uses only the symbol vector probabilities corresponding to the kth user. This results in a decentralized implementation of the receiver. Clearly this partitioned receiver is equivalent to a single-user space–time coded system, except for a different noise variance value. This leads to the following upper bound on the pairwise error probability of the decorrelator-based partitioned space–time multi-user receiver

$$\mathrm{P}_e^{k,(d)}[\mathbf{D}_k \to \hat{\mathbf{D}}_k] \leq \frac{1}{\prod_{n=1}^{r_k}\lambda_{k,n}(\mathbf{E}_k)}\left(\frac{A_k^2/M_T}{4N_0(\overline{\mathbf{R}}^{-1})_{kk}}\right)^{-r_k},$$

where r_k is the rank of the codeword error matrix $\mathbf{E}_k = \mathbf{D}_k - \hat{\mathbf{D}}_k$ and $\lambda_{k,n}(\mathbf{E}_k)$, for $n = 1, \ldots, r_k$, are the non-zero eigenvalues of the $M_T \times M_T$ matrix $\mathbf{E}_k^{\mathsf{T}}\mathbf{E}_k$.

Linear MMSE-based partitioned space–time multi-user receiver

As is well-known, the decorrelator performance degrades when the background noise is dominant, since it completely ignores the presence of background noise [38]. A better compromise between suppressing the multiple-access interference (MAI) and the background noise is obtained by employing a linear MMSE filter at the first stage of the space–time receiver. The linear MMSE multi-user detector output at symbol time i is given by [38]

$$\hat{\mathbf{z}}[i] = \mathbf{A}^{-1}(\overline{\mathbf{R}} + N_0\mathbf{A}^{-2})^{-1}\mathbf{z}[i].$$

The decision statistic corresponding to the kth user can then be written as

$$\begin{aligned}\hat{z}_k[i] &= \frac{A_k}{M_T}\sum_{j=1}^{K}\mathbf{M}_{kj}A_j\mathbf{b}_j^{\mathsf{T}}[i]\mathbf{h}_j + \hat{\eta}_k[i] \\ &= \frac{A_k^2}{M_T}\mathbf{M}_{kk}\mathbf{b}_k^{\mathsf{T}}[i]\mathbf{h}_k + \frac{A_k}{M_T}\sum_{j\neq k}\mathbf{M}_{kj}A_j\mathbf{b}_j^{\mathsf{T}}[i]\mathbf{h}_j + \hat{\eta}_k[i]\end{aligned} \tag{6.81}$$

where we have defined $\mathbf{M} = (\mathbf{A}^2 + N_0\overline{\mathbf{R}}^{-1})^{-1}$ and $\hat{\eta}_k[i] \sim \mathcal{N}(0, \frac{A_k^2}{M_T}N_0(\mathbf{M}\overline{\mathbf{R}}^{-1}\mathbf{M})_{kk})$.

In order to compute the soft output *a posteriori* probabilities at the end of the first stage, we make the assumption that the noise at the output of an MMSE multi-user detector (residual MAI plus the background noise) can be modeled as being Gaussian [32]. Therefore, we may model (6.81) as

$$\hat{z}_k[i] = \frac{A_k^2}{M_T}\mathbf{M}_{kk}\mathbf{b}_k^{\mathsf{T}}[i]\mathbf{h}_k + \tilde{\eta}_k[i], \tag{6.82}$$

with $\tilde{\eta}_k[i] \sim \mathcal{N}(0, \nu_k^2[i])$. It can be shown that

$$\nu_k^2[i] = 4\frac{A_k^2}{M_T}\left[\sum_{j\neq k}\frac{A_j^2}{M_T}\mathbf{M}_{kj}^2|\mathbf{h}_j[i]|^2 + N_0(\mathbf{M}\overline{\mathbf{R}}^{-1}\mathbf{M})_{kk}\right]. \tag{6.83}$$

Using this model, the soft output *a posteriori* probabilities at the output of the linear MMSE multi-user stage can be written as

$$\begin{aligned}p_{k,l}[i] &= \mathrm{P}\left[\mathbf{b}_k[i] = \mathbf{s}_l|\hat{\mathbf{z}}[i], \mathbf{h}\right] \\ &= C_2\exp\left(-\frac{1}{\nu_k^2[i]}|\hat{z}_k[i] - \frac{A_k^2}{M_T}\mathbf{M}_{kk}\mathbf{s}_l^{\mathsf{T}}\mathbf{h}_k|^2\right),\end{aligned}$$

where C_2 is a normalizing constant.

The second stage of this receiver operates exactly the same way as that in the decorrelator-based partitioned receiver.

Figure 6.9 shows the FER performance of partitioned space–time multi-user receivers based on linear first-stage multi-user detectors and ML single-user decoders, in a four-user system with each having two transmit antennas. As before, we make use of the full diversity BPSK space–time trellis code with constraint length $\nu = 5$ and based on

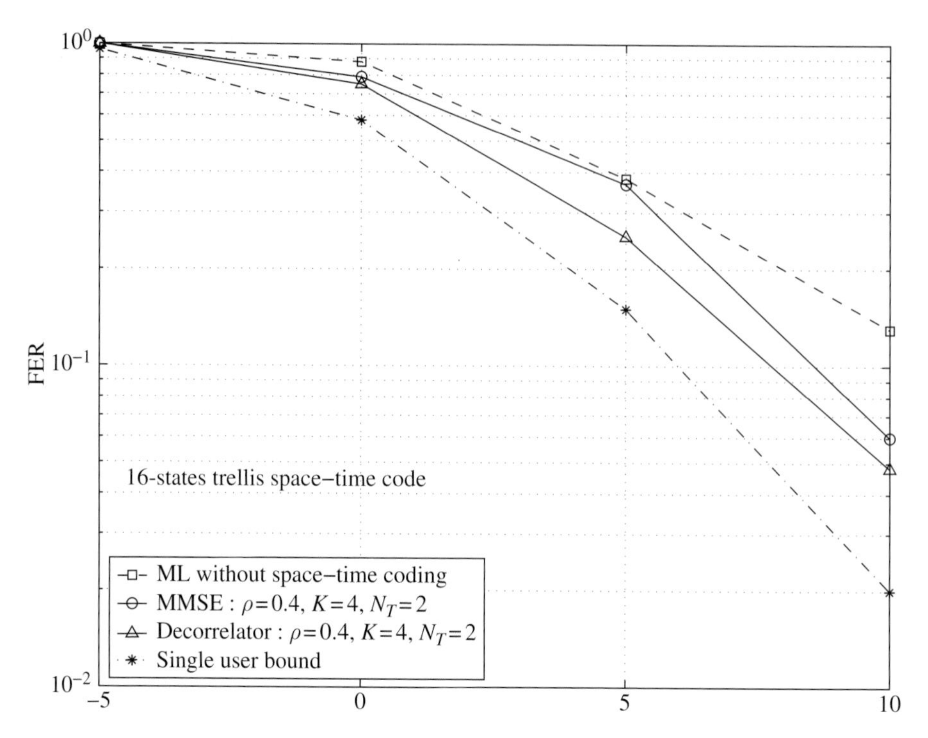

Figure 6.9. FER performance versus E_b/N_0 (in dB) of the linear first stage-based partitioned space–time multi-user detectors: $K = 4$, $\rho = 0.4$, and $M_T = 2$.

the underlying rate-1/2 convolutional code with octal generators (46, 72) [16]. User cross-correlations are assumed to be $\rho_{jk} = 0.4$ for all $k \neq j$.

From Fig. 6.9 it can be seen that the linear first stage-based partitioned space–time receivers may offer some diversity gain over single-antenna systems, though they fail to capture the full gains achievable with space–time coding. This is clear from the large performance gap between that of linear first stage-based partitioned receivers and the single-user bound in Fig. 6.9. This performance degradation becomes severe with increasing user cross-correlations, as one would expect. These results also justify our iterative approach, which is capable of providing near single-user performance even in severe MAI environments (as we will see below). We observe that for the given cross-correlation values, the MMSE first stage performance is no better than that with a decorrelator first stage. Of course in the case of smaller MAI than what we have simulated, the MMSE first stage would outperform the decorrelator-based receiver, since in this case the background noise would be the dominant noise source. In either case, these linear detectors fail to exploit the large performance gains available with space–time coding.

Iterative MUD with interference cancellation for space–time coded CDMA

In this section we present a simple iterative receiver structure based on interference cancellation and the turbo principle. Suppose that at the first stage of the receiver, we have available *a priori* probabilities of all users' transmitted symbol vectors, $p_{k,l}[i]_2^p = \mathrm{P}[\mathbf{b}_k[i] = \mathbf{s}_l]$, for $l = 1, \ldots, 2^{M_T}$, $k = 1, \ldots, K$ and $i = 0, \ldots, B-1$. Note that the subscript 2 and superscript p indicate that these *a priori* probabilities were, in fact, generated by the second stage of the receiver (i.e., the single-user space–time decoders) at the previous iteration. Using these *a priori* probabilities $p_{k,l}[i]_2^p$, the interference-canceling multi-user detector at the first stage of the receiver computes soft estimates of the transmitted symbol vectors of all the users as

$$\hat{\mathbf{b}}_k[i] = \sum_{l=1}^{2^{M_T}} \mathbf{s}_l p_{k,l}[i]_2^p. \tag{6.84}$$

These soft estimates are used to cancel the multiple-access interference at the output of the kth user's matched filter. The interference cancelled output corresponding to the kth user is obtained as the kth component of the vector

$$\hat{\mathbf{z}}_k[i] = \hat{\mathbf{z}}[i] - \overline{\mathbf{R}}\mathbf{A}\hat{\mathbf{D}}_k[i]\mathbf{h}, \tag{6.85}$$

where $\hat{\mathbf{D}}_k[i] = \mathrm{diag}(\hat{\mathbf{b}}_1[i], \ldots, \hat{\mathbf{b}}_{k-1}[i], \mathbf{0}, \hat{\mathbf{b}}_{k+1}[i], \ldots, \hat{\mathbf{b}}_K[i])$. From (6.85), with $\hat{z}_k[i]$ denoting the kth element of $\hat{\mathbf{z}}_k[i]$, we have that

$$\hat{z}_k[i] = \frac{A_k}{\sqrt{M_T}} \mathbf{b}_k^{\mathsf{T}} \mathbf{h}_k + \sum_{j \neq k} \rho_{kj} \frac{A_j}{\sqrt{M_T}} (\mathbf{b}_j - \hat{\mathbf{b}}_j)^{\mathsf{T}} \mathbf{h}_j + \eta_k[i]. \tag{6.86}$$

Since $\eta_k[i] \sim \mathcal{N}(0, N_0)$, assuming all the previous estimates of the symbol vectors were correct, the iterative interference-canceling space–time multi-user detector (IC-ST-MUD) computes the soft output *a posteriori* probabilities of the transmitted symbol vectors of user k, for $k = 1, \ldots, K$, as

$$\begin{aligned} \mathrm{P}\left[\mathbf{b}_k[i] = \mathbf{s}_l | \mathbf{z}[i], \{\hat{\mathbf{b}}_j\}_{j=1, j\neq k}^{K}\right] &= C_3 \exp\left[-\frac{1}{N_0} |\hat{z}_k[i] - \frac{A_k}{\sqrt{M_T}} \mathbf{s}_l^{\mathsf{T}} \mathbf{h}_k|^2\right] p_{k,l}[i]_2^p \\ &= p_{k,l}[i]_1 p_{k,l}[i]_2^p, \end{aligned}$$

where C_3 is a normalizing constant.

Following turbo decoding terminology, the term $p_{k,l}[i]_1$ is called the extrinsic *a posteriori* probability as computed by the space–time multi-user detector. These extrinsic *a posteriori* probabilities, $p_{k,l}[i]_1$, are de-interleaved and passed on to a bank of K single-user soft-input/soft-output space–time MAP decoders, described in Section 6.4.4 below (for a more general factor graph interpretation refer to Chapter 5). The kth user's SISO space–time MAP decoder computes *a posteriori* probabilities of the transmitted symbol vectors for all the symbols in a given frame [44]. The extrinsic components of these symbol vector APPs, $p_{k,l}[i]_2$, are then interleaved and fed back to the first stage of the IC-ST-MUD, to be used as the *a priori* probabilities $p_{k,l}[i]_2^p$, in the next iteration. At the final iteration, the space–time MAP decoders output hard decisions on the information symbols. A block diagram of this iterative, interference-canceling space–time multi-user detector is shown in Fig. 6.10.

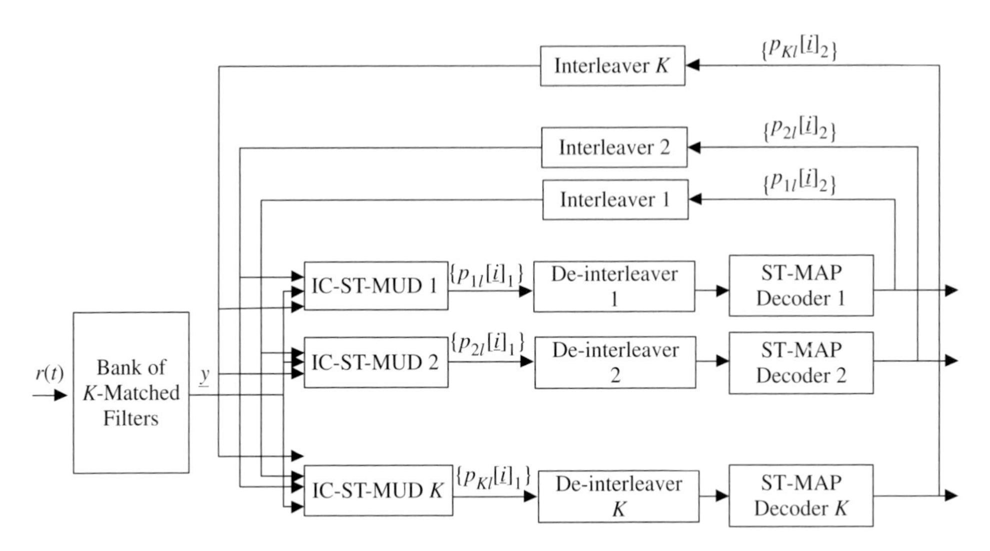

Figure 6.10. Iterative, interference-canceling, space–time multi-user detector.

FER performance of the iterative receiver based on interference cancellation is shown in Fig. 6.11 for the same four-equal-power-user system in which each user has two antennas considered in Fig. 6.9. From Fig. 6.11 we observe that with about four iterations we can achieve most of the gain available from the iterative decoding process. Significantly, we see that for medium values of ρ, this simple interference cancellation scheme can achieve near single-user performance with few iterations, which is not possible with linear first stages as we observed earlier.

However, this simple interference-cancellation-based iterative detector fails when the cross-correlations between users are increased. In this case, the performance becomes almost insensitive to the number of iterations since when the user cross-correlations are high our estimates at the end of the initial iteration are very poor (which of course is the same as a system employing a single-user matched filter front-end), and thus the subsequent iterations will be based on these poor estimates.

The conventional matched filter complexity is $\mathcal{O}(1)$. At each iteration, the first stage of the receiver needs to compute 2^{M_T} symbol vector *a posteriori* probabilities. Hence, the computational complexity of this partitioned receiver is $\mathcal{O}(2^{M_T}+2^{\nu})$ per user per iteration. Note that even though both MAP and ML decoding have the same $\mathcal{O}(2^{\nu})$ complexity order, the MAP decoding in general requires more computations compared to the ML decoding. It has been shown that MAP decoding can be done with a complexity roughly four times that of ML decoding [40].

Iterative MUD with interference cancellation and instantaneous MMSE filtering for space–time coded multi-user systems

As we mentioned above, the performance of the iterative IC-ST-MUD receiver, proposed in the previous section degrades considerably for medium to large cross-correlation values.

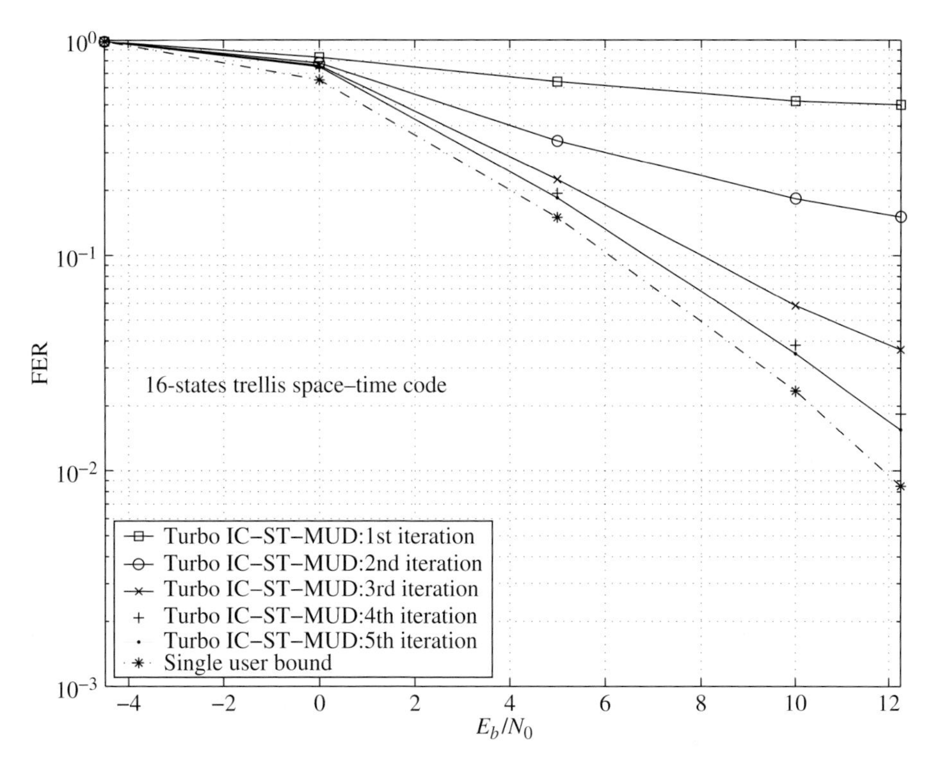

Figure 6.11. FER performance versus E_b/N_0 (in dB) of the partitioned iterative space–time receiver based on interference-canceling multi-user detection: $K = 4$, $\rho = 0.4$, and $M_T = 2$.

Especially when the user cross-correlations are high, the soft estimates at the initial iteration can be very poor and thus the performance does not improve significantly on subsequent iterations. In order to overcome this shortcoming, in this section we modify the iterative receiver proposed in the previous section by the addition of an instantaneous filter. This becomes similar to the iterative decoder proposed in [44] for a convolutionally coded CDMA channel.

Specifically, we choose a linear MMSE filter that minimizes the mean square error between the interference-suppressed output and the kth user's fading-modulated transmitted symbol vector. Clearly, when the soft estimates of the multiple-access interference are very poor or they are not available at all (as in the case of the first iteration), this filtering helps the receiver to still maintain an acceptable performance level, as we will see by the simulation results.

The kth user's linear MMSE filter at symbol time i applies weights $\mathbf{w}_k[i]$ to the interference-suppressed output $\hat{\mathbf{z}}_k[i]$ of (6.85), where $\mathbf{w}_k[i]$ is designed so that

$$\mathbf{w}_k[i] = \arg\min_{\mathbf{w}} \mathrm{E}\left[\|\mathbf{b}_k^{\mathsf{T}}[i]\mathbf{h}_k - \mathbf{w}^{\mathsf{H}}\hat{\mathbf{z}}_k[i]\|^2\right]. \tag{6.87}$$

It can easily be shown that the solution to (6.87) is given by

$$\mathbf{w}_k[i] = \mathrm{E}\left[\hat{\mathbf{z}}_k[i]\hat{\mathbf{z}}_k^{\mathsf{H}}[i]\right]^{-1}\mathrm{E}\left[\hat{\mathbf{z}}_k[i]\mathbf{b}_k^{\mathsf{T}}[i]\mathbf{h}_k\right], \tag{6.88}$$

with

$$\mathrm{E}\left[\hat{\mathbf{z}}_k[i]\hat{\mathbf{z}}_k^{\mathsf{H}}[i]\right] = \overline{\mathbf{R}}\mathbf{V}_k[i]\overline{\mathbf{R}} + N_0\overline{\mathbf{R}},$$

and

$$\mathrm{E}\left[\hat{\mathbf{z}}_k[i]\mathbf{b}_k^{\mathsf{T}}[i]\mathbf{h}_k\right] = \frac{A_k}{\sqrt{M_T}}|\mathbf{h}_k|^2\overline{\mathbf{R}}\mathbf{e}_k,$$

where we have defined the matrix $\mathbf{V}_k[i]$ as

$$\mathbf{V}_k[i] = \mathrm{diag}\left(\frac{A_1^2}{M_T}\sum_{m=1}^{M_T}(1-\hat{b}_{1,m}^2)|h_{1,m}|^2, \ldots, \frac{A_k^2}{M_T}|\mathbf{h}_k|^2, \ldots, \frac{A_K^2}{M_T}\sum_{m=1}^{M_T}(1-\hat{b}_{K,m}^2)|h_{K,m}|^2\right),$$

and $\mathbf{e}_k$ is the kth unit vector. Denoting the matrix $\left(\overline{\mathbf{R}}\mathbf{V}_k[i]\overline{\mathbf{R}} + N_0\overline{\mathbf{R}}\right)^{-1}$ by $\mathbf{M}_k[i]$, we can write the instantaneous linear MMSE filter corresponding to the kth user at symbol time i as

$$\begin{aligned}\mathbf{w}_k[i] &= \frac{A_k}{\sqrt{M_T}}|\mathbf{h}_k|^2\left(\overline{\mathbf{R}}\mathbf{V}_k[i]\overline{\mathbf{R}} + N_0\overline{\mathbf{R}}\right)^{-1}\overline{\mathbf{R}}\mathbf{e}_k \\ &= \frac{A_k}{\sqrt{M_T}}|\mathbf{h}_k|^2\mathbf{M}_k[i]\overline{\mathbf{R}}\mathbf{e}_k. \end{aligned} \tag{6.89}$$

We again model the residual noise at the linear MMSE filter output as having a Gaussian distribution [32, 44]. Thus, we have the following model for $v_k[i]$, the output of the linear MMSE filter corresponding to the kth user at symbol time i:

$$v_k[i] = \mathbf{w}_k^{\mathsf{H}}[i]\hat{\mathbf{z}}_k[i] = \mu_k[i]\mathbf{b}_k^{\mathsf{T}}[i]\mathbf{h}_k + u_k[i], \tag{6.90}$$

where $u_k[i] \sim \mathcal{N}(0, \nu_k^2[i])$. It can be shown that

$$\mu_k[i] = \frac{A_k^2}{M_T}|\mathbf{h}_k|^2(\mathbf{M}_k[i])_{k,k} \tag{6.91}$$

and

$$\nu_k^2[i] = |\mathbf{h}_k|^2\left(\mu_k[i] - \mu_k^2[i]\right). \tag{6.92}$$

The soft-output interference-canceling multi-user detector with instantaneous MMSE filtering makes use of the model in (6.90) in order to compute the *a posteriori* probabilities of the transmitted symbol vectors corresponding to the kth user:

$$\begin{aligned}\mathrm{P}\left[\mathbf{b}_k[i] = \mathbf{s}_l \mid \mathbf{z}[i], \{\hat{\mathbf{b}}_j\}_{j=1, j\neq k}^{K}\right] &= C_4\exp\left[-\frac{|v_k[i] - \mu_k[i]\mathbf{s}_l\mathbf{h}_k|^2}{\nu_k^2[i]}\right]p_{k,l}[i]_2^p \\ &= p_{k,l}[i]_1 p_{k,l}[i]_2^p,\end{aligned}$$

where C_4 is a normalizing constant.

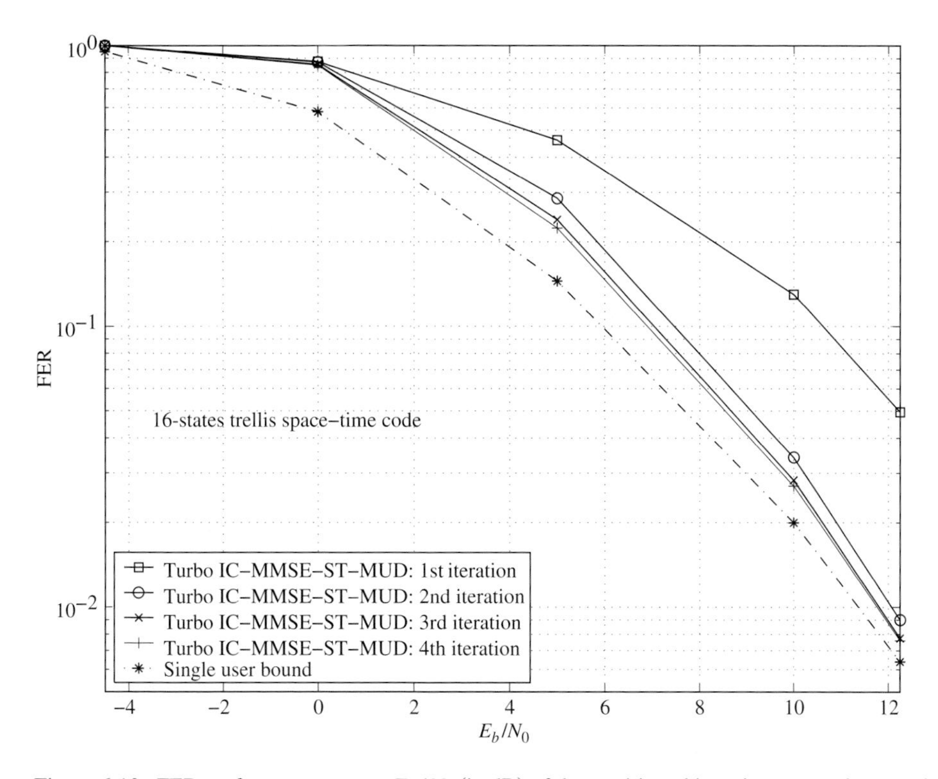

Figure 6.12. FER performance versus E_b/N_0 (in dB) of the partitioned iterative space–time receiver based on interference canceling and linear MMSE filtering multi-user detection stage. $K = 4$, $\rho = 0.75$, and $M_T = 2$.

The second stage of this modified iterative receiver is a SISO space–time MAP decoder which operates in exactly the same way as the receiver described in the previous section. This decoder is described briefly in the following section.

Figure 6.12 shows the FER performance of the interference-canceling space–time multi-user receiver with instantaneous linear MMSE filtering assuming the same four-user system but with the cross-correlation between any pair of users being equal to 0.75. We observe that this modified iterative receiver provides excellent performance and is able to achieve near single-user performance with only a few iterations (two to three iterations), even in the presence of considerable MAI.

The complexity of this MMSE-based interference-canceling partitioned receiver is roughly about $\mathcal{O}(K^2 + 2^{M_T} + 2^{\nu})$ per user per iteration. Note also that this iterative receiver does not rely on spatial diversity for interference suppression but exploits the multi-user signal structure, which is likely to be available at a base-station receiver.

6.4.4 Single-user soft-input soft-output space–time map decoder

In the following we briefly outline the single-user soft-input soft-output space–time MAP decoder assumed in the iterative receivers above. The space–time encoder of each user is assumed to append zero bits to a given information bit block of size B', so that the trellis is always terminated in the zero state. Thus, the actual space–time code block length is $B = B' + \nu - 1$ (since we assume that the rate of the space–time code is 1), where ν is the constraint length of the underlying convolutional code. In this section, we use the MAP decoding algorithm [2] to compute the *a posteriori* probabilities of all the symbol vectors and the information bits.

Similarly to the notation in [44], we will denote the state of the space–time trellis at time i by a $(\nu - 1)$-tuple, as $S_i = (s_i^1, \ldots, s_i^{\nu-1}) = (d_{i-1}, \ldots, d_{i-\nu-1})$, where d_i is the input information bit to the space–time encoder at time i. The corresponding output code symbol vector is denoted by $\mathbf{b}_i$. (Note that here we are using the subscripts to denote the time index.) Let $d(s', s)$ be the input information bit that causes the state transition from $S_{i-1} = s'$ to $S_i = s$ and $\mathbf{b}(s', s)$ be the corresponding output bit vector, which is of length M_T.

Define the forward and backward recursions [2] as

$$\alpha_i(s) = \sum_{s'} \alpha_{i-1}(s')\mathrm{P}[\mathbf{b}_i(s', s)], \qquad i = 1, \ldots, B, \tag{6.93}$$

and

$$\beta_i(s) = \sum_{s'} \beta_{i+1}(s')\mathrm{P}[\mathbf{b}_{i+1}(s', s)], \qquad i = B - 1, \ldots, 0, \tag{6.94}$$

where $\mathrm{P}[\mathbf{b}_i(s', s)] = \mathrm{P}[\mathbf{b}_i = \mathbf{b}(s', s)]$. Initial conditions for (6.93) and (6.94) are given as $\alpha_0(\mathbf{0}) = 1$, $\alpha_0(s \neq \mathbf{0}) = 0$, $\beta_B(\mathbf{0}) = 1$, and $\beta_B(s \neq \mathbf{0}) = 0$. The summations are over all the states s' where the state transition (s', s) is allowed in the code trellis. Normalization of forward and backward variables is done as in [44] to avoid numerical instabilities, though we do not elaborate them here.

Let $\mathcal{S}^l$ denote the set of state pairs (s', s) such that the output symbol vector corresponding to this transition is $\mathbf{s}_l$. The SISO ST MAP decoder of user k updates the *a posteriori* symbol vector probabilities as

$$\begin{aligned}
\mathrm{P}[\mathbf{b}_k[i] = \mathbf{s}_l | \{p_{k,l'}[i]_1\}_{i=0}^{B-1}, l' = 1, \ldots, L] &= \sum_{(s',s)\in\mathcal{S}^l} \alpha_{i-1}(s')\beta_i(s)\mathrm{P}[\mathbf{b}_i(s', s)] \\
&= \left(\sum_{(s',s)\in\mathcal{S}^l} \alpha_{i-1}(s')\beta_i(s) \right) \mathrm{P}[\mathbf{b}_k[i] = \mathbf{s}_l] \\
&= p_{k,l}[i]_2 p_{k,l}[i]_1.
\end{aligned} \tag{6.95}$$

The extrinsic part of the above *a posteriori* symbol vector probability, $p_{k,l}[i]_2$, is interleaved and fed back to the interference-canceling space–time multi-user detector, to be used as the *a priori* probability $p_{k,l}[i]_2^p$, in the next iteration.

In the final iteration the SISO ST MAP decoder also computes the *a posteriori* log-likelihood ratio (LLR) of the information bits. Again, similarly to the notation in [44], let

$\mathcal{U}^+$ denote the set of state pairs (s', s) such that the corresponding input information bit is $+1$. $\mathcal{U}^-$ is defined similarly. Then we have that

$$\begin{aligned}\Lambda[d_k[i]] &= \frac{\mathrm{P}[d_k[i] = +1]}{\mathrm{P}[d_k[i] = -1]} \\ &= \log \frac{\sum_{\mathcal{U}^+} \alpha_{i-1}(s')\beta_i(s)\mathrm{P}[\mathbf{b}_i(s', s)]}{\sum_{\mathcal{U}^-} \alpha_{i-1}(s')\beta_i(s)\mathrm{P}[\mathbf{b}_i(s', s)]}.\end{aligned}$$

Based on these *a posteriori* log-likelihood ratios, the decoder outputs a final hard decision on the information bit $d_k[i]$ for $i = 1, \ldots, B' - 1$, at the last iteration.

6.4.5 *Summary*

In this section, we have considered space–time coding for multiple-access systems in the presence of quasi-static Rayleigh fading. We first obtained the joint ML receiver for a space–time coded CDMA multi-user channel. This joint ML receiver can be shown to achieve full-diversity advantage for each user if the individual space–time codes are of full-diversity. A better trade-off between performance and computational complexity at the receiver can be obtained by partitioning the multi-user detection and space–time decoding into two stages at the receiver. In particular, a nonlinear iterative receiver based on interference cancellation and instantaneous MMSE filtering is capable of capturing most of the gains available with space–time coding in multiple-access channels, with only a few iterations.

6.5 Adaptive linear space–time multi-user detection

We now turn to the situation in which some of the parameters of the model of (6.1) are not known, and thus the receiver must adapt itself to the environment. To examine this situation, two linear multi-user MIMO reception strategies are presented: diversity and space–time multi-user detection. Citing advantages of the space–time technique, *linear adaptive* implementations, including batch and sequential-adaptive algorithms for synchronous CDMA in flat-fading channels, are then developed. The section concludes with extensions to asynchronous CDMA in multi-path fading. Portions of this work first appeared in [35].

6.5.1 *Diversity multi-user detection versus space–time multi-user detection*

We consider a K-user code division multiple-access (CDMA) system with processing gain N, operating in flat-fading with M_R receiver antennas and M_T transmitter antennas. For simplicity of exposition, we will consider only $M_T = M_R = 2$ and BPSK modulation in this section. Extensions to other antenna configurations and modulation techniques are straightforward. When two antennas are employed at the transmitter, we must first specify how the information symbols are transmitted across the two antennas. Here we adopt the

Alamouti space–time block coding scheme [1, 36] discussed in Chapter 1. Specifically, for each user k, two information symbols $b_{k,1}$ and $b_{k,2}$ are transmitted over two symbol intervals. At the first time interval, the symbol pair $(b_{k,1}, b_{k,2})$ is transmitted across the two transmitter antennas; and at the second time interval, the symbol pair $(-b_{k,2}, b_{k,1})$ is transmitted. After chip-matched filtering with respect to $\psi(t)$ and chip-rate sampling, the received signals at antenna 1 during the two symbol intervals are[4]

$$\mathbf{r}_{1,1} = \sum_{k=1}^{K} \left[h_{k,1,1}b_{k,1} + h_{k,2,1}b_{k,2}\right]\mathbf{s}_k + \mathbf{n}_{1,1} \tag{6.96}$$

and

$$\mathbf{r}_{2,1} = \sum_{k=1}^{K} \left[-h_{k,1,1}b_{k,2} + h_{k,2,1}b_{k,1}\right]\mathbf{s}_k + \mathbf{n}_{2,1}, \tag{6.97}$$

and the corresponding signals received at antenna 2 are

$$\mathbf{r}_{1,2} = \sum_{k=1}^{K} \left[h_{k,1,2}b_{k,1} + h_{k,2,2}b_{k,2}\right]\mathbf{s}_k + \mathbf{n}_{1,2} \tag{6.98}$$

and

$$\mathbf{r}_{2,2} = \sum_{k=1}^{K} \left[-h_{k,1,2}b_{k,2} + h_{k,2,2}b_{k,1}\right]\mathbf{s}_k + \mathbf{n}_{2,2}, \tag{6.99}$$

where $h_{k,i,j}$, $i, j \in \{1, 2\}$ is the complex channel response between transmitter antenna i and receiver antenna j for user k and $\mathbf{s}_k = [c_k^{(0)} c_k^{(1)} \cdots c_k^{(N-1)}]^T \in \{\pm 1/\sqrt{N}\}^N$ is the spreading code assigned to user k, as discussed previously in this chapter.

The noise vectors $\mathbf{n}_{1,1}$, $\mathbf{n}_{1,2}$, $\mathbf{n}_{2,1}$, and $\mathbf{n}_{2,2}$ are assumed to be independent and identically distributed with distribution $\mathcal{N}_c(\mathbf{0}, \sigma^2\mathbf{I}_N)$.

Linear diversity multi-user detector

Denote

$$\mathbf{S} \triangleq [\mathbf{s}_1 \cdots \mathbf{s}_K]$$

and

$$\overline{\mathbf{R}} \triangleq \mathbf{S}^T\mathbf{S}.$$

Suppose that user 1 is the user of interest. The combining weights for the linear decorrelating detector [38] for user 1 can be written as

$$\mathbf{w}_1 = \mathbf{S}\overline{\mathbf{R}}^{-1}\mathbf{e}_1, \tag{6.100}$$

[4] In this section, we assume complex signaling waveforms and channel coefficients.

where $\mathbf{e}_1$ denotes the first unit vector in $\mathbb{R}^K$. Our first detection strategy, which we call *linear diversity multi-user detection*, applies the linear multi-user detector $\mathbf{w}_1$ in (6.100) to each of the four received signals $\mathbf{r}_{1,1}$, $\mathbf{r}_{1,2}$, $\mathbf{r}_{1,2}$, and $\mathbf{r}_{2,2}$ and then performs space–time decoding. Specifically, denote the filter outputs as

$$z_{1,1} \triangleq \mathbf{w}_1^T \mathbf{r}_{1,1} = h_{1,1,1} b_{1,1} + h_{1,2,1} b_{1,2} + u_{1,1}, \tag{6.101}$$

$$z_{2,1} \triangleq \left(\mathbf{w}_1^T \mathbf{r}_{2,1}\right)^* = -h_{1,1,1}^* b_{1,2} + h_{1,2,1}^* b_{1,1} + u_{2,1}^*, \tag{6.102}$$

$$z_{1,2} \triangleq \mathbf{w}_1^T \mathbf{r}_{1,2} = h_{1,1,2} b_{1,1} + h_{1,2,2} b_{1,2} + u_{1,2}, \tag{6.103}$$

$$z_{2,2} \triangleq \left(\mathbf{w}_1^T \mathbf{r}_{2,2}\right)^* = -h_{1,1,2}^* b_{1,2} + h_{1,2,2}^* b_{1,1} + u_{2,2}^*, \tag{6.104}$$

with

$$u_{i,j} \triangleq \mathbf{w}_1^T \mathbf{n}_{i,j} \sim \mathcal{N}_c\left(\mathbf{0}, \frac{\sigma^2}{\eta_1^2}\right), \qquad i, j = 1, 2 \tag{6.105}$$

where $\eta_1^2 \triangleq 1/\left[\overline{\mathbf{R}}^{-1}\right]_{1,1}$.

We define the following quantities:

$$\begin{aligned}
\mathbf{z} &\triangleq [z_{1,1} z_{2,1} z_{1,2} z_{2,2}]^T \\
\mathbf{u} &\triangleq [u_{1,1} u_{2,1}^* u_{1,2} u_{2,2}^*]^T \\
\mathbf{h}_{1,1} &\triangleq [h_{1,1,1} h_{1,2,1}]^H \\
\bar{\mathbf{h}}_{1,1} &\triangleq [h_{1,2,1} - h_{1,1,1}]^T \\
\mathbf{h}_{1,2} &\triangleq [h_{1,1,2} h_{1,2,2}]^H \\
\bar{\mathbf{h}}_{1,2} &\triangleq [h_{1,2,2} - h_{1,1,2}]^T.
\end{aligned}$$

Then (6.101)–(6.105) can be written as

$$\mathbf{z} = \underbrace{\left[\mathbf{h}_{1,1} \bar{\mathbf{h}}_{1,1} \mathbf{h}_{1,2} \bar{\mathbf{h}}_{1,2}\right]^H}_{\mathbf{H}_1^H} \begin{bmatrix} b_{1,1} \\ b_{2,1} \end{bmatrix} + \mathbf{u}, \tag{6.106}$$

with

$$\mathbf{u} \sim \mathcal{N}_c\left(\mathbf{0}, \frac{\sigma^2}{\eta_1^2} \cdot \mathbf{I}_4\right). \tag{6.107}$$

It is readily verified that

$$\mathbf{H}_1 \mathbf{H}_1^H = \begin{bmatrix} E_1 & 0 \\ 0 & E_1 \end{bmatrix}, \tag{6.108}$$

$$E_1 \triangleq \left|h_{1,1,1}\right|^2 + \left|h_{1,1,2}\right|^2 + \left|h_{1,2,1}\right|^2 + \left|h_{1,2,2}\right|^2. \tag{6.109}$$

To form the maximum-likelihood decision statistic, we premultiply $\mathbf{z}$ by $\mathbf{H}_1$ and obtain

$$\begin{bmatrix} d_{1,1} \\ d_{1,2} \end{bmatrix} \triangleq \mathbf{H}_1 \mathbf{z} = E_1 \begin{bmatrix} b_{1,1} \\ b_{1,2} \end{bmatrix} + \mathbf{v}, \tag{6.110}$$

with

$$\mathbf{v} \sim \mathcal{N}_c \left(\mathbf{0}, \frac{E_1 \sigma^2}{\eta_1^2} \cdot \mathbf{I}_2 \right). \tag{6.111}$$

The corresponding symbol estimates are given by

$$\begin{bmatrix} \hat{b}_{1,1} \\ \hat{b}_{1,2} \end{bmatrix} = \text{sign}\left(\Re \left\{ \begin{bmatrix} d_{1,1} \\ d_{1,2} \end{bmatrix} \right\} \right). \tag{6.112}$$

The bit error probability is then given by

$$\begin{aligned} P_1^{\mathrm{D}}(e) &= P\left(\Re\{d_{1,1}\} < 0 \mid b_{1,1} = +1 \right) \\ &= P\left[E_1 + \mathcal{N}\left(0, \frac{E_1 \sigma^2}{2\eta_1^2} \right) < 0 \right] = Q\left(\frac{\sqrt{2E_1}}{\sigma} \cdot \eta_1 \right), \end{aligned} \tag{6.113}$$

which fully exploits the available antenna diversity.

Linear space–time multi-user detector

Now consider the quantities:

$$\tilde{\mathbf{r}} \triangleq \begin{bmatrix} \mathbf{r}_{1,1} \\ \mathbf{r}_{2,1}^* \\ \mathbf{r}_{1,2} \\ \mathbf{r}_{2,2}^* \end{bmatrix}, \quad \tilde{\mathbf{n}} \triangleq \begin{bmatrix} \mathbf{n}_{1,1} \\ \mathbf{n}_{2,1}^* \\ \mathbf{n}_{1,2} \\ \mathbf{n}_{2,2}^* \end{bmatrix}, \quad \mathbf{h}_k \triangleq \begin{bmatrix} h_{k,1,1} \\ h_{k,2,1}^* \\ h_{k,1,2} \\ h_{k,2,2}^* \end{bmatrix}, \quad \bar{\mathbf{h}}_k \triangleq \begin{bmatrix} h_{k,2,1} \\ -h_{k,1,1}^* \\ h_{k,2,2} \\ -h_{k,1,2}^* \end{bmatrix}. \tag{6.114}$$

Then (6.96)–(6.99) may be written as

$$\tilde{\mathbf{r}} = \sum_{k=1}^{K} \left(b_{k,1} \mathbf{h}_k \otimes \mathbf{s}_k + b_{k,2} \bar{\mathbf{h}}_k \otimes \mathbf{s}_k \right) + \tilde{\mathbf{n}} = \tilde{\mathbf{S}} \mathbf{b} + \tilde{\mathbf{n}}, \tag{6.115}$$

where

$$\tilde{\mathbf{S}} \triangleq \left[\mathbf{h}_1 \otimes \mathbf{s}_1, \bar{\mathbf{h}}_1 \otimes \mathbf{s}_1, \ldots, \mathbf{h}_K \otimes \mathbf{s}_K, \bar{\mathbf{h}}_K \otimes \mathbf{s}_K \right]_{4N \times 2K} \tag{6.116}$$

$$\mathbf{b} \triangleq \left[b_{1,1} b_{1,2} b_{2,1} b_{2,2} \cdots b_{K,1} b_{K,2} \right]^T. \tag{6.117}$$

Since $\mathbf{h}_k^H \bar{\mathbf{h}}_k = 0$ it is easy to show that the decorrelating detector for detecting the symbol $b_{1,1}$ based on $\tilde{\mathbf{r}}$ is given by

$$\tilde{\mathbf{w}}_{1,1} = \frac{\mathbf{h}_1 \otimes \mathbf{w}_1}{\|\mathbf{h}_1\|^2}, \tag{6.118}$$

which we call *linear space–time multi-user detection*. Hence the output of the linear space–time detector in this case is given by

$$\tilde{z}_1 = \tilde{\mathbf{w}}_{1,1}^H \tilde{\mathbf{r}} = b_{1,1} + u_1 \tag{6.119}$$

with

$$u_1 \triangleq \tilde{\mathbf{w}}_{1,1}^H \tilde{\mathbf{n}} \sim \mathcal{N}_c\left(0, \sigma^2 \|\tilde{\mathbf{w}}_{1,1}\|^2\right) \tag{6.120}$$

where

$$\|\tilde{\mathbf{w}}_{1,1}\|^2 = \frac{\|\mathbf{w}_1\|^2}{\|\mathbf{h}_1\|^2} = \frac{1}{E_1 \eta_1^2}. \tag{6.121}$$

Therefore the probability of error is given by

$$\begin{aligned} P_1^{\mathrm{ST}}(e) &= P\left(\Re\{\tilde{z}_1\} < 0 \mid b_{1,1} = +1\right) \\ &= P\left[1 + \mathcal{N}\left(0, \frac{1}{2E_1 \eta_1^2}\right) < 0\right] = Q\left(\frac{\sqrt{2E_1}}{\sigma} \cdot \eta_1\right). \end{aligned} \tag{6.122}$$

Comparing (6.122) with (6.113) it is seen that when two transmitter antennas and two receiver antennas are employed and the signals are transmitted in the form of a space–time block code, then the linear diversity receiver and the linear space–time receiver have identical performance. What, then, are the benefits of the space–time detection technique? They include the following.

1. The user capacity for CDMA systems is limited by correlations among composite signature waveforms. This multiple-access interference will tend to decrease as the dimension of the vector space in which the signature waveforms reside increases. The signature waveforms for linear diversity detection are of length N, i.e., they reside in $\mathbb{C}^N$. Since the received signals are stacked for space–time detection, these signature waveforms reside in $\mathbb{C}^{2N}$ for two transmit and one receive antenna or $\mathbb{C}^{4N}$ for two transmit and two receive antennas. As a result, the space–time structure can support more users than linear diversity detection for a given performance threshold. A specific example of this phenomenon is discussed in Section 6.5.3.
2. For adaptive configurations, linear diversity multi-user detection requires four independent subspace trackers operating simultaneously since the receiver performs detection on each of the four received signals, and each has a different signal subspace. The space–time structure requires only one subspace tracker.

6.5.2 Adaptive linear space–time multi-user detection for flat-fading CDMA

Signal model

Motivated by the above discussion, we now discuss *adaptive* space–time multi-user detection algorithms for systems with two transmitter antennas and two receiver antennas.

These algorithms are also blind, in the sense that the receiver requires knowledge only of the signature waveform of the user of interest, i.e., neither *a priori* channel knowledge nor the spreading codes of the interfering users are necessary for detection. As before, the Alamouti space–time block code is used for transmission, so that during the first symbol interval of block i, user k transmits $(b_{k,1}[i], b_{k,2}[i])$ from the two transmit antennas. During the second symbol interval, user k transmits $(-b_{k,2}[i], b_{k,1}[i])$. Note that inherent to any blind receiver in multiple transmitter antenna systems is an ambiguity issue. That is, if the same spreading waveform is used for a user at both transmitter antennas, the blind receiver cannot distinguish which symbol is transmitted from which antenna. To resolve this ambiguity, we use two different spreading waveforms for each user, i.e., $\mathbf{s}_{k,j}$, $j \in \{1, 2\}$ is the spreading code for user k for the transmission of symbol $b_{k,j}[i]$. The discrete-time received N-vector at base-station antenna 1 during the two symbol periods for block i is

$$\mathbf{r}_{1,1}[i] = \sum_{k=1}^{K} \left(h_{k,1,1} b_{k,1}[i] \mathbf{s}_{k,1} + h_{k,2,1} b_{k,2}[i] \mathbf{s}_{k,2}\right) + \mathbf{n}_{1,1}[i] \tag{6.123}$$

and

$$\mathbf{r}_{2,1}[i] = \sum_{k=1}^{K} \left(-h_{k,1,1} b_{k,2}[i] \mathbf{s}_{k,2} + h_{k,2,1} b_{k,1}[i] \mathbf{s}_{k,1}\right) + \mathbf{n}_{2,1}[i], \tag{6.124}$$

and the corresponding signals received at antenna 2 are

$$\mathbf{r}_{1,2}[i] = \sum_{k=1}^{K} \left(h_{k,1,2} b_{k,1}[i] \mathbf{s}_{k,1} + h_{k,2,2} b_{k,2}[i] \mathbf{s}_{k,2}\right) + \mathbf{n}_{1,2}[i] \tag{6.125}$$

and

$$\mathbf{r}_{2,2}[i] = \sum_{k=1}^{K} \left(-h_{k,1,2} b_{k,2}[i] \mathbf{s}_{k,2} + h_{k,2,2} b_{k,1}[i] \mathbf{s}_{k,1}\right) + \mathbf{n}_{2,2}[i]. \tag{6.126}$$

We stack the received signal vectors and denote

$$\tilde{\mathbf{r}}[i] \triangleq \begin{bmatrix} \mathbf{r}_{1,1}[i] \\ \mathbf{r}^*_{2,1}[i] \\ \mathbf{r}_{1,2}[i] \\ \mathbf{r}^*_{2,2}[i] \end{bmatrix}, \qquad \tilde{\mathbf{n}}[i] \triangleq \begin{bmatrix} \mathbf{n}_{1,1}[i] \\ \mathbf{n}^*_{2,1}[i] \\ \mathbf{n}_{1,2}[i] \\ \mathbf{n}^*_{2,2}[i] \end{bmatrix},$$

$$\mathbf{h}_k \triangleq \begin{bmatrix} h_{k,1,1} \\ h^*_{k,2,1} \\ h_{k,1,2} \\ h^*_{k,2,2} \end{bmatrix}, \qquad \bar{\mathbf{h}}_k \triangleq \begin{bmatrix} h_{k,2,1} \\ -h^*_{k,1,1} \\ h_{k,2,2} \\ -h^*_{k,1,2} \end{bmatrix}. \tag{6.127}$$

Then we have

$$\tilde{\mathbf{r}}[i] = \sum_{k=1}^{K} \left(b_{k,1}[i] \mathbf{h}_k \otimes \mathbf{s}_{k,1} + b_{k,2}[i] \bar{\mathbf{h}}_k \otimes \mathbf{s}_{k,2}\right) + \tilde{\mathbf{n}}[i] \tag{6.128}$$

$$= \tilde{\mathbf{S}} \mathbf{b}[i] + \tilde{\mathbf{n}}[i], \tag{6.129}$$

where

$$\tilde{\mathbf{S}} \triangleq \left[\mathbf{h}_1 \otimes \mathbf{s}_{1,1}, \bar{\mathbf{h}}_1 \otimes \mathbf{s}_{1,2}, \ldots, \mathbf{h}_K \otimes \mathbf{s}_{K,1}, \bar{\mathbf{h}}_K \otimes \mathbf{s}_{K,2}\right]_{4N\times 2K}$$

$$\mathbf{b}[i] \triangleq \left[b_{1,1}[i]b_{1,2}[i]b_{2,1}[i]b_{2,2}[i]\cdots b_{K,1}[i]b_{K,2}[i]\right]^T_{2K\times 1}$$

and where $\otimes$ denotes the Kronecker product. The auto-correlation matrix of the stacked signal $\tilde{\mathbf{r}}[i]$, $\mathbf{C}$, and its eigendecomposition are given by

$$\mathbf{C} = \mathsf{E}\left[\tilde{\mathbf{r}}[i]\tilde{\mathbf{r}}[i]^H\right] = \tilde{\mathbf{S}}\tilde{\mathbf{S}}^H + \sigma^2\mathbf{I}_{4N} \tag{6.130}$$

$$= \mathbf{U}_s\boldsymbol{\Lambda}_s\mathbf{U}_s^H + \sigma^2\mathbf{U}_n\mathbf{U}_n^H, \tag{6.131}$$

where $\boldsymbol{\Lambda}_s = \text{diag}\{\lambda_1, \lambda_2, \ldots, \lambda_{2K}\}$ contains the largest $(2K)$ eigenvalues of $\mathbf{C}$, the columns of $\mathbf{U}_s$ are the corresponding eigenvectors; and the columns of $\mathbf{U}_n$ are the $(4N-2K)$ eigenvectors corresponding to the smallest eigenvalue σ^2.

The blind linear space–time MMSE filter for joint suppression of multiple access interference and space–time decoding for symbol $[\mathbf{b}[i]]_1 = b_{1,1}[i]$ is given by the solution to the optimization problem

$$\mathbf{w}_{1,1} \triangleq \arg\min_{\mathbf{w}\in\mathbb{C}^{4N}} \mathsf{E}\left[\left|b_{1,1}[i] - \mathbf{w}^H\tilde{\mathbf{r}}[i]\right|^2\right]. \tag{6.132}$$

It has been shown in [43, 46] that a scaled version of the solution can be written in terms of the signal subspace components as

$$\mathbf{w}_{1,1} = \mathbf{U}_s\boldsymbol{\Lambda}_s^{-1}\mathbf{U}_s^H\left(\mathbf{h}_1 \otimes \mathbf{s}_{1,1}\right), \tag{6.133}$$

and the decision is made according to

$$z_{1,1}[i] = \mathbf{w}_{1,1}^H\tilde{\mathbf{r}}[i], \tag{6.134}$$

$$\hat{b}_{1,1}[i] = \text{sign}\left[\Re\left(z_{1,1}[i]\right)\right] \quad \text{(coherent detection)}, \tag{6.135}$$

and

$$\hat{\beta}_{1,1}[i] = \text{sign}\left[\Re\left(z_{1,1}[i-1]^* z_{1,1}[i]\right)\right] \quad \text{(differential detection)}. \tag{6.136}$$

Before we address specific batch and sequential adaptive algorithms, we note that these algorithms can also be implemented using linear *group-blind* multi-user detectors [41] which, in contrast to their blind counterparts, are constructed with knowledge of the spreading codes of a subset of the active users. They would be appropriate, for example, in cellular uplink environments in which the receiver has knowledge of the signature waveforms of all of the users in its cell, but not those of interfering users outside the cell. Specifically, we may re-write (6.129) as

$$\tilde{\mathbf{r}}[i] = \breve{\mathbf{S}}\breve{\mathbf{b}}[i] + \bar{\mathbf{S}}\bar{\mathbf{b}}[i] + \tilde{\mathbf{n}}[i], \tag{6.137}$$

where we have separated the users into two groups. The signature sequences of the known users are the columns of $\breve{\mathbf{S}}$. The unknown users' sequences are the columns of $\bar{\mathbf{S}}$. Then the group-blind linear hybrid detector for symbol $b_{1,1}[i]$ is given by [41]

$$\mathbf{w}_{1,1}^{GB} = \mathbf{U}_s \mathbf{\Lambda}_s^{-1} \mathbf{U}_s^H \breve{\mathbf{S}} \left[\breve{\mathbf{S}}^H \mathbf{U}_s \mathbf{\Lambda}_s^{-1} \mathbf{U}_s^H \breve{\mathbf{S}} \right]^{-1} \left(\mathbf{h}_1 \otimes \mathbf{s}_{1,1} \right). \tag{6.138}$$

This detector offers a significant performance improvement over blind implementations of (6.133) for environments in which the signature sequences of some of the interfering users are known.

Batch blind linear space–time multi-user detection

Implementation of (6.133) requires knowledge of the signal subspace components and the channel. The subspace components can be estimated blindly from the received signal using the sample auto-correlation matrix of the received signal. In order to obtain an estimate of $\mathbf{h}_1$ we make use of the orthogonality between the signal and noise subspaces, i.e., the fact that $\mathbf{U}_n^H \left(\mathbf{h}_1 \otimes \mathbf{s}_{1,1} \right) = \mathbf{0}$. In particular, we have

$$\begin{aligned}
\hat{\mathbf{h}}_1 &= \arg \min_{\mathbf{h} \in \mathbb{C}^4} \left\| \mathbf{U}_n^H \left(\mathbf{h} \otimes \mathbf{s}_{1,1} \right) \right\|^2 \\
&= \arg \max_{\mathbf{h} \in \mathbb{C}^4} \left\| \mathbf{U}_s^H \left(\mathbf{h} \otimes \mathbf{s}_{1,1} \right) \right\|^2 \\
&= \arg \max_{\mathbf{h} \in \mathbb{C}^4} \left(\mathbf{h}^H \otimes \mathbf{s}_{1,1}^T \right) \mathbf{U}_s \mathbf{U}_s^H \left(\mathbf{h} \otimes \mathbf{s}_{11} \right) \\
&= \arg \max_{\mathbf{h} \in \mathbb{C}^4} \mathbf{h}^H \underbrace{\left[\left(\mathbf{I}_4 \otimes \mathbf{s}_{1,1}^T \right) \mathbf{U}_s \mathbf{U}_s^H \left(\mathbf{I}_4 \otimes \mathbf{s}_{1,1} \right) \right]}_{\mathbf{Q}} \mathbf{h} && (6.139) \\
&= \text{principal eigenvector of } \mathbf{Q}. && (6.140)
\end{aligned}$$

In (6.140), $\hat{\mathbf{h}}_1$ specifies $\mathbf{h}_1$ up to an arbitrary complex scale factor α, i.e., $\hat{\mathbf{h}}_1 = \alpha \mathbf{h}_1$, but this ambiguity can be circumvented using differential modulation and detection. The following is the summary of a batch blind space–time multi-user detection algorithm for the two transmitter antenna/two receiver antenna configuration. The channel is assumed to be constant for at least the duration of the batch size M.

Algorithm 1 (Batch blind linear space–time multi-user detector: synchronous CDMA, two transmitter antennas and two receiver antennas)

- *Estimate the signal subspace:*

$$\begin{aligned}
\hat{\mathbf{C}} &= \frac{1}{M} \sum_{i=0}^{M-1} \tilde{\mathbf{r}}[i] \tilde{\mathbf{r}}[i]^H, && (6.141) \\
&= \hat{\mathbf{U}}_s \hat{\mathbf{\Lambda}}_s \hat{\mathbf{U}}_s^H + \hat{\mathbf{U}}_n \hat{\mathbf{\Lambda}}_n \hat{\mathbf{U}}_n^H. && (6.142)
\end{aligned}$$

- *Estimate the channels:*

$$\hat{\mathbf{Q}}_1 = \left(\mathbf{I}_4 \otimes \mathbf{s}_{1,1}^T\right) \hat{\mathbf{U}}_s \hat{\mathbf{U}}_s^H \left(\mathbf{I}_4 \otimes \mathbf{s}_{1,1}\right), \tag{6.143}$$

$$\hat{\mathbf{Q}}_2 = \left(\mathbf{I}_4 \otimes \mathbf{s}_{1,2}^T\right) \hat{\mathbf{U}}_s \hat{\mathbf{U}}_s^H \left(\mathbf{I}_4 \otimes \mathbf{s}_{1,2}\right), \tag{6.144}$$

$$\hat{\mathbf{h}}_1 = \text{principal eigenvector of } \hat{\mathbf{Q}}_1, \tag{6.145}$$

$$\hat{\hat{\mathbf{h}}}_1 = \text{principal eigenvector of } \hat{\mathbf{Q}}_2. \tag{6.146}$$

- *Form the detectors:*

$$\hat{\mathbf{w}}_{1,1} = \hat{\mathbf{U}}_s \hat{\boldsymbol{\Lambda}}_s^{-1} \hat{\mathbf{U}}_s^H \left(\hat{\mathbf{h}}_1 \otimes \mathbf{s}_{1,1}\right), \tag{6.147}$$

$$\hat{\mathbf{w}}_{1,2} = \hat{\mathbf{U}}_s \hat{\boldsymbol{\Lambda}}_s^{-1} \hat{\mathbf{U}}_s^H \left(\hat{\hat{\mathbf{h}}}_1 \otimes \mathbf{s}_{1,2}\right). \tag{6.148}$$

- *Perform differential detection:*

$$z_{1,1}[i] = \hat{\mathbf{w}}_{1,1}^H \tilde{\mathbf{r}}[i], \tag{6.149}$$

$$z_{1,2}[i] = \hat{\mathbf{w}}_{1,2}^H \tilde{\mathbf{r}}[i], \tag{6.150}$$

$$\hat{\beta}_{1,1}[i] = \text{sign}\left(\Re\left\{z_{1,1}[i] z_{1,1}[i-1]^*\right\}\right), \tag{6.151}$$

$$\hat{\beta}_{1,2}[i] = \text{sign}\left(\Re\left\{z_{1,2}[i] z_{1,2}[i-1]^*\right\}\right), \tag{6.152}$$

$$i = 0, \ldots, M-1.$$

A batch group-blind space–time multi-user detector algorithm can be implemented with simple modifications to (6.147) and (6.148).

Adaptive blind linear space–time multi-user detection

To form a sequential blind adaptive receiver, we need adaptive algorithms for sequentially estimating the channel and the signal subspace components $\mathbf{U}_s$ and $\boldsymbol{\Lambda}_s$. First, we address sequential adaptive channel estimation. Denote by $\mathbf{z}[i]$ the projection of the stacked signal $\tilde{\mathbf{r}}[i]$ onto the noise subspace, i.e.,

$$\mathbf{z}[i] = \tilde{\mathbf{r}}[i] - \mathbf{U}_s \mathbf{U}_s^H \tilde{\mathbf{r}}[i] \tag{6.153}$$

$$= \mathbf{U}_n \mathbf{U}_n^H \tilde{\mathbf{r}}[i]. \tag{6.154}$$

Since $\mathbf{z}[i]$ lies in the noise subspace, it is orthogonal to any signal in the signal subspace, and in particular, it is orthogonal to $(\mathbf{h}_1 \otimes \mathbf{s}_{1,1})$. Hence $\mathbf{h}_1$ is the solution to the following constrained optimization problem:

$$\min_{\mathbf{h}_1 \in \mathbb{C}^4} \mathsf{E}\left[\left\| \mathbf{z}[i]^H (\mathbf{h}_1 \otimes \mathbf{s}_{1,1}) \right\|^2 \right]$$
$$= \min_{\mathbf{h}_1 \in \mathbb{C}^4} \mathsf{E}\left[\left\| \mathbf{z}[i]^H (\mathbf{I}_4 \otimes \mathbf{s}_{1,1}) \mathbf{h}_1 \right\|^2 \right]$$
$$= \min_{\mathbf{h}_1 \in \mathbb{C}^4} \mathsf{E}\left[\left\| \left[\left(\mathbf{I}_4 \otimes \mathbf{s}_{1,1}^T\right) \mathbf{z}[i] \right]^H \mathbf{h}_1 \right\|^2 \right] \text{s.t.} \|\mathbf{h}_1\| = 1. \tag{6.155}$$

In order to obtain a sequential algorithm to solve the above optimization problem, we write it in the following (trivial) state space form

$$\mathbf{h}_1[i+1] = \mathbf{h}_1[i], \qquad \text{state equation}$$
$$0 = \left[\left(\mathbf{I}_4 \otimes \mathbf{s}_{1,1}^T\right) \mathbf{z}[i] \right]^H \mathbf{h}_1[i], \qquad \text{observation equation.}$$

The standard Kalman filter can then be applied to the above system as follows. Denote $\mathbf{x}[i] \triangleq \left(\mathbf{I}_4 \otimes \mathbf{s}_{1,1}^T\right) \mathbf{z}[i]$:

$$\mathbf{k}[i] = \mathbf{\Sigma}[i-1]\mathbf{x}[i] \left(\mathbf{x}[i]^H \mathbf{\Sigma}[i-1]\mathbf{x}[i]\right)^{-1}, \tag{6.156}$$
$$\mathbf{h}_1[i] = \left(\mathbf{h}_1[i-1] - \mathbf{k}[i]\left(\mathbf{x}[i]^H \mathbf{h}_1[i-1]\right)\right) / \left\| \mathbf{h}_1[i-1] - \mathbf{k}[i]\left(\mathbf{x}[i]^H \mathbf{h}_1[i-1]\right) \right\|, \tag{6.157}$$
$$\mathbf{\Sigma}[i] = \mathbf{\Sigma}[i-1] - \mathbf{k}[i]\mathbf{x}[i]^H \mathbf{\Sigma}[i-1]. \tag{6.158}$$

Once we have obtained channel estimates at block i, we can combine them with estimates of the signal subspace components to form the detector in (6.133). Subspace tracking algorithms of various complexities exist in the literature. Since we are stacking received signal vectors and subspace tracking complexity increases at least linearly with the signal subspace dimension, it is imperative that we choose an algorithm with minimal complexity. The best existing low-complexity algorithm for this purpose appears to be noise-averaged Hermitian–Jacobi fast subspace tracking (NAHJ-FST) [34]. This algorithm has the lowest complexity of any algorithm used for similar purposes and has performed well when used for signal subspace tracking in multi-path fading environments. Since the size of $\mathbf{U}_s$ is $4N \times 2K$, the complexity is $40 \times 4N \times 2K + 3 \times 4N + 7.5(2K)^2 + 7 \times 2K$ floating point operations per iteration. The algorithm and a multi-user detection application are presented in [34]. The application to the current tracking problem is straightforward and will not be discussed in detail.

Algorithm 2 (Blind adaptive linear space–time multi-user detector: synchronous CDMA, two transmitter antennas, and two receiver antennas)

- *Using a suitable signal subspace tracking algorithm, e.g., NAHJ-FST, update the signal subspace components* $\mathbf{U}_s[i]$ *and* $\mathbf{\Lambda}_s[i]$ *at each block i.*
- *Track the channel* $\mathbf{h}_1[i]$ *and* $\bar{\mathbf{h}}_1[i]$ *according to the following:*

$$\mathbf{z}[i] = \tilde{\mathbf{r}}[i] - \mathbf{U}_s[i]\mathbf{U}_s[i]^H\tilde{\mathbf{r}}[i], \tag{6.159}$$

$$\mathbf{x}[i] = \left(\mathbf{I}_4 \otimes \mathbf{s}_{1,1}^T\right)\mathbf{z}[i], \tag{6.160}$$

$$\bar{\mathbf{x}}[i] = \left(\mathbf{I}_4 \otimes \mathbf{s}_{1,2}^T\right)\mathbf{z}[i], \tag{6.161}$$

$$\mathbf{k}[i] = \mathbf{\Sigma}[i-1]\mathbf{x}[i]\left(\mathbf{x}[i]^H\mathbf{\Sigma}[i-1]\mathbf{x}[i]\right)^{-1}, \tag{6.162}$$

$$\bar{\mathbf{k}}[i] = \bar{\mathbf{\Sigma}}[i-1]\bar{\mathbf{x}}[i]\left(\bar{\mathbf{x}}[i]^H\bar{\mathbf{\Sigma}}[i-1]\bar{\mathbf{x}}[i]\right)^{-1}, \tag{6.163}$$

$$\mathbf{h}_1[i] = \left(\mathbf{h}_1[i-1] - \mathbf{k}[i]\left(\mathbf{x}[i]^H\mathbf{h}_1[i-1]\right)\right) / \left\|\mathbf{h}_1[i-1] - \mathbf{k}[i]\left(\mathbf{x}[i]^H\mathbf{h}_1[i-1]\right)\right\|, \tag{6.164}$$

$$\bar{\mathbf{h}}_1[i] = \left(\bar{\mathbf{h}}_1[i-1] - \bar{\mathbf{k}}[i]\left(\bar{\mathbf{x}}[i]^H\bar{\mathbf{h}}_1[i-1]\right)\right) / \left\|\bar{\mathbf{h}}_1[i-1] - \bar{\mathbf{k}}[i]\left(\bar{\mathbf{x}}[i]^H\bar{\mathbf{h}}_1[i-1]\right)\right\|, \tag{6.165}$$

$$\mathbf{\Sigma}[i] = \mathbf{\Sigma}[i-1] - \mathbf{k}[i]\mathbf{x}[i]^H\mathbf{\Sigma}[i-1], \tag{6.166}$$

$$\bar{\mathbf{\Sigma}}[i] = \bar{\mathbf{\Sigma}}[i-1] - \bar{\mathbf{k}}[i]\bar{\mathbf{x}}[i]^H\bar{\mathbf{\Sigma}}[i-1]. \tag{6.167}$$

- *Form the detectors:*

$$\hat{\mathbf{w}}_{1,1}[i] = \mathbf{U}_s[i]\mathbf{\Lambda}_s^{-1}[i]\mathbf{U}_s[i]^H\left(\mathbf{h}_1[i] \otimes \mathbf{s}_{1,1}\right), \tag{6.168}$$

$$\hat{\mathbf{w}}_{1,2}[i] = \mathbf{U}_s[i]\mathbf{\Lambda}_s^{-1}[i]\mathbf{U}_s[i]^H\left(\bar{\mathbf{h}}_1[i] \otimes \mathbf{s}_{1,2}\right). \tag{6.169}$$

- *Perform differential detection:*

$$z_{1,1}[i] = \hat{\mathbf{w}}_{1,1}[i]^H\tilde{\mathbf{r}}[i], \tag{6.170}$$

$$z_{1,2}[i] = \hat{\mathbf{w}}_{1,2}[i]^H\tilde{\mathbf{r}}[i], \tag{6.171}$$

$$\hat{\beta}_{1,1}[i] = \text{sign}\left(\Re\left\{z_{1,1}[i]z_{1,1}[i-1]^*\right\}\right), \tag{6.172}$$

$$\hat{\beta}_{1,2}[i] = \text{sign}\left(\Re\left\{z_{1,2}[i]z_{1,2}[i-1]^*\right\}\right). \tag{6.173}$$

A group-blind sequential adaptive space–time multi-user detector can be implemented similarly. The adaptive receiver structure is illustrated in Fig. 6.13.

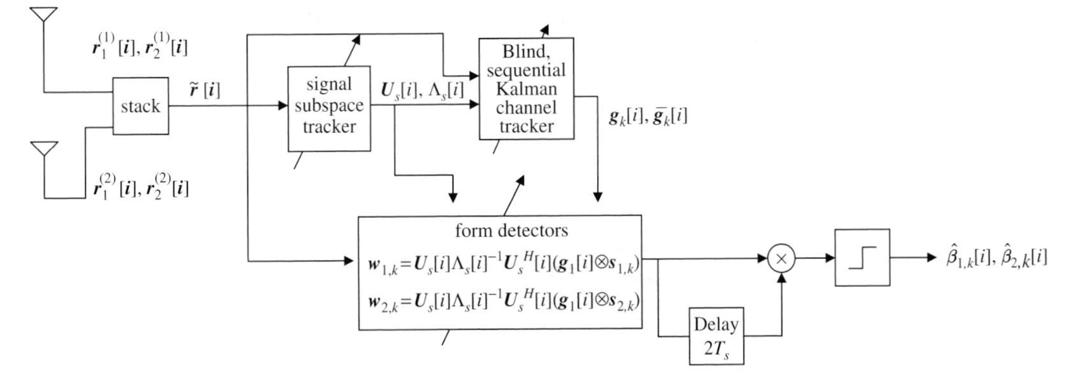

Figure 6.13. Adaptive receiver structure for linear space–time multi-user detectors.

6.5.3 *Blind adaptive space–time multi-user detection for asynchronous CDMA in fading multi-path channels*

Signal model

To extend the previous development to asynchronous multi-path channels, we must begin with a continuous-time baseband signal model. The signal transmitted from antennas 1 and 2 due to the kth user for time interval $i \in \{0, 1, \ldots, M-1\}$ is given by

$$x_{k,1}(t) = \sum_{i=0}^{M-1} \left[b_{k,1}[i] s_{k,1}(t - 2iT_s) - b_{k,2}[i] s_{k,2}(t - (2i+1)T_s) \right] \tag{6.174}$$

$$x_{k,2}(t) = \sum_{i=0}^{M-1} \left[b_{k,2}[i] s_{k,2}(t - 2iT_s) + b_{k,1}[i] s_{k,1}(t - (2i+1)T_s) \right] \tag{6.175}$$

where M denotes the length of the data frame, T_s denotes the information symbol interval, and $\{b_k[i]\}_i$ is the symbol stream of user k. Although this is an asynchronous system, we have, for notational simplicity, suppressed the delay associated with each user's transmitted signal and incorporated it into the path delays in (6.3). We assume that for each k, the symbol stream, $\{b_k[i]\}_i$, is a collection of independent random variables that take on values of $+1$ and -1 with equal probability. Furthermore, we assume that the symbol streams of different users are independent. The transmitted signature waveforms $\{s_{k,m}(t)\}$ are described in (6.26). The kth user's space–time coded signals, $x_{k,1}(t)$ and $x_{k,2}(t)$, propagate from transmitter to receiver through the multi-path fading channel described by (6.3), where $\tau_{k,m,p,l}$, satisfying $\tau_{k,m,p,1} \leq \tau_{k,m,p,2} \leq \cdots \leq \tau_{k,m,p,L}$, is the sum of the corresponding path delay and the initial transmission delay of user k. It is assumed that the channel is slowly varying, so that the path gains and delays remain constant over the duration of one signal frame (MT_s).

The received signal component due to the transmission of $x_{k,1}(t)$ and $x_{k,2}(t)$ through the channel at receiver antennas 1 and 2 is given by

$$y_{k,1}(t) = x_{k,1}(t) \star h_{k,1,1}(t) + x_{k,2}(t) \star h_{k,2,1}(t), \tag{6.176}$$

$$y_{k,2}(t) = x_{k,1}(t) \star h_{k,1,2}(t) + x_{k,2}(t) \star h_{k,2,2}(t). \tag{6.177}$$

The total received signal at receiver antenna $b \in \{1, 2\}$ is given by

$$r_b(t) = \sum_{k=1}^{K} y_{k,b}(t) + n_b(t). \tag{6.178}$$

At the receiver, the received signal is match filtered to the chip waveform and sampled at the chip rate, i.e., the sampling interval is T_c, N is the total number of samples per symbol interval, and $2N$ is the total number of samples per time slot. The nth matched filter output during the ith time slot is given by

$$r_b[i, n] \triangleq \int_{2iT_s+nT_c}^{2iT_s+(n+1)T_c} r_b(t) \psi(t - 2iT_s - nT_c) \mathrm{d}t. \tag{6.179}$$

Denote the maximum delay (in symbol intervals) by

$$\iota_{k,m,p} \triangleq \left\lceil \frac{\tau_{k,m,p,L} + T_c}{T_s} \right\rceil \quad \text{and} \quad \iota \triangleq \max_{k,m,p} \iota_{k,m,p}. \tag{6.180}$$

Closed-form expressions for the matched filter outputs $r_b[i, n]$ are provided in [35].

To fully exploit available diversity, we stack the matched filter outputs from both receive antennas, forming the vector

$$\underline{r}[i] \triangleq \begin{bmatrix} \underline{r}_1[i] \\ \underline{r}_2[i] \end{bmatrix}_{4N\times 1}, \tag{6.181}$$

where, for $b \in \{1, 2\}$,

$$\underline{r}_b[i] \triangleq \begin{bmatrix} r_b[i, 0] \\ \vdots \\ r_b[i, 2N-1] \end{bmatrix}_{2N\times 1}. \tag{6.182}$$

Stacking $\bar{m}$ successive sample vectors, we form

$$\mathbf{r}[i] \triangleq \begin{bmatrix} \underline{r}[i] \\ \vdots \\ \underline{r}[i+m-1] \end{bmatrix}_{4N\bar{m}\times 1} \tag{6.183}$$

$$= \mathbf{H}\mathbf{b}[i] + \mathbf{n}[i], \tag{6.184}$$

where $\mathbf{H}$ is a function of the spreading codes, the channel conditions, and the chip waveform (see [35] for details), $\mathbf{n}[i]$ is additive white Gaussian noise, and where

$$\mathbf{b}[i] \triangleq \begin{bmatrix} \underline{b}[i - \lceil \iota/2 \rceil] \\ \vdots \\ \underline{b}[i+m-1] \end{bmatrix}_{r\times 1}, \quad \underline{b}[i] \triangleq \begin{bmatrix} b_{1,1}[i] \\ \vdots \\ b_{K,1}[i] \\ b_{1,2}[i] \\ \vdots \\ b_{K,2}[i] \end{bmatrix}_{2K\times 1}, \tag{6.185}$$

and $r \triangleq 2K(\bar{m} + \lceil \iota/2 \rceil)$.

We will see on page 285 that the smoothing factor, $\bar{m}$, is chosen such that

$$\bar{m} \geq \left\lceil \frac{N(\iota+1) + K\lceil \iota/2 \rceil + 1}{2N - K} \right\rceil \tag{6.186}$$

for channel identifiability. Note that the columns of $\mathbf{H}$ (the composite signature vectors) contain information about both the timings and the complex path gains of the multi-path channel of each user. Hence an estimate of these waveforms eliminates the need for separate estimates of the timing information $\{\tau_{k,m,p,l}\}$.

Blind MMSE space–time multi-user detection

Since the ambient noise is white, i.e., $\mathsf{E}[\mathbf{n}[i]\mathbf{n}[i]^H] = \sigma^2 \mathbf{I}_{4N\bar{m}}$, the auto-correlation matrix of the received signal in (6.184) is

$$\mathbf{C}_{\mathbf{r}} \stackrel{\triangle}{=} \mathsf{E}[\mathbf{r}[i]\mathbf{r}[i]^H] = \mathbf{H}\mathbf{H}^H + \sigma^2 \mathbf{I}_{4N\bar{m}} \tag{6.187}$$

$$= \mathbf{U}_s \boldsymbol{\Lambda}_s \mathbf{U}_s^H + \sigma^2 \mathbf{U}_n \mathbf{U}_n^H, \tag{6.188}$$

where (6.188) is the eigendecomposition of $\mathbf{C}_{\mathbf{r}}$. Note that $\mathbf{U}_s$ has size $4N\bar{m} \times r$ and $\mathbf{U}_n$ has size $4N\bar{m} \times (4N\bar{m} - r)$.

The joint MMSE multi-user detector and space–time decoder with corresponding symbol estimate for $b_{k,a}[i]$, $a \in \{1, 2\}$ are given by

$$\mathbf{w}_{k,a}[i] \stackrel{\triangle}{=} \arg \min_{\mathbf{w} \in \mathbb{C}^{4P\bar{m}}} \mathsf{E}\left[\left|b_{k,a}[i] - \mathbf{w}^H \mathbf{r}[i]\right|^2\right], \tag{6.189}$$

$$\hat{b}_{k,a}[i] = \text{sign}\left[\text{Re}\left\{\mathbf{w}_{k,a}[i]^H \mathbf{r}[i]\right\}\right]. \tag{6.190}$$

The solution to (6.189) can be written in terms of the signal subspace components as [42]

$$\mathbf{w}_{k,a}[i] = \mathbf{U}_s \boldsymbol{\Lambda}_s^{-1} \mathbf{U}_s^H \mathbf{h}_{k,a}, \tag{6.191}$$

where $\mathbf{h}_{k,a} \stackrel{\triangle}{=} \mathbf{H}\mathbf{e}_{K(2\lceil\iota/2\rceil+a-1)+k}$ is the composite signature waveform of user k for symbol $a \in \{1, 2\}$. As for the synchronous case, this detector can be implemented in blind mode, requiring knowledge only of the signature sequence of the user of interest and a (blind) estimate of the channel.

Blind sequential Kalman channel estimation

The full details of the discrete-time channel model for the asynchronous multi-path case appear in [35]. In summary, the composite signature waveform of user k for symbol a can be written as

$$\mathbf{h}_{k,a} = \overline{\mathbf{C}}_{k,a} \mathbf{f}_{k,a} \tag{6.192}$$

where $\overline{\mathbf{C}}_{k,a}$ is a matrix of size $4N(\lceil\iota/2\rceil + 1) \times (2N(\iota + 1) + 2)$ that is constructed from the ath spreading code assigned to user k. The vector $\mathbf{f}_{k,a}$, with size $(2N(\iota+1)+2) \times 1$, is a function of the channel state information for user k and is also defined in [35]. The blind channel estimation problem involves the estimation of $\mathbf{f}_{k,a} (1 \le k \le K, a = 1, 2)$ from the received signal $\mathbf{r}[i]$. As we did for the synchronous case, we will exploit the orthogonality between the signal subspace and the noise subspace. Specifically, since $\mathbf{U}_n$ is orthogonal to the columnspace of $\mathbf{H}$, we have

$$\mathbf{U}_n^H \mathbf{h}_{k,a} = \mathbf{U}_n^H \overline{\mathbf{C}}_{k,a} \mathbf{f}_{k,a} = \mathbf{0}. \tag{6.193}$$

Denote by $\mathbf{z}[i]$ the projection of the received signal $\mathbf{r}[i]$ onto the noise subspace, i.e.,

$$\mathbf{z}[i] = \mathbf{r}[i] - \mathbf{U}_s\mathbf{U}_s^H\mathbf{r}[i] \tag{6.194}$$

$$= \mathbf{U}_n\mathbf{U}_n^H\mathbf{r}[i]. \tag{6.195}$$

Using (6.193) we have

$$\mathbf{f}_{k,a}^H\overline{\mathbf{C}}_{k,a}^H\mathbf{z}[i] = 0. \tag{6.196}$$

Our channel estimation problem, then, involves the solution of the optimization problem

$$\hat{\mathbf{f}}_{k,a} = \arg\min_{\mathbf{f}} \mathsf{E}\left[\left|\mathbf{f}^H\overline{\mathbf{C}}_{k,a}^H\mathbf{z}[i]\right|^2\right] \tag{6.197}$$

subject to the constraint $\|\mathbf{f}\| = 1$. If we denote $\mathbf{x}[i] \triangleq \overline{\mathbf{C}}_{k,a}^H\mathbf{z}[i]$ then we can use the Kalman-type algorithm described in (6.156)–(6.158) where $\mathbf{h}_1[i]$ is replaced with $\mathbf{f}_{k,a}[i]$.

Note that a necessary condition for the channel estimate to be unique is that the matrix $\mathbf{U}_n^H\overline{\mathbf{C}}_{k,a}$ is tall, i.e., $4N\bar{m} - 2K(\bar{m} + \lceil\iota/2\rceil) \geq 2N(\iota+1) + 2$. Therefore we choose the smoothing factor, $\bar{m}$, such that

$$\bar{m} \geq \left\lceil \frac{N(\iota+1) + K\lceil\iota/2\rceil + 1}{2N - K} \right\rceil. \tag{6.198}$$

Using the same constraint, we find that for a fixed m, the maximum number of users that can be supported is

$$\min\left\{\left\lfloor \frac{N(2\bar{m} - \iota - 1) - 1}{\bar{m} + \lceil\iota/2\rceil} \right\rfloor, \left\lfloor \frac{N}{2} \right\rfloor\right\}. \tag{6.199}$$

Notice that for reasonable choices of $\bar{m}$ and ι, (6.199) is larger than the maximum number of users for the linear diversity receiver structure, given by

$$\left\lfloor \frac{N(\bar{m} - \iota)}{2(\bar{m} + \iota)} \right\rfloor. \tag{6.200}$$

This represents a quantitative example of the user capacity benefit of space–time multi-user detection discussed in Section 6.5.1.

Once an estimate of the channel state, $\hat{\mathbf{f}}_{k,a}$, is obtained, the composite signature vector of the kth user for symbol a is given by (6.192). Note that there is an arbitrary phase ambiguity in the estimated channel state, which necessitates differential encoding and decoding of the transmitted data.

Algorithm summary

Algorithm 3 (Blind adaptive linear space–time multi-user detector: asynchronous multi-path CDMA, two transmitter antennas and two receiver antennas)

- *Stack matched filter outputs in (6.179) to create* $\mathbf{r}[i]$.
- *Create* $\overline{\mathbf{C}}_{k,a}$.
- *Using a suitable signal subspace tracking algorithm, e.g., NAHJ-FST, update the signal subspace components* $\mathbf{U}_s[i]$ *and* $\mathbf{\Lambda}_s[i]$ *at each time slot i.*
- *Track the channel* $\mathbf{f}_{k,a}(1 \leq k \leq K, a = 1, 2)$ *according to the following:*

$$\mathbf{z}[i] = \mathbf{r}[i] - \mathbf{U}_s[i]\mathbf{U}_s[i]^H\mathbf{r}[i], \tag{6.201}$$

$$\mathbf{x}[i] = \overline{\mathbf{C}}_{k,a}^H\mathbf{z}[i], \tag{6.202}$$

$$\mathbf{k}[i] = \mathbf{\Sigma}[i-1]\mathbf{x}[i]\left(\mathbf{x}[i]^H\mathbf{\Sigma}[i-1]\mathbf{x}[i]\right)^{-1}, \tag{6.203}$$

$$\mathbf{f}_{k,a}[i] = \left(\mathbf{f}_{k,a}[i-1] - \mathbf{k}[i]\left(\mathbf{x}[i]^H\mathbf{f}_{k,a}[i-1]\right)\right) / \left\|\mathbf{f}_{k,a}[i-1] - \mathbf{k}[i]\left(\mathbf{x}[i]^H\mathbf{f}_{k,a}[i-1]\right)\right\|, \tag{6.204}$$

$$\mathbf{\Sigma}[i] = \mathbf{\Sigma}[i-1] - \mathbf{k}[i]\mathbf{x}[i]^H\mathbf{\Sigma}[i-1]. \tag{6.205}$$

- *Form the detectors:*

$$\mathbf{w}_{k,a}[i] = \mathbf{U}_s[i]\mathbf{\Lambda}_s^{-1}[i]\mathbf{U}_s[i]^H\overline{\mathbf{C}}_{k,a}\mathbf{f}_{k,a}[i]. \tag{6.206}$$

- *Perform differential detection:*

$$z_{k,a}[i] = \mathbf{w}_{k,a}[i]^H\mathbf{r}[i], \tag{6.207}$$

$$\hat{\beta}_{k,a}[i] = \text{sign}\left(\Re\left\{z_{k,a}[i]z_{k,a}[i-1]^*\right\}\right). \tag{6.208}$$

6.5.4 *Simulation results*

In this section, we present simulation results to illustrate the performance of blind adaptive space–time multi-user detection. We first look at the synchronous flat-fading case; then we consider the asynchronous multi-path-fading scenario. For all simulations we use the two-transmit/two-receive antenna configuration. m-sequences of length 15 and their shifted versions are employed as user spreading sequences. The chip pulse is a raised cosine with a roll-off factor of 0.5. For the multi-path case, each user has $L = 3$ paths. The delay of each path is uniformly distributed on $[0, T_s]$. Hence, the maximum delay spread is one symbol interval, i.e., $\iota = 1$. The fading gain for each user's channel is generated from a complex Gaussian distribution and is fixed for all simulations. The path gains in each users' channel are normalized so that all user's signals arrive at the receiver with the same power. The smoothing factor is $\bar{m} = 2$ and the forgetting factor for the subspace tracking algorithm for all simulations is 0.995. The performance measures are the bit-error probability and the signal-to-interference-plus-noise ratio, defined by $\text{SINR} \triangleq E^2\{\mathbf{w}^H\mathbf{r}\}/\text{Var}\{\mathbf{w}^H\mathbf{r}\}$, where the expectation is with respect to the data symbols of interfering users and the ambient noise. In the simulations, the expectation operation is replaced by time averaging. SINR is a particularly appropriate figure of merit for MMSE detectors since it has been shown [32] that the output of an MMSE detector is approximately Gaussian distributed. Hence, the SINR values (approximately) translate directly and simply to bit-error probabilities, i.e., $\Pr(e) \approx Q\left(\sqrt{\text{SINR}}\right)$. The labeled horizontal lines on the SINR plot represent bit-error-probability thresholds. For the SINR plots, the number of users for the first 1500 iterations is four. At iteration 1501, three users are added so that the system is fully loaded. At iteration 3001, five users are removed.

Figure 6.14 illustrates the adaptation performance for the synchronous, flat-fading case. The SNR is fixed at 8 dB. Figure 6.15 shows the adaptation performance for the asynchronous multi-path case. The SNR for this simulation is 11 dB. Notice that in both cases the bit-error probability does not drop below tolerable levels even during transitions, when users enter or leave the system. Convergence of the SINR to its maximum value is almost instantaneous when users leave the system, and requires fewer than 500 iterations when users are added to the system.

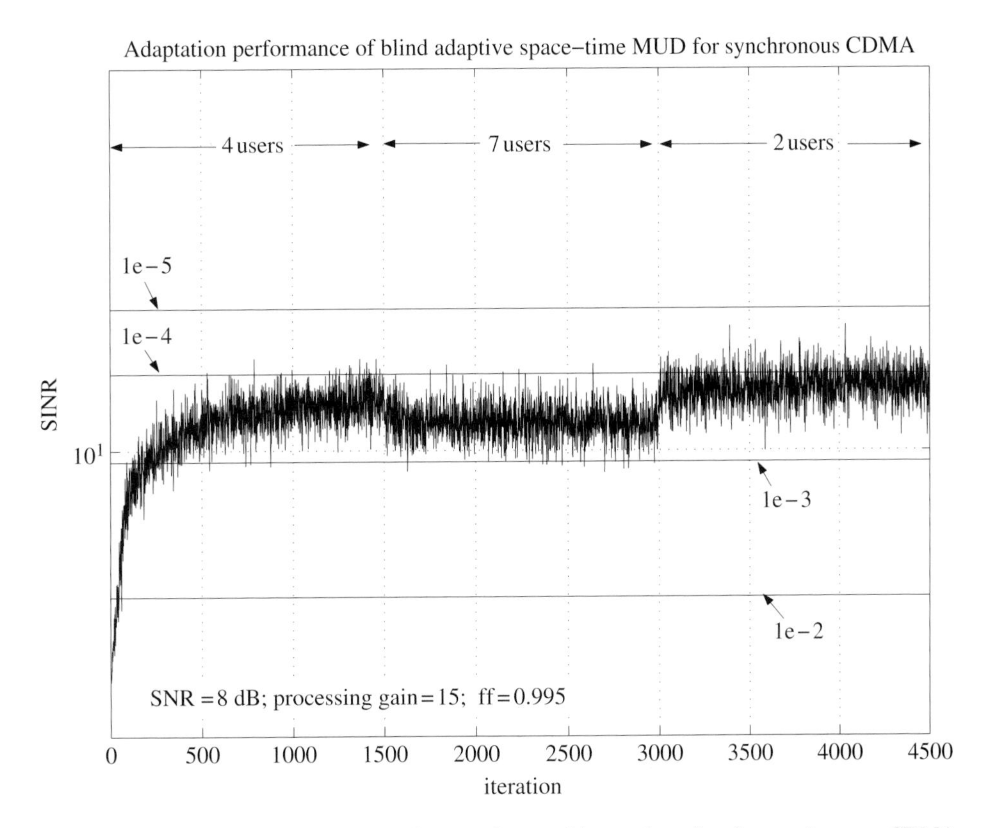

Figure 6.14. Adaptation performance of space–time multi-user detection for synchronous CDMA. The labeled horizontal lines represent bit-error probability thresholds.

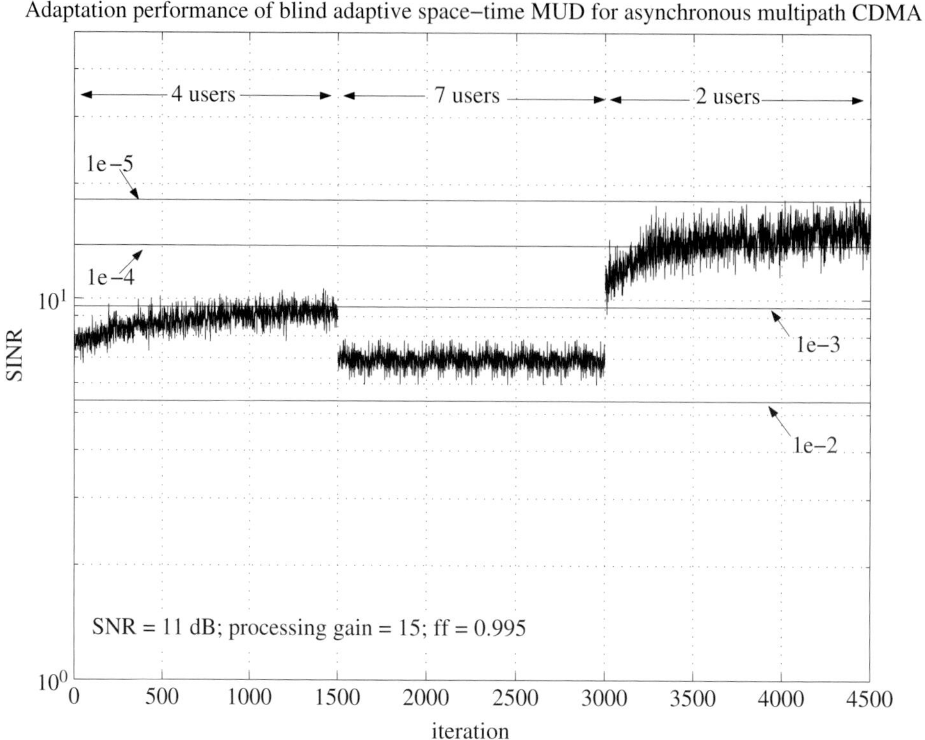

Figure 6.15. Adaptation performance of space–time multi-user detection for asynchronous multipath CDMA. The labeled horizontal lines represent bit-error probability thresholds.

6.6 Summary

In this chapter, we have taken the work of the preceding chapters in several directions. In Section 6.2, we introduced a general model for multiple-access signaling in MIMO channels, and used this model to derive canonical receiver structures for multi-user MIMO systems. This development ties the MIMO multi-user channel models discussed in Chapter 2 together with receiver designs described in Chapters 3 and 5, and then extends the latter to the multiple-access, frequency-selective channels arising in many applications. Section 6.3 also echoes the detection problems discussed in Chapters 3 and 5, notably by re-emphasizing the importance of iterative algorithms in complexity reduction for MIMO receivers. Section 6.4 describes how the structure imposed by space–time coding techniques of Chapter 4 apply can be exploited, together with the turbo-style iterative methods of Chapter 5, and can be used to significantly enhance the overall receiver performance with little attendant increase in complexity. Although most of the techniques in this chapter apply to general interference-type channels (multi-access, inter-symbol, and inter-antenna), the focus has been on the direct-sequence CDMA channels introduced in Chapter 1. Section 6.5 specifically deals with such channels, which are particularly amenable to adaptive implementation. Moreover, Section 6.5 exploits the Alamouti space–time coding structure described in Chapters 1 and 4 in its adaptation algorithms.

6.7 Bibliographical notes

As noted in Section 6.2, the methods discussed in this chapter have been developed over a period of several decades. Early work on receiver design for channel-coded systems and inter-symbol-interference channels dates from the 1960s and 1970s, respectively, while the techniques for multiple-access and inter-antenna interference channels began largely in the 1980s and 1990s, respectively. A review of these developments is found in [30]. Complexity reduction through iterative algorithms and through adaptation have been major issues throughout this development, with turbo-style algorithms gaining significant interest in the 1990s. (An overview of iterative techniques is found in [31].) The current decade has seen a number of developments, particularly in the development of new analytical tools using methods of statistical physics, and in refinement, analysis and understanding of adaptive and iterative methods. However, all of these areas are still areas of active research, and new developments continue today. Perhaps the most critical open issue lies in the transition of these methods into more widespread practice. Although current wireless standards and systems do incorporate some of the ideas explored in this chapter, there is still considerable opportunity for further practical development. The iterative and adaptive methods are, of course, directed at precisely this goal.

For further additional reading on the subject matter of this chapter, the reader is referred to the books by Verdú [38], Wang and Poor [46], and Comaniciu *et al.* [5]. The first of these three books contains an excellent exposition of the fundamentals of multi-user detection, while the second contains further elaboration and additional examples illustrating the model of Section 6.2, as well as considerable discussion of various methods of adaptive and iterative receiver design. Issues not treated in this chapter, such as fast-fading and OFDM systems, are also considered there. Finally, the impact of these methods on higher-layer networking issues, such as resource allocation, quality-of-service provision, and network performance, is discussed in the third of these three books.

References

[1] S. M. Alamouti, "A simple transmit diversity technique for wireless communications," *IEEE J. Select. Areas. Commun.*, vol. 16, no. 8, pp. 1451–1458, 1998.

[2] L. R. Bahl, J. Cocke, F. Jelinek, and J. Raviv, "Optimum decoding of linear codes for minimizing symbol error rate," *IEEE Trans. Inform. Theory*, vol. 20 , no. 3, pp. 284–287, 1974.

[3] C. Berrou and A. Glavieux, "Near optimum error-correcting coding and decoding: turbo codes," *IEEE Trans. Commun.*, vol. 44, no. 10, pp. 1261–1271, 1996.

[4] C. Berrou, A. Glavieux, and P. Thitimajshima, "Near Shanon limit error-correcting coding and decoding: turbo codes," *Proc. 1993 IEEE Intl. Conf. Commun.*, Geneva, Switzerland, vol. 2, pp. 1064–1070, 1993.

[5] C. Comaniciu, N. Mandayam, and H. V. Poor, "Wireless Networks: Multiuser Detection in Cross-layer Design." New York: Springer, 2005.

[6] H. Dai, A. F. Molisch, and H. V. Poor, "Downlink capacity of interference-limited MIMO systems with joint detection," *IEEE Trans. Wireless Commun.*, vol. 3, no. 2, pp. 442–453, 2004.

[7] H. Dai and H. V. Poor, "Sample-by-sample adaptive space–time processing for multiuser detection in multipath CDMA systems," *Proc. 2001 Fall IEEE Vehicular Technology Conf.*, Atlantic City, NJ, Oct. 2001.

[8] H. Dai and H. V. Poor, "Iterative space–time processing for multiuser detection in multipath CDMA channels," *IEEE Trans. Signal Processing*, vol. 50 , no. 9, pp. 2116–2127, 2002.

[9] M. O. Damen, A. Safavi, and K. Abed-Meriam, "On CDMA with space–time codes over multipath fading channels," *IEEE Trans. Wireless Commun.*, vol. 2, pp. 11–19, 2003.

[10] A. P. Dempster, N. M. Laird, and D. B. Rubin, "Maximum-likelihood from incomplete data via the EM algorithm," *J. Royal Statist. Soc. B*, vol. 39, pp. 1–38, 1977.

[11] J. A. Fessler and A. O. Hero, "Space-alternating generalized EM algorithm," *IEEE Trans. Signal Processing*, vol. 42, no. 10, pp. 2664–2677, 1994.

[12] T. R. Giallorenzi and S. G. Wilson, "Multiuser ML sequence estimator for convolutionally coded asynchronous DS-CDMA systems," *IEEE Trans. Commun.*, vol. 44, no. 8, pp. 997–1008, 1996.

[13] T. R. Giallorenzi and S. G. Wilson, "Suboptimum multiuser receivers for convolutionally coded asynchronous DS-CDMA systems," *IEEE Trans. Commun.*, vol. 44 no. 9, pp. 1183–1196, 1996.

[14] G. H. Golub and C. F. Van Loan, "Matrix Computation." Baltimore, MD: Johns Hopkins University Press, 1996.

[15] J.-C. Guey, M. P. Fitz, M. R. Bell, and W.-Y. Kuo, "Signal design for transmitter diversity wireless communication systems over Rayleigh fading channels," *Proc. IEEE Vehicular Technol. Conf.*, Atlanta, GA, pp. 136–140, 1996.

[16] A. R. Hammons and H. El Gamal, "On the theory of space–time codes for PSK modulation," *IEEE Trans. Inform. Theory*, vol. 46, no. 2, pp. 524–542, 2000.

[17] B. Hochwald, T. L. Marzetta, and C. B. Papadias, "A novel space–time spreading scheme for wireless CDMA systems," *Proc. 37th Ann. Allerton Conf. Commun., Contr., Comput.*, Monticello, IL, Sept. pp. 22–24, 1999.

[18] B. Hochwald, T. L. Marzetta, and C. B. Papadias, "A transmitter diversity scheme for wideband CDMA systems based on space–time spreading," *IEEE J. Select. Areas. Commun.*, vol. 19, pp. 48–60, 2001.

[19] J. Hou, J. E. Smee, H. D. Pfister, and S. Tomasin, "Implementing interference cancellation to increase the EV-DO REV A link capacity," *IEEE Commun., Magazine*, vol. 44, no. 2, pp. 96–102, 2006.

[20] S. K. Jayaweera and H. V. Poor, "Iterative multiuser detection for space–time coded synchronous CDMA," *Proc. IEEE Vehicular Technol. Conf.*, Atlantic City, NJ, vol. 4, Fall 2001, pp. 2736–2739.

[21] S. K. Jayaweera and H. V. Poor, "Low complexity receiver structures for space-time coded multiple-access systems," *EURASIP J. Appl. Signal Processing* (special issue on space–time coding), vol. 2002, pp. 275–288, 2002.

[22] P. Lancaster and M. Tismenetsky, *The Theory of Matrices with Applications*. Orlando, FL: Academic Press, Inc., 1985.

[23] H. Li, X. Lu, and G. B. Giannakis, "Capon multiuser receiver for CDMA systems with space–time coding," *IEEE Trans. Sig. Processing*, vol. 50, pp. 1193–1204, 2002.

[24] J. Liu, J. Li, H. Li, and E. G. Larsson, "Differential space–time modulatio for interference suppression," *IEEE Trans. Sig. Processing*, vol. 49, pp. 1786–1795, 2001.

[25] Z. Liu, G. B. Giannakis, B. Muquet, and S. Zhou, "Space–time coding for broadband wireless communications," *Wireless Syst. Mobile Comput.*, vol. 1, pp. 35–53, 2001.

[26] B. Lu and X. Wang, "Iterative receivers for multiuser space–time coding systems," *IEEE J. Select. Areas Commun.*, vol. 18, no. 11, pp. 2322–2335, 2000.

[27] A. Naguib and N. Seshadri, "Combined interference cancellation and ML decoding of space–time block codes," *Proc. 7th Commun. Theory Mini-conference at Globecom'98*, Sydney, Australia, 1998.

[28] I. Oppermann, "CDMA space–time coding using an LMMSE receiver," *Proc. Intl. Conf. Commun. (ICC'99)*, Vancouver, BC, Canada, pp. 182–186, 1999.

[29] H. V. Poor, *An Introduction to Signal Detection and Estimation*. New York: Springer-Verlag, 1994.

[30] H. V. Poor, "Dynamic programming in digital communications: Viterbi decoding to turbo multiuser detection," *J. Optimiz. Theory Applic.* vol. 115, no. 3, pp. 629–657, 2002.

[31] H. V. Poor, "Iterative multiuser detection," *IEEE Signal Processing Magazine*, vol. 21, no. 1, pp. 81–88, 2004.

[32] H. V. Poor and S. Verdú, "Probability of error in MMSE multiuser detection," *IEEE Trans. Inform. Theory*, pp. 858–871, 1997.

[33] J. Proakis, *Digital Communications*, 4th edn. New York: McGraw-Hill, 2000.

[34] D. Reynolds and X. Wang, "Adaptive group-blind multiuser detection based on a new subspace tracking algorithm," *IEEE Trans. Commun.*, vol. 49, no. 7, pp. 1135–1141, 2001.

[35] D. Reynolds, X. Wang, and H. V. Poor, "Blind adaptive space–time multiuser detection with multiple transmitter and receiver antennas," *IEEE Trans. Signal Processing*, vol. 50, no. 6, pp. 1261–1276, 2002.

[36] V. Tarokh, H. Jafarkhani, and A. R. Calderbank, "Space–time block codes from orthogonal designs," *IEEE Trans. Inform. Theory*, vol. 45, no. 5, pp. 1456–1467, 1999.

[37] V. Tarokh, N. Seshadri, and A. R. Calderbank, "Space–time codes for high rate wireless communication: performance criterion and code construction," *IEEE Trans. Inform. Theory*, vol. 44, no. 2, pp. 744–765, 1998.

[38] S. Verdú, *Multiuser Detection*. Cambridge: Cambridge University Press, 1998.

[39] S. Verdú and H. V. Poor, "Abstract dynamic programming models under commutativity conditions," *SIAM J. Control and Optimiz.*, vol. 25, no. 4, pp. 990–1006, 1987.

[40] A. J. Viterbi, "An intuitive justification and a simplified implementation of the MAP decoder for convolutional codes," *IEEE J. Select. Areas Commun.*, vol. 16, no. 2, pp. 260–264, 1998.

[41] X. Wang and A. Host-Madsen, "Group-blind multiuser detection for uplink CDMA," *IEEE J. Select. Areas Commun.*, vol. 17, no. 11, pp. 1971–1984, 1999.

[42] X. Wang and H. V. Poor, "Blind equalization and multiuser detection for CDMA communications in dispersive channels," *IEEE Trans. Commun.*, vol. 46, no. 1, pp. 91–103, 1998.

[43] X. Wang and H. V. Poor, "Blind multiuser detection: a subspace approach", *IEEE Trans. Inform. Theory*, vol. 44, no. 2, pp. 677–691, 1998.

[44] X. Wang and H. V. Poor, "Iterative (turbo) soft interference cancellation and decoding for coded CDMA," *IEEE Trans. Commun.*, vol. 47, no. 7, pp. 1046–1061, 1999.

[45] X. Wang and M. V. Poor, "Space–time multiuser detection in multipath CDMA channels," *IEEE Trans. Signal Processing*, vol. 47, no. 9, pp. 2356–2374, 1999.

[46] X. Wang and M. V. Poor, *Wireless Communication Systems: Advanced Techniques for Signal Reception*. Upper Saddle River, NJ: Prentice-Hall, 2002.

[47] Y. Zhang and R. S. Blum, "Multistage multiuser detection for CDMA with space–time coding," *Proc. 10th IEEE Workshop on Statistical Signal and Array Processing,* Poconos, PA, pp. 1–5, Aug. 2000.

[48] R. E. Ziemer, R. L. Peterson, and D. E. Borth, *Introduction to Spread Spectrum Communications*. Upper Saddle River, NJ: Prentice-Hall, 1995.

Bibliography

Abe, T. and Matsumoto, T. (2003). Space–time turbo equalization in frequency-selective MIMO channels. *IEEE Trans. Vehic. Technol.*, **52**(3), 469–475.

Aftas, D., Bacha, M., Evans, J., and Hanly, S. (2004). On the sum capacity of multiuser MIMO channels. In *Proc. Intl. Symp. on Information Theory and its Applications.*

Agarwal, D., Tarokh, V., Naguib, A., and Seshadri, N. (1998). Space–time coded OFDM for high data rate wireless communication over wideband channels. In *Proc. IEEE VTC*, vol. 3, pp. 2232–2236.

Agrell, E., Eriksson, T., Vardy, A., and Zeger, K. (2002). Closest point search in lattices. *IEEE Trans. Inform. Theory*, **48**(8), 2201–2214.

Ahlswede, R. (1973). Multi-way communication channels. In *Proc. 2nd Intl. Symp. Information Theory*, pp. 23–52.

Al-Dhahir, N. (2001). Single-carrier frequency-domain equalization for space–time block-coded transmissions over frequency-selective fading channels. *IEEE Commun. Lett.*, **5**(7), 304–306.

(2002). Overview and comparison of equalization schemes for space–time-coded signals with application to EDGE. *IEEE Trans. Signal Processing.*, **50**(10), 2477–2488.

Al-Dhahir, N., Fragouli, C., Stamoulis, A., Younis, Y., and Calderbank, A. R. (2002a). Space–time processing for broadband wireless access. *IEEE Commun. Mag.*, **40**(9), 136–142.

Al-Dhahir, N., Giannakis, G., Hochwald, B., Hughes, B., and Marzetta, T. (2002b). Guest editorial. *Trans. Signal Processing*, **50**(10), 2381–2384.

Alamouti, S. M. (1998). A simple transmit diversity technique for wireless communications. *IEEE J. Select. Areas Commun.*, **16**(8), 1451–1458.

Ariyavisitakul, S. L. (2000). Turbo space–time processing to improve wireless channel capacity. *IEEE Trans. Commun.*, **48**(8), 1347–1359.

Azarian, K., Gamal, H. E., and Schniter, P. (2005). On the achievable diversity–multiplexing tradeoff in half-duplex cooperative channels. *IEEE Trans. Inform. Theory*, **51**(12), 4152–4172.

Bahl, L. R., Cocke, J., Jelinek, F., and Raviv, J. (1974). Optimal decoding of linear codes for minimizing symbol error rate. *IEEE Trans. Inform. Theory*, **20**(2), 284–287.

Balaban, P. and Salz, J. (1992). Optimum diversity combining and equalization in digital data transmission with applications to cellular mobile radio – Part I: Theoretical considerations. *IEEE Trans. Commun.*, **40**(5), 885–894.

Balakrishnan, H., Padmanabhan, V. N., Seshan, S., and Katz, R. H. (1997). A comparison of mechanisms for improving TCP performance over wireless links. *IEEE/ACM Trans. Network.*, **5**(6), 756–769.

Bauch, G. and Al-Dhahir, N. (2002). Reduced-complexity space–time turbo equalization for frequency-selective MIMO channels. *IEEE Trans. Wireless Commun.*, **1**(4), 819–828.

Baum, D. S., Gore, D., Nabar, R., Panchanathan, S., Hari, K. V. S., Erceg, V., and Paulraj, A. J. (2000). Measurement and characterization of broadband MIMO fixed wireless channels at 2.5 GHz. In *Proc. IEEE ICPWC*, Hyderabad, pp. 203–206.

Belfiore, J.-C., Rekaya, G., and Viterbo, E. (2005). The Golden code: a 2×2 full-rate space–time code with nonvanishing determinants. *IEEE Trans. Inform. Theory*, **51**(4), 1432–1436.

Benedetto, S. and Montorsi, G. (1996). Unveiling turbo codes: some results on parallel concatenated coding schemes. *IEEE Trans. Inform. Theory*, **42**(2), 409–428.

Bengtsson, M. and Ottersten, B. (2001). Optimal and suboptimal transmit beamforming. In *Handbook of Antennas in Wireless Communications*. Boca Raton, FL: CRC Press.

Berrou, C. and Glavieux, A. (1996). Near optimum error correcting coding and decoding: turbo-codes. *IEEE Trans. Commun.*, **44**(10), 1261–1271.

Berrou, C., Glavieux, A., and Thitimajshima, P. (1993). Near Shannon limit error-correcting coding and decoding: turbo codes. In *Proc. 1993 Intl. Conf. on Communications*, Geneva, vol. 2, pp. 1064–1070.

Biglieri, E. (2005). *Coding for Wireless Channels*. New York: Springer-Verlag.

Biglieri, E., Nordio, A., and Taricco, G. (2004a). Doubly-iterative decoding of space–time turbo codes with a large number of antennas. *IEEE Intl. Conf. Commun. (ICC 2004)*, Paris, June 20–24.

(2004b). Iterative receivers for coded MIMO signaling. *Wireless Commun. Mob. Comput.*, **4**(7), 697–710.

(2005). MIMO doubly-iterative receivers: pre- vs. post-cancellation filtering. *IEEE Commun. Lett.*, **9**(2), 106–108.

Biglieri, E., Proakis, J., and Shamai, S. (1998). Fading channels: information-theoretic and communications aspects. *IEEE Trans. Inform. Theory*, **44**(6), 2619–2692.

Biglieri, E., Taricco, G., and Tulino, A. (2002a). Decoding space–time codes with BLAST architectures. *IEEE Trans. Signal Processing.*, **50**(10), 2547–2552.

(2002b). Performance of space–time codes for a large number of antennas. *IEEE Trans. Inform. Theory*, **48**(7), 1794–1803.

Blackwell, D., Breiman, L., and Thomasian, A. J. (1959). The capacity of a class of channels. *Ann. Math. Stat.*, **30**, 1229–1241.

Bliss, D., Chan, A., and Chang, N. (2004). MIMO wireless communication channel phenomenology. *IEEE Trans. Antennas Propagation*, **52**(8), 2073–2082.

Boche, H. and Jorswieck, E. (2004). Outage probability of multiple antenna systems: optimal transmission and impact of correlation. *Intl. Zurich Seminar.*

Boche, H. and Schubert, M. (2002). A general duality theory for uplink and downlink beamforming. In *Proc. IEEE Vehicular Technology Conf.*

Bölcskei, H., Gesbert, D., and Paulraj, A. J. (2002a). On the capacity of OFDM-based spatial multiplexing systems. *IEEE Trans. Commun.*, **50**(2), 225–234.

(2002b). On the capacity of OFDM-based spatial multiplexing systems. *IEEE Trans. Commun.*, **50**(2), 225–234.

Bölcskei, H., Nabar, R., Oyman, O., and Paulraj, A. (2006). Capacity scaling laws in MIMO relay networks. *IEEE Trans. Wireless Commun.*, **5**(6), 1433–1444.

Bölcskei, H. and Paulraj, A. J. (2000). Space–frequency coded broadband OFDM systems. In *Proc. IEEE WCNC*, Chicago, IL, vol. 1, pp. 1–6.

Borst, S. and Whiting, P. (2001). The use of diversity antennas in high-speed wireless systems: capacity gains, fairness issues, multi-user scheduling. *Bell Labs. Tech. Mem.*, download available at http://mars.bell-labs.com

Boullé, K. and Belfiore, J. C. (1992). Modulation schemes designed for the Rayleigh channel. In *Proc. Conf. on Information Science Systems (CISS '92)*, pp. 288–293.

Bourdoux, A., Come, B., and Khaled, N. (2003). Non-reciprocal transceivers in OFDM/SDMA systems: impact and mitigation. In *Proc. Radio and Wireless Conf.*, pp. 183–186.

Boutros, J. and Caire, G. (2002). Iterative multiuser joint detection: unified framework and asymptotic analysis. *IEEE Trans. Inform. Theory*, **48**(7), 1772–1793.

Boutros, J. and Viterbo, E. (1998). Signal space diversity: a power and bandwidth efficient diversity technique for the Rayleigh fading channel. *IEEE Trans. Inform. Theory*, **44**(4), 1453–1467.

Boyd, S. and Vandenberghe, L. (2003). *Convex Optimization*. Cambridge: Cambridge University Press. Available: http://www.stanford.edu/~boyd/cvxbook.html

Burg, A., Borgmann, M., Wenk, M., Zellweger, M., Fichtner, W., and Bölcskei, H. (2005). VLSI implementation of MIMO detection using the sphere decoding algorithm. *IEEE J. Solid-State Circuits*, **40**(7), 1566–1577.

Bäro, S. (2005). *Iterative Detection for Coded MIMO Systems*. Fortschritt-Berichte VDI, Reihe 10, Nr. 752. Düsseldorf: VDI Verlag.

Caire, G. and Shamai, S. (1999). On the capacity of some channels with channel state information. *IEEE Trans. Inform. Theory*, **45**(6), 2007–2019.

(2003). On the achievable throughput of a multiantenna Gaussian broadcast channel. *IEEE Trans. Inform. Theory*, **49**(7), 1691–1706.

Caire, G., Taricco, G., and Biglieri, E. (1999). Optimum power control over fading channels. *IEEE Trans. Inform. Theory*, **45**(5), 1468–1489.

Calderbank, A. R., Das, S., Al-Dhahir, N., and Diggavi, S. (2005). Construction and analysis of a new quaternionic space–time code for 4 transmit antennas. *Commun. Inform. Syst.*, **5**(1), 97–121.

Calderbank, A. R., Diggavi, S. N., and Al-Dhahir, N. (2004). Space–time signaling based on kerdock and Delsarte–Goethals codes. In *Proc. ICC*, pp. 483–487.

Calderbank, A. R. and Naguib, A. F. (2001). Orthogonal designs and third generation wireless communication. In *Surveys in Combinatorics 2001, London Mathematical Society Lecture Note Series 288*, ed. J. W. P. Hirschfeld. Cambridge: Cambridge University Press, pp. 75–107.

Carleial, A. B. (1975). A case where interference does not reduce capacity. *IEEE Trans. Inform. Theory*, **21**(5), 569–570.

Catreux, S., Driessen, P., and Greenstein, L. (2000). Simulation results for an interference-limited multiple-input multiple-output cellular system. *IEEE Commun. Lett.*, **4**(11), 334–336.

(2001). Attainable throughput of an interference-limited multiple-input multiple-output (MIMO) cellular system. *IEEE Trans. Commun.*, **49**(8), 1307–1311.

Chiang, M., Low, S. H., Doyle, J. C., and Calderbank, A. R. (2006). Layering as optimization decomposition. *Proc. IEEE*, to appear.

Chiani, M., Win, M. Z., and Zanella, Z. (2003). On the capacity of spatially correlated MIMO Rayleigh-fading channels. *IEEE Trans. Inform. Theory*, **49**(10), 2363–2371.

Chizhik, D., Ling, J., Wolniansky, P. W., Valenzuela, R. A., Costa, N., and Huber, K. (2002). Multiple input multiple output measurements and modeling in Manhattan. In *Proc. IEEE Vehicular Technology Conf.*

(2003). Multiple-input–multiple-output measurements and modeling in Manhattan. *IEEE J. Select. Areas Commun.*, **23**(3), 321–331.

Chu, D. (1972). Polyphase codes with good periodic correlation properties. *IEEE Trans. Inform. Theory*, **18**, 531–532.

Chuah, C., Tse, D., Kahn, J., and Valenzuela, R. (2002). Capacity scaling in MIMO wireless systems under correlated fading. *IEEE Trans. Inform. Theory*, **48**(3), 637–650.

Clifford, W. K. (1878). Applications of Grassman's extensive algebra. *Amer. J. Math.*, **1**, 350–358.

Comaniciu, C., Mandayam, N., and Poor, H. V. (2005). *Wireless Networks: Multiuser Detection in Cross-layer Design*. New York: Springer-Verlag.

Costa, M. (1983). Writing on dirty paper. *IEEE Trans. Inform. Theory*, **29**(3), 439–441.

Costa, M. and El Gamal, A. (1987). The capacity region of the discrete memoryless interference channel with strong interference. *IEEE Trans. Inform. Theory*, **33**, 710–711.

Cover, T. and El Gamal, A. (1979). Capacity theorems for the relay channel. *IEEE Trans. Inform. Theory*, **25**(5), 572–584.

Cover, T. and Thomas, J. (1991). *Elements of Information Theory*. New York: Wiley.

Csiszár, I. (1992). Arbitrarily varying channels with general alphabets and states. *IEEE Trans. Inform. Theory*, **38**(6), 1725–1742.

Csiszár, I. and Körner, J. (1997). *Information Theory: Coding Theorems for Discrete Memoryless Systems*. New York: Academic Press.

Dai, H., Molisch, A. F., and Poor, H. V. (2004). Downlink capacity of interference-limited MIMO systems with joint detection. *IEEE Trans. Wireless Commun.*, **3**(2), 442–453.

Dai, H. and Poor, H. V. (2001). Sample-by-sample adaptive space–time processing for multiuser detection in multipath CDMA systems. In *Proc. 2001 Fall IEEE Vehicular Technology Conf.*, Atlantic City, NJ.

(2002). Iterative space–time processing for multiuser detection in multipath CDMA channels. *IEEE Trans. Signal Processing.*, **50**(9), 2116–2127.

Damen, M. O., Abed-Meriam, K., and Belfiore, J.-C. (2002). Diagonal algebraic space–time block codes. *IEEE Trans. Inform. Theory*, **48**(3), 628–636.

Damen, M. O., Chkeif, A., and Belfiore, J.-C. (2000). Lattice codes decoder for space–time codes. *IEEE Commun. Lett.*, **4**, 161–163.

Damen, M. O., El Gamal, H., and Beaulieu, N. (2003). On optimal linear space–time constellations. In *Intl. Conf. on Communications (ICC)*.

Damen, M. O., El Gamal, H., and Caire, G. (2003). On maximum-likelihood detection and the search for the closest lattice point. *IEEE Trans. Inform. Theory*, **49**(10), 2389–2402.

Damen, M. O., Safavi, A., and Abed-Meriam, K. (2003). On CDMA with space–time codes over multipath fading channels. *IEEE Trans. Wireless Commun.*, **2**, 11–19.

Dempster, A. P., Laird, N. M., and Rubin, D. B. (1977). Maximum likelihood from incomplete data via the EM algorithm. *J. R. Stat. Soc.* B, **39**, 1–38.

Diggavi, S. N., Al-Dhahir, N., and Calderbank, A. R. (2003a). Diversity embedded space–time codes. In *IEEE Global Communications Conference (GLOBECOM)*, pp. 1909–1914.

(2003b). Algebraic properties of space–time block codes in intersymbol interference multiple access channels. *IEEE Trans. Inform. Theory*, **49**(10), 2403–2414.

(2004). Diversity embedding in multiple antenna communications. In Gupta, P., Kramer, G., and van Wijngaarden, A. J., eds. (2003). Network information theory. In *AMS Series on Discrete Mathematics and Theoretical Computer Science*, vol. 66, pp. 285–302. Appeared as a part of *DIMACS Workshop on Network Information Theory*.

Diggavi, S. N., Al-Dhahir, N., Stamoulis, A., and Calderbank, A. R. (2002). Differential space–time coding for frequency-selective channels. *IEEE Commun. Lett.*, **6**(6), 253–255.

(2004). Great expectations: the value of spatial diversity in wireless networks. *Proc. IEEE*, **92**(2), 219–270.

Diggavi, S. N., Dusad, S., Calderbank, A. R., and Al-Dhahir, N. (2005). On embedded diversity codes. In *Allerton Conf. on Communication, Control, and Computing*.

Diggavi, S. N., Grossglauser, M., and Tse, D. (2002). Even one-dimensional mobility increases ad hoc wireless capacity. In *Proc. Intl. Symp. on Information Theory*.

Diggavi, S. N. and Tse, D. N. C. (2004). On successive refinement of diversity. In *Allerton Conf. on Communication, Control, and Computing*.

(2005). Fundamental limits of diversity-embedded codes over fading channels. In *IEEE Intl. Symp. on Information Theory (ISIT)*, pp. 510–514.

Dong, B. and Wang, X. (2005). Sampling-based soft equalization for frequency-selective MIMO channels. *IEEE Trans. Commun.*, **53**(2), 278–288.

Dong, B., Wang, X., and Doucet, A. (2003). A new class of soft MIMO demodulation algorithms. *IEEE Trans. Signal Processing.*, **51**(11), 2752–2763.

El Gamal, A. and Cover, T. (1980). Multiple user information theory. *Proc. IEEE*, **68**(12), 1466–1483.

El Gamal, A. Mammen, J., Prabhakar, B., and Shah, D. (2004). Throughput-delay trade-off in energy constrained wireless networks. In *Proc. Intl. Symp. on Information Theory*.

El Gamal, H. and Damen, M. O. (2003). Universal space–time coding. *IEEE Trans. Inform. Theory*, **49**(5), 1097–1119.

El Gamal, H. and Hammons, A. R. (2001). A new approach to layered space–time coding and signal processing. *IEEE Trans. Inform. Theory*, **47**(6), 2321–2334.

El Gamal, H., Caire, G., and Damen, O. (2004). Lattice coding and decoding achieve the optimal diversity–multiplexing of MIMO channels. *IEEE Trans. Inform. Theory*, **50**(6), 968–985.

Elia, P., Kumar, K. R., Pawar, S. A., Kumar, P. V., and Lu, H.-F. (2004). Explicit minimum-delay space–time codes achieving the diversity–multiplexing gain tradeoff. Submitted.

Erez, U., Shamai, S., and Zamir, R. (2000). Capacity and lattice strategies for cancelling known interference. In *Proc. Intl. Symp. on Information Theory and its Applications*, pp. 681–684.

Erez, U. and ten Brink, S. (2003). Approaching the dirty paper limit for cancelling known interference. In *Proc. 41st Annual Allerton Conf. on Communications, Control and Computing*.

Fessler, J. A. and Hero, A. O. (1994). Space-alternating generalized EM algorithm. *IEEE Trans. Signal Processing.*, **42**(10), 2664–2677.

Fischer, R., Stierstorfer, C., and Huber, J. (2004). Precoding for point-to-multipoint transmission over MIMO ISI channels. In *Proc. Intl. Zurich Seminar on Communications*, pp. 208–211.

Forney, G. D., Jr. (2001). Codes on graphs: normal realizations. *IEEE Trans. Inform. Theory*, **47**(2), 520–548.

Foschini, G. J. (1996). Layered space–time architecture for wireless communication in a fading environment when using multi-element antennas. *Bell Labs Tech. J.*, **1**(2), 41–59.

Foschini, G. J. and Gans, M. J. (1998). On limits of wireless communications in a fading environment when using multiple antennas. *Wireless Personal Commun.*, **6**, 311–335.

Fragouli, C., Al-Dhahir, N., and Turin, W. (2002). Effect of spatio-temporal channel correlation on the performance of space–time codes. In *ICC*, vol. 2, pp. 826–830.

(2003). Training-based channel estimation for multiple-antenna broadband transmissions. *IEEE Trans. Wireless Commun.*, **2**(2), 384–391.

Frey, B. J. and Kschischang, F. R. (1998). Early detection and trellis splicing: reduced-complexity iterative decoding. *IEEE J. Select. Areas Commun.*, **16**(2), 153–159.

Gallager, R. G. (1963). *Low Density Parity Check Codes*. Cambridge, MA: MIT Press. Available at http://justice.mit.edu/people/gallager.html

(1968). *Information Theory and Reliable Communication*. New York: Wiley.

(1994). An inequality on the capacity region of multiaccess fading channels. *Communication and Cryptography: Two Sides of One Tapestry*, ed. R. E. Blahut, D. J. Costello, and T. Mittelholzer. Boston, MA: Kluwer, pp. 129–139.

Gans, M. J. *et al.* (2002). Outdoor BLAST measurement system at 2.44 GHz: calibration and initial results. *IEEE J. Select. Areas Commun.*, **20**(3), 570–581.

Geramita, A. V. and Seberry, J. (1979). *Orthogonal Designs, Quadratic Forms and Hadamard Matrices. Lecture Notes in Pure and Applied Mathematics*, vol. 43. New York: Marcel Dekker.

Gesbert, D., Bölcskei, H., Gore, D. A., and Paulraj, A. J. (2002). Outdoor MIMO wireless channels: models and performance prediction. *IEEE Trans. Commun.*, **50**(12), 1926–1934.

Gesbert, D., Shafi, M., Shiu, D., Smith, P. J., and Naguib, A. (2003). From theory to practice: an overview of IMO space–time coded wireless systems. *IEEE J. Select. Areas Commun.*, **21**(3), 281–302.

Giallorenzi, T. R. and Wilson, S. G. (1996a). Multiuser ML sequence estimator for convolutionally coded asynchronous DS-CDMA systems. *IEEE Trans. Commun.*, **44**(8), 997–1008.

(1996b). Suboptimum multiuser receivers for convolutionally coded asynchronous DS-CDMA systems. *IEEE Trans. Commun.*, **44**(9), 1183–1196.

Giese, J. and Skoglund, M. (2005). Space–time constellation design for partial CSI at the receiver. In *Proc. IEEE Intl. Symp. on Information Theory (ISIT)*, pp. 2213–2217.

Godara, L. (1997a). Applications of antenna arrays to mobile communications. Part I. Performance improvement, feasibility, and system considerations. *Proc. IEEE*, **85**, 1031–1060.

(1997b). Applications of antenna arrays to mobile communications. Part II. Beamforming and direction-of-arrival considerations. *Proc. IEEE*, **85**, 1195–1245.

Goeckel, D. (1999). Adaptive coding for time-varying channels using outdated fading estimates. *IEEE Trans. Commun.*, **47**(6), 844–855.

Goldsmith, A. J. (2005). *Wireless Communications*. Cambridge: Cambridge University Press.

Goldsmith, A. J., Jafar, S., Jindal, N., and Vishwanath, S. (2003). Capacity limits of MIMO channels. *IEEE J. Select. Areas Commun.*, **21**(3), 684–702.

Goldsmith, A. J. and Varaiya, P. (1997). Capacity of fading channels with channel side information. *IEEE Trans. Inform. Theory*, **43**(6), 1986–1992.

Golub, G. H. and Van Loan, C. F. (1996). *Matrix Computation*. Baltimore, MD: Johns Hopkins University Press.

Graham, A. (1981). *Kronecker Products and Matrix Calculus with Application*. Chichester: Ellis Horwood.

Grossglauser, M. and Tse, D. (2002). Mobility increases the capacity of ad hoc wireless networks. *IEEE/ACM Trans. Network.*, **10**(4), 477–486.

Guess, T. and Varanasi, M. K. (1996). Multiuser decision-feedback receivers for the general Gaussian multiple-access channel. In *Proc. Allerton Conf. on Communications, Control, and Computing*.

Guey, J.-C., Fitz, M. P., Bell, M. R., and Kuo, W.-Y. (1996). Signal design for transmitter diversity wireless communication systems over Rayleigh fading channels. In *Proc. IEEE Vehicular Technology Conf.*, Atlanta, GA, pp. 136–140.

(1999). Signal design for transmitter diversity wireless communication systems over Rayleigh fading channels. *IEEE Trans. Commun.*, **47**(4), 527–537.

Gupta, P. and Kumar, P. R. (2000). The capacity of wireless networks. *IEEE Trans. Inform. Theory*, **46**(2), 388–404.

Hammons, A. R. and El Gamal, H. (2000). On the theory of space–time codes for PSK modulation. *IEEE Trans. Inform. Theory*, **46**(2), 524–542.

Hanly, S. and Tse, D. (1998). Multiaccess fading channels – Part II: Delay-limited capacities. *IEEE Trans. Inform. Theory*, **44**(7), 2816–2831.

Hanly, S. V. and Whiting, P. (1992). Information theory and the design of multi-receiver networks. In *IEEE 2nd Intl. Symp. on Spread Spectrum Technological Applications (ISSTA)*, pp. 103–106.

Hassibi, B. and Hochwald, B. (2002). High-rate codes that are linear in space and time. *IEEE Trans. Inform. Theory*, **48**(7), 1804–1824.

Hassibi, B. and Hochwald, B. (2003). How much training is needed in multiple-antenna wireless links? *IEEE Trans. Inform. Theory*, **49**(4), 951–963.

(2005). On sphere decoding algorithm. I. Expected complexity. *IEEE Trans. Signal Processing.*, **53**(8), August, 2806–2818.

Haustein, T. and Boche, H. (2003). Optimal power allocation for MSE and bit-loading in MIMO systems and the impact of correlation. In *Proc. IEEE Intl. Conf. on Acoustics, Speech, and Signal Processing*, vol. 4, pp. 405–408.

Haykin, S. (1991). *Adaptive Filter Theory*, 2nd edn. Upper Saddle River, NJ: Prentice-Hall.

Haykin, S., Sellathurai, M., de Jong, Y., and Willink, T. (2004). Turbo-MIMO for wireless communications. *IEEE Commun. Mag.*, **42**(10), 48–53.

Heath, R. W. and Paulraj, A. (2005). Switching between diversity and multiplexing in MIMO systems. *IEEE Trans. Commun.*, **53**, 962–968.

Hermosilla, C. and Szczeciński, L. (2003). EXIT charts for turbo receivers in MIMO systems. In *Proc. 7th Intl. Symp. Signal Processing and its Applications (ISSPA 2003)*, July 1–4, pp. 209–212.

Hochwald, B. M. and Marzetta, T. L. (1999). Capacity of a mobile multiple-antenna communication link in Rayleigh flat fading. *IEEE Trans. Inform. Theory*, **45**(1), 139–157.

(2000). Unitary space–time modulation for multiple antenna communications in Rayleigh fading. *IEEE Trans. Inform. Theory*, **46**, 543–564.

Hochwald, B. M., Marzetta, T. L., Richardson, T., Sweldens, W., and Urbanke, R. (2000). Systematic design of unitary space–time constellations. *IEEE Trans. Inform. Theory*, **46**, 1962–1973.

Hochwald, B. M., Marzetta, T. L., and Papadias, C. B. (1999). A novel space–time spreading scheme for wireless CDMA systems. In *Proc. 37th Ann. Allerton Conf. on Communications, Control, Computing*, Monticello, IL.

(2001). A transmitter diversity scheme for wideband CDMA systems based on space–time spreading. *IEEE J. Select. Areas. Commun.*, **19**, 48–60.

Hochwald, B. M., Marzetta, T. L., and Tarokh, V. (2004). Multi-antenna channel-hardening and its implications for rate feedback and scheduling. *IEEE Trans. Inform. Theory*, **50**(9), 1893–1909.

Hochwald, B. M. and Sweldens, W. (2000). Differential unitary space–time modulation. *IEEE Trans. Commun.*, **48**(12), 2041–2052.

Hochwald, B. M. and Vishwanath, S. (2002). Space–time multiple access: linear growth in sum rate. In *Proc. 40th Annual Allerton Conf. on Communications, Control and Computing*.

Hoesli, D., Kim, Y.-H., and Lapidoth, A. (2005). Monotonicity results for coherent MIMO Rician channels. *IEEE Trans. Inform. Theory*, **51**(12), 4334–4339.

Hoesli, D. and Lapidoth, A. (2004). The capacity of a MIMO Ricean channel is monotonic in the singular values of the mean. In *Proc. 5th Intl. ITG Conf. on Source and Channel Coding (SCC)*, Erlangen, Nuremberg.

Horn, R. A. and Johnson, C. R. (1991). *Matrix Analysis*. Cambridge: Cambridge University Press.

Host-Madsen, A. (2004). On the achievable rate for receiver cooperation in ad-hoc networks. In *Proc. IEEE Intl. Symp. on Information Theory*, p. 272.

Host-Madsen, A. and Yang, Z. (2005). Interference and cooperation in multi-source wireless networks. In *IEEE Communication Theory Workshop*.

Hottinen, A., Tirkkonen, O., and Wichman, R. (2003). *Multi-antenna Transceiver Techniques for 3G and Beyond*. New York: Wiley.

Hou, J., Smee, J. E., Pfister, H. D., and Tomasin, S. (2006). Implementing interference cancellation to increase the EV-DO REV A link capacity. *IEEE Commun. Mag.*, **44**(2), 96–102.

Huang, H. and Venkatesan, S. (2004). Asymptotic downlink capacity of coordinated cellular networks. In *Proc. Asilomar Conf. on Signals, Systems, and Computing*, pp. 850–855.

Hughes, B. L. (2000). Differential space–time modulation. *IEEE Trans. Inform. Theory*, **46**(7), 2567–2578.

Hunter, T. E. and Nosratinia, A. (2002). Cooperative diversity through coding. In *Proc. IEEE ISIT*, Lausanne, p. 220.

Hösli, D. and Lapidoth, A. (2004). The capacity of a MIMO Ricean channel is monotonic in the singular values of the mean. In *Proc. 5th Intl. ITG Conf. on Source and Channel Coding*.

I.S. 802.11a (1999). Part 11: Wireless LAN medium access control (MAC) and physical layer (PHY) specifications high-speed physical layer in the 5 GHz band. IEEE Standards.

I.S. 802.16e (2005). Part 16: Air interface for fixed and mobile broadband wireless access systems. IEEE Standards.

Jafar, S. (2005). Degrees of freedom in distributed MIMO communications. In *IEEE Communication Theory Workshop*.

Jafar, S. and Goldsmith, A. (2001a). On optimality of beamforming for multiple antenna systems with imperfect feedback. In *Proc. Intl. Symp. on Information Theory*, p. 321.

(2001b). Vector MAC capacity region with covariance feedback. In *Proc. Intl. Symp. on Information Theory*, p. 321.

(2002). Transmitter optimization for multiple antenna cellular systems. In *Proc. Intl. Symp. on Information Theory*, p. 50.

(2004). Transmitter optimization and optimality of beamforming for multiple antenna systems with imperfect feedback. *IEEE Trans. Wireless Commun.*, **3**(4), 1165–1175.

(2005a). Isotropic fading vector broadcast channels: the scalar upperbound and loss in degrees of freedom. *IEEE Trans. Inform. Theory*, **51**(3), 848–857.

(2005b). Multiple-antenna capacity in correlated Rayleigh fading with channel covariance information. *IEEE Trans. Wireless Commun.*, **4**(3), 990–997.

Jafar, S., Vishwanath, S., and Goldsmith, A. (2001). Channel capacity and beamforming for multiple transmit and receive antennas with covariance feedback. In *Proc. Intl. Conf. on Communications*, vol. 7, pp. 2266–2270.

Jafarkhani, H. (2001). A quasi-orthogonal space time block code. *IEEE Trans. Commun.*, **49**(1), 1–4.

Jafarkhani, H. and Tarokh, V. (2001). Multiple transmit antenna differential detection from generalized orthogonal designs. *IEEE Trans. on Inform. Theory*, **47**(6), 2626–2631.

Jakes, W. (1994). *Microwave Mobile Communications*, 2nd edn. New York: IEEE Press.

Jayaweera, S. K. and Poor, H. V. (2001). Iterative multiuser detection for space–time coded synchronous CDMA. In *Proc. IEEE Vehicular Technology Conf.*, Atlantic City, NJ, vol. 4, pp. 2736–2739.

(2002). Low complexity receiver structures for space–time coded multiple-access systems. Special issue on space–time coding. *EURASIP J. Appl. Signal Processing.*, **2002**, 275–288.

(2003). Capacity of multiple-antenna systems with both receiver and transmitter channel state information. *IEEE Trans. Inform. Theory*, **49**(10), 2697–2709.

Jindal, N. (2005a). High SNR analysis of MIMO broadcast channels. In *Proc. Intl. Symp. on Information Theory*.

(2005b). MIMO broadcast channels with finite rate feedback. In *Proc. IEEE Globecom*.

Jindal, N. and Goldsmith, A. (2004). Optimal power allocation for parallel broadcast channels with independent and common information. In *Proc. Intl. Symp. on Information Theory*.

(2005). Dirty paper coding vs. TDMA for MIMO broadcast channels. *IEEE Trans. Inform. Theory*, **51**(5), 1783–1794.

Jindal, N., Mitra, U., and Goldsmith, A. (2004). Capacity of ad-hoc networks with no decooperation. In *Proc. Intl. Symp. on Information Theory*.

Jindal, N., Rhee, W., Vishwanath, S., Jafar, S., and Goldsmith, A. (2005). Sum power iterative water-filling for multi-antenna Gaussian broadcast channels. *IEEE Trans. Inform. Theory*, **51**(4), 1570–1580.

Jindal, N., Vishwanath, S., and Goldsmith, A. (2004). On the duality of Gaussian multiple-access and broadcast channels. *IEEE Trans. Inform. Theory*, **50**(5), 768–783.

Jorswieck, E. and Boche, H. (2003). Optimal transmission with imperfect channel state information at the transmit antenna array. *Wireless Personal Commun.*, **27**, 33–56.

(2004a). Channel capacity and capacity-range of beamforming in MIMO wireless systems under correlated fading with covariance feedback. *IEEE Trans. Wireless Commun.*, **3**(5), 1543–1553.

(2004b). Optimal transmission strategies and impact of correlation in multi-antenna systems with different types of channel state information. *IEEE Trans. Signal Processing.*, **52**(12), 3440–3453.

Jorswieck, E., Sezgin, A., Boche, H., and Costa, E. (2004). Optimal transmit strategies in MIMO Ricean channels with MMSE receiver. In *Proc. Vehicular Technology Conf.*

Jöngren, G., Skoglund, M., and Ottersten, B. (2002). Combining beamforming and orthogonal space–time block coding. *IEEE Trans. Inform. Theory*, **48**(3), 611–627.

Kailath, T., Sayed, A., and Hassibi, B. (2000). *Linear Estimation*. Englewood Cliffs, NJ: Prentice-Hall.

Kermoal, J. P., Schumacher, L., Mogensen, P. E., and Pedersen, K. I. (2000). Experimental investigation of correlation properties of MIMO radio channels for indoor picocell scenarios. In *Proc. IEEE VTC*, vol. 1, pp. 14–21.

Kermoal, J. P., Schumacher, L., Pedersen, K., Mogensen, P., and Frederiksen, F. (2002). A stochastic MIMO radio channel model with experimental validation. *IEEE J. Select. Areas Commun.*, **20**(6), 1211–1226.

Kerpez, K. J. (1993). Constellations for good diversity performance. *IEEE Trans. Commun.*, **41**(9), 1412–1421.

Khisti, A., Erez, U., and Wornell, G. (2004). A capacity theorem for co-operative multicasting in large wireless networks. In *Proc. Allerton Conf. on Communications, Control, and Computing*.

Kim, T., Jöngren, G., and Skoglund, M. (2004). Weighted space–time bit-interleaved coded modulation. In *Proc IEEE Information Theory Workshop*, pp. 375–380.

Komninakis, C., Fragouli, C., Sayed, A., and Wesel, R. (2002). Multi-input multi-output fading channel tracking and equalization using Kalman estimation. *IEEE Trans. Signal Processing.*, **50**(5), 1065–1076.

Kramer, G., Gastpar, M., and Gupta, P. (2005). Cooperative strategies and capacity theorems for relay networks. *IEEE Trans. Inform. Theory*, **51**(9), 3037–3063.

Kschischang, F. R., Frey, B. J., and Loeliger, H.-A. (2001). Factor graphs and the sum–product algorithm. *IEEE Trans. Inform. Theory*, **47**(2), 498–519.

Kumar, P. and Gamal, H. E. (2006). On the throughput–delay tradeoff in cellular multicast. *IEEE Trans. Inform. Theory.*, submitted.

Kuo, W. and Fitz, M. P. (1997). Design and analysis of transmitter diversity using intentional frequency offset for wireless communications. *IEEE Trans. Vehicular. Technol.*, **46**(4), 871–881.

Kyritsi, P. (2002). Capacity of multiple input–multiple output wireless systems in an indoor environment. Ph.D. thesis, Stanford University.

Lan, T. and Yu, W. (2004). Input optimization for multi-antenna broadcast channels and per-antenna power constraints. In *Proc. IEEE Globecom*.

Lancaster, P. and Tismenetsky, M. (1985). *The Theory of Matrices with Applications*. Orlando, FL: Academic Press.

Laneman, J. N., Tse, D. N. C., and Wornell, G. W. (2004). Cooperative diversity in wireless networks: efficient protocols and outage behavior. *IEEE Trans. Inform. Theory*, **50**(12), 3062–3080.

Laneman, J. N. and Wornell, G. W. (2003). Distributed space–time-coded protocols for exploiting cooperative diversity in wireless networks. *IEEE Trans. Inform. Theory*, **49**(10), 2415–2425.

Lapidoth, A. and Moser, S. M. (2003). Capacity bounds via duality with applications to multi-antenna systems on flat fading channels. *IEEE Trans. Inform. Theory*, **49**(10), 2426–2467.

Lapidoth, A., Shamai, S., and Wigger, M. (2005). On the capacity of fading MIMO broadcast channels with imperfect transmitter side-information. In *Proc. 43rd Annual Allerton Conf. on Communications, Control and Computing*.

Lee, S.-J., Singer, A. C., and Shanbhag, N. R. (2003). Analysis of linear turbo equalizer via EXIT chart. In *Proc. IEEE Global Telecomm. Conf. (GLOBECOM 2003)*, vol. 4, pp. 2237–2242.

Li, H., Lu, X., and Giannakis, G. B. (2002). Capon multiuser receiver for CDMA systems with space–time coding. *IEEE Trans. Signal Processing.*, **50**, 1193–1204.

Li, L. and Goldsmith, A. (2001). Capacity and optimal resource allocation for fading broadcast channels – Part I: Ergodic capacity. *IEEE Trans. Inform. Theory*, **47**(3), 1083–1102.

Li, L., Jindal, N., and Goldsmith, A. (2005). Outage capacities and optimal power allocation for fading multiple-access channels. *IEEE Trans. Inform. Theory*, **51**(4), 1326–1347.

Liao, H. (1972). Multiple access channels. Ph.D. dissertation, Dept. of Electrical Engineering, University of Hawaii.

Lindskog, E. and Paulraj, A. (2000). A transmit diversity scheme for delay spread channels. In *Intl. Conf. on Communications (ICC)*, pp. 307–311.

Liu, J., Li, J., Li, H., and Larsson, E. G. (2001). Differential space–time modulation for interference suppression. *IEEE Trans. Signal Processing.*, **49**, 1786–1795.

Liu, Y., Fitz, M. P., and Takeshita, O. (2001a). Space–time codes performance criteria and design for frequency selective fading channels. In *Intl. Conf. on Communications (ICC)*, vol. 9, pp. 2800–2804.

(2001b). Full rate space–time turbo codes. *IEEE J. Select. Areas Commun.*, **19**(5), 969–980.

Liu, Z., Giannakis, G. B., Muquet, B., and Zhou, S. (2001). Space–time coding for broadband wireless communications. *Wireless Syst. Mobile Comput.*, **1**, 35–53.

Liu, Z., Giannakis, G. B., Scaglione, A., and Barbarossa, S. (1999). Decoding and equalization of unknown multipath channels based on block precoding and transmit-antenna diversity. In *Asilomar Conf. on Signals, Systems, and Computers*, pp. 1557–1561.

Loeliger, H.-A. (2004). An introduction to factor graphs. *IEEE Signal Processing. Mag.*, **21**(1), 28–41.

Love, D. J. and Heath, R., Jr. (2005a). Limited feedback unitary precoding for orthogonal space–time block codes. *IEEE Trans. Signal Processing.*, **53**(1), 64–73.

(2005b). Limited feedback unitary precoding for spatial multiplexing. *IEEE Trans. Inform. Theory*, **51**(8), 2967–2976.

Love, D. J., Heath, R. W. Jr., and Strohmer, T. (2003). Grassmannian beamforming for multiple-input multiple-output wireless systems. *IEEE Trans. Inform. Theory*, **49**, 2735–2747.

Lozano, A. and Tulino, A. (2002). Capacity of multiple-transmit multiple-receive antenna architectures. *IEEE Trans. Inform. Theory*, **48**(12), 3117–3128.

Lozano, A., Tulino, A., and Verdú, S. (2003). Multiple-antenna capacity in the low-power regime. *Trans. Inform. Theory*, **49**(10), 2527–2544.

(2005). High-SNR power offset in multi-antenna communication. *Trans. Inform. Theory*, **51**(12), 4134–4151.

(2006). Multiantenna capacity: myths and realities. In *Space–Time Wireless Systems: From Array Processing to MIMO Communications*, ed. H. Bölcskei, D. Gesbert, C. Papadias, and A. J. van der Veen. Cambridge: Cambridge University Press.

Lu, B. and Wang, X. (2000). Iterative receivers for multiuser space–time coding systems. *IEEE J. Select. Areas. Commun.*, **18**(11), 2322–2335.

Lu, B., Wang, X., and Narayanan, K. R. (2002). LDPC-based space–time coded OFDM systems over correlated fading channels: performance analysis and receiver design. *IEEE Trans. Commun.*, **50**(1), 74–88.

Lu, H. F. and Kumar, P. V. (2003). Rate–diversity trade-off of space–time codes with fixed alphabet and optimal constructions for PSK modulation. *IEEE Trans. Inform. Theory*, **49**(10), 2747–2752.

(2005). A unified construction of space–time codes with optimal rate–diversity tradeoff. *IEEE Trans. Inform. Theory*, **51**(5), 1709–1730.

Marshall, A. and Olkin, I. (1979). *Inequalities: Theory of Majorization and its Applications*. New York: Academic.

Marzetta, T. L. and Hochwald, B. M. (1999). Capacity of a mobile multiple-antenna communication link in Rayleigh flat fading. *IEEE Trans. Inform. Theory*, **45**(1), 139–157.

(2000). Unitary space–time modulation for multiple-antenna communications in Rayleigh flat fading. *IEEE Trans. Inform. Theory*, **46**(2), 543–564.

McEliece, R. J., MacKay, D. J. C., and Cheng, J.-F. (1998). Turbo decoding as an instance of Pearl's 'belief propagation' algorithm. *IEEE J. Select. Areas Commun.*, **16**(2), 140–152.

Minn, H. and Al-Dhahir, N. (2005a). PAR-constrained training signal designs for MIMO OFDM channel estimation in the presence of frequency offsets. In *Vehicular Technology Conf.*

(2005b). Training signal design for MIMO OFDM channel estimation in the presence of frequency offsets. In *Wireless Communications and Networking Conf.*

Molisch, A., Stienbauer, M., Toeltsch, M., Bonek, E., and Thoma, R. S. (2002). Capacity of MIMO systems based on measured wireless channels. *IEEE J. Select. Areas Commun.*, **20**(3), 561–569.

Moustakas, A. and Simon, S. (2003). Optimizing multiple-input single-output (MISO) communication systems with general Gaussian channels: nontrivial covariance and nonzero mean. *IEEE Trans. Inform. Theory*, **49**(10), 2770–2780.

Mukkavilli, K., Sabharwal, A., Erkip, E., and Aazhang, B. (2003). On beamforming with finite rate feedback in multiple antenna systems. *IEEE Trans. Inform. Theory*, **49**, 2562–2579.

Nabar, R. U., Bölcskei, H., and Kneubühler, F. W. (2004). Fading relay channels: performance limits and space–time signal design. *IEEE J. Select. Areas Commun.*, **22**(6), 1099–1109.

Naguib, A. and Seshadri, N. (1998). Combined interference cancellation and ML decoding of space–time block codes. In *Proc. 7th Communications Theory Mini-conference at Globecom'98*, Sydney.

Naguib, A., Tarokh, V., Seshadri, N., and Calderbank, A. R. (1998). A space–time coding modem for high-data-rate wireless communications. *IEEE J. Select. Areas Commun.*, **16**(8), 1459–1477.

Narula, A., Lopez, M., Trott, M., and Wornell, G. (1998). Efficient use of side information in multiple-antenna data transmission over fading channels. *IEEE J. Select. Areas Commun.*, **16**(8), 1423–1436.

Narula, A., Trott, M., and Wornell, G. (1999). Performance limits of coded diversity methods for transmitter antenna arrays. *IEEE Trans. Inform. Theory*, **45**(7), 2418–2433.

Ng, C. and Goldsmith, A. (2004). Transmitter cooperation in ad-hoc wireless networks: does dirty-paper coding beat relaying?. In *Proc. IEEE Information Theory Workshop*, pp. 277–282.

Oppermann, I. (1999). CDMA space–time coding using an LMMSE receiver. In *Proc. Intl. Conf. Commun. (ICC'99)*, Vancouver, BC, pp. 182–186.

Oyman, Ö., Nabar, R. U., Bölcskei, H., and Paulraj, A. J. (2003). Characterizing the statistical properties of mutual information in MIMO channels. *IEEE Trans. Signal Processing.*, **51**(11), 2784–2795.

Ozarow, L. H., Shamai, S., and Wyner, A. D. (1994). Information theoretic considerations for cellular mobile radio. *IEEE Trans. Vehicular Technol.*, **43**(2), 359–378.

Parkvall, S., Karlsson, M., Samuelsson, M., Hedlund, L., and Göransson, B. (2000). Transmit diversity in WCDMA: link and system level results. In *Vehicular Technology Conf.*, pp. 864–868.

Paulraj, A. J. and Kailath, T. (1994). *Increasing Capacity in Wireless Broadcast Systems using Distributed Transmission/directional Reception*, U.S. Patent, no. 5,345,599.

Paulraj, A. J., Nabar, R., and Gore, D. (2003). *Introduction to Space–Time Wireless Communications*. Cambridge: Cambridge University Press.

Peel, C., Hochwald, B., and Swindlehurst, L. (2005a). A vector-perturbation technique for near-capacity multiantenna multiuser communication – Part I: Channel inversion and regularization. *IEEE Trans. Commun.*, **53**(1), 195–202.

(2005b). A vector-perturbation technique for near-capacity multiantenna multiuser communication – Part II: Perturbation. *IEEE Trans. Commun.*, **53**(3), 537–544.

Poor, H. V. (1994). *An Introduction to Signal Detection and Estimation*. New York: Springer-Verlag.

(2002). Dynamic programming in digital communications: Viterbi decoding to turbo multiuser detection. *J. Optimiz. Theory Applic.*, **115**(3), 629–657.

(2004). Iterative multiuser detection. *IEEE Signal Processing. Mag.*, **21**(1), 81–88.

Poor, H. V. and Verdú, S. (1997). Probability of error in MMSE multiuser detection. *IEEE Trans. Inform. Theory*, **43**(3), 858–871.

Prasad, N. and Varanasi, M. K. (2004). Diversity and multiplexing tradeoff bounds for cooperative diversity schemes. In *Proc. IEEE Intl. Symp. Inform. Theory*, p. 268.

Proakis, J. (2000). *Digital Communications*, 4th edn. New York: McGraw-Hill.

Raleigh, G. G. and Cioffi, J. M. (1998). Spatio-temporal coding for wireless communication. *IEEE Trans. Commun.*, **46**(3), 357–366.

Rappaport, T. (1996) *Wireless Communications: Principles and Practice*. Englewood Cliffs, NJ: Prentice-Hall.

Rashid-Farrokhi, F., Liu, K. R., and Tassiulas, L. (1998). Transit beamforming and power control for cellular wireless systems. *IEEE J. Select. Areas Commun.*, **16**(8), 1437–1450.

Rashid-Farrokhi, F., Lozano, A., Foschini, G. J., and Valenzuela, R. A. (2000). Spectral efficiency of wireless systems with multiple transmit and receive antennas. In *Proc. IEEE Intl. Symp. on PIMRC*, London, vol. 1, pp. 373–377.

Rekaya, G. and Belfiore, J.-C. (2006). Complexity of ML lattice decoders for the decoding of linear full-rate space–time codes. *IEEE Trans. Wireless Commun.*, to be published.

Reynolds, D. and Wang, X. (2001). Adaptive group-blind multiuser detection based on a new subspace tracking algorithm. *IEEE Trans. Commun.*, **49**(7), 1135–1141.

Reynolds, D., Wang, X., and Poor, H. V. (2002). Blind adaptive space–time multiuser detection with multiple transmitter and receiver antennas. *IEEE Trans. Signal Processing.*, **50**(6), 1261–1276.

Richardson, T. J. and Urbanke, R. L. (2001). The capacity of low-density parity-check codes under message-passing decoding. *IEEE Trans. Inform. Theory*, **47**(2), 599–618.

Sampath, H. and Paulraj, A. (2002). Linear precoding for space–time coded systems with known fading correlations. *IEEE Commun. Lett.*, **6**(6), 239–241.

Sari, H., Karam, G., and Jeanclaude, I. (1995). Transmission techniques for digital terrestrial TV broadcasting. *IEEE Commun. Mag.*, **33**(2), 100–109.

Sato, H. (1981). The capacity of Gaussian interference channel under strong interference (corresp.). *IEEE Trans. Inform. Theory*, pp. 786–788.

Sayeed, A. (2002). Deconstructing multiantenna fading channels. *IEEE Trans. Signal Processing.*, **50**(10), 2563–2579.

Sellathurai, M. and Haykin, S. (2002). TURBO-BLAST for wireless communications: theory and experiments. *IEEE Trans. Signal Processing.*, **50**(10), 2538–2546.

Sendonaris, A., Erkip, E., and Aazhang, B. (1994). Increasing uplink capacity via user cooperation diversity. In *Proc. Intl. Symp. on Information Theory*, p. 156.

(2003a). User cooperation diversity – Part I: System description. *IEEE Trans. Commun.*, **51**(11), 1927–1938.

(2003b). User cooperation diversity – Part II: Implementation aspects and performance analysis. *IEEE Trans. Commun.*, **51**(11), 1939–1948.

Seshadri, N. and Winters, J. H. (1993). Two signaling schemes for improving the error performance of frequency-division-duplex (FDD) transmission systems using transmitter antenna diversity. In *Vehicular Technology Conf. (VTC)*, pp. 508–511.

(1994). Two signaling schemes for improving the error performance of frequency-division-duplex (FDD) transmission systems using transmitter antenna diversity. *Intl. J. Wireless Inform. Networks*, **1**, 49–60.

Sethuraman, B., Sundar Rajan, B., and Shashidhar, V. (2003). Full-diversity, high-rate space–time block codes from division algebras. *IEEE Trans. Inform. Theory*, **49**, 2596–2616.

Shamai, S. and Marzetta, T. L. (2002). Multiuser capacity in block fading with no channel state information. *IEEE Trans. Inform. Theory*, **48**(4), 938–942.

Shamai, S. and Wyner, A. D. (1997). Information-theoretic considerations for symmetric, cellular, multiple-access fading channels: Part I. *IEEE Trans. Inform. Theory*, **43**, 1877–1894.

Shamai, S. and Zaidel, B. M. (2001). Enhancing the cellular downlink capacity via co-processing at the transmitting end. In *Proc. IEEE Vehicular Technology Conf.*, pp. 1745–1749.

Shannon, C. (1948). A mathematical theory of communication. *Bell Sys. Tech. J.*, **27**, 379–423, 623–656.

(1949). Communications in the presence of noise. In *Proc. IRE*, **37**, 10–21.

(1958). Channels with side information at the transmitter. *IBM J. Res. Devel.*, **2**(4), 289–293.

Shannon, C. and Weaver, W. (1949). *The Mathematical Theory of Communication*. Urbana, IL: University of Illinois Press.

Sharif, M. and Hassibi, B. (2005). On the capacity of MIMO broadcast channels with partial side information. *IEEE Trans. Inform. Theory*, **51**(2), 506–522.

Shin, H. and Lee, J. (2003). Capacity of multiple-antenna fading channels: spatial fading correlation, double scattering and keyhole. *IEEE Trans. Inform. Theory*, **49**(10), 2636–2647.

Shiu, D., Foschini, G., Gans, M., and Kahn, J. (2000). Fading correlation and its effect on the capacity of multielement antenna systems. *IEEE Trans. Commun.*, **48**(3), 502–513.

Sidiropoulos, N., Davidson, T., and Luo, Z. Q. (2006). Transmit beamforming for physical layer multicasting. *IEEE Trans. Signal Processing.*, **54**(6), 2239–2251.

Simon, S. and Moustakas, A. (2003). Optimizing MIMO antenna systems with channel covariance feedback. *IEEE J. Select. Areas Commun.*, **21**(3), 406–417.

Siwamogsatham, S., Fitz, M. P., and Grimm, J. H. (2002). A new view of performance analysis of transmit diversity schemes in correlated Rayleigh fading. *IEEE Trans. Inform. Theory*, **48**(4), 950–956.

Skoglund, M. and Jöngren, G. (2003). On the capacity of a multiple-antenna communication link with channel side information. *IEEE J. Select. Areas Commun.*, **21**(3), 395–405.

Smith, P. J. and Shafi, M. (2002). On a Gaussian approximation to the capacity of wireless MIMO systems. In *Proc. Intl. Conf. on Communications*, pp. 406–410.

Soni, R., Buehrer, M., and Benning, R. (2002). Intelligent antenna system for cdma2000. *IEEE Signal Processing. Mag.*, **19**(4), 54–67.

Spencer, Q., Swindlehurst, L., and Haardt, M. (2004). Zero-forcing methods for downlink spatial multiplexing in multiuser MIMO channels. *IEEE Trans. Signal Processing.*, **52**(2), 461–471.

Srinivasa, S. and Jafar, S. (2005). Vector channel capacity with quantized feedback. In *Proc. IEEE Intl. Conf. on Communications (ICC)*.

Stamoulis, A. and Al-Dhahir, N. (2003). Impact of space–time block codes on 802.11 network throughput. *IEEE Trans. Wireless Commun.*, **2**(5), 1029–1039.

Stefanov, A. and Erkip, E. (2002). Cooperative coding for wireless networks. In *Proc. Intl. Workshop on Mobile and Wireless Communications Networks*, Stockholm, pp. 273–277.

Stoica, P. and Lindskog, E. (2001). Space–time block coding for channels with intersymbol interference. In *Proc. Asilomar Conf. on Signals, Systems and Computers*, Pacific Grove, CA, vol. 1, pp. 252–256.

Stridh, R., Ottersten, B., and Karlsson, P. (2000). MIMO channel capacity of a measured indoor radio channel at 5.8 GHz. In *Proc. Asilomar Conf. on Signals, Systems and Computers*, vol. 1, pp. 733–737.

Suard, B., Xu, G., and Kailath, T. (1998). Uplink channel capacity of space-division-multiple-access schemes. *IEEE Trans. Inform. Theory*, **44**(4), 1468–1476.

Swindlehurst, A. L., German, G., Wallace, J., and Jensen, M. (2001). Experimental measurements of capacity for MIMO indoor wireless channels. In *Proc. IEEE Signal Processing Workshop on Signal Processing Advances in Wireless Communications*, Taoyuan, Taiwan, pp. 30–33.

T.I. (1998). Space–time block coded transmit antenna diversity for WCDMA. Texas Instruments SMG2 document 581/98.

Taricco, G. and Biglieri, E. (2005). Space–time decoding with imperfect channel estimation. *IEEE Trans. on Wireless Commun.*, **4**(4), 1874–1888.

Tarokh, V. and Jafarkhani, H. (2000). A differential detection scheme for transmit diversity. *IEEE J. Select. Areas Commun.*, **18**(7), 1169–1174.

Tarokh, V., Jafarkhani, H., and Calderbank, A. R. (1999a). Space–time block codes from orthogonal designs. *IEEE Trans. Inform. Theory*, **45**(5), 1456–1467.

Tarokh, V., Naguib, A., Seshadri, N., and Calderbank, A. R. (1999b). Space–time codes for high data rate wireless communication: performance criteria in the presence of channel estimation errors, mobility, and multiple paths. *IEEE Trans. Commun.*, **47**(2), 199–207.

Tarokh, V., Seshadri, N., and Calderbank, A. R. (1998). Space–time codes for high data rate wireless communication: performance criterion and code construction. *IEEE Trans. Inform. Theory*, **44**(2), 744–765.

Tavildar, S. and Viswanath, P. (2006). Approximately universal codes over slow-fading channels. *IEEE Trans. Inform. Theory*, **52**(7), 3233–3258. See also http://www.ifp.uiuc.edu/~pramodv/pubs.html

Telatar, I. E. (1995). Capacity of multi-antenna Gaussian channels. *Bell Laboratories Technical Memorandum*, http://mars.bell-labs.com/papers/proof/

(1999). Capacity of multi-antenna Gaussian channels. *European Trans. Tel.*, **10**(6), 585–595.

ten Brink, S. (2001). Convergence behavior of iteratively decoded parallel concatenated codes. *IEEE Trans. Commun.*, **49**(10), 1727–1737.

TIA (1998). The CDMA 2000 candidate submission. Draft of TIA 45.5 Subcommittee.

Tirkkonen, O., Boariu, A., and Hottinen, A. (2000). Minimal non-orthogonality rate 1 space–time block code for 3+ tx antennas. In *Proc. IEEE ISSSTA2000*, vol. 2, pp. 429–432.

Tolhuizen, L. M. G. M. (2005). Soft-decision sphere decoding for systems with more transmit antennas than receive antennas. *12th Annual Symp. of the IEEE/CVT*, Enschede, The Netherlands.

Tonello, A. (2003). MIMO MAP equalization and turbo decoding in interleaved space–time coded systems. *IEEE Trans. Commun.*, **51**(2), 155–160.

Tong, L. and Perreau, S. (1998). Multichannel blind identification: from subspace to maximum likelihood methods. *Proc. IEEE*, **86**(10), 1951–1968.

Tse, D. and Hanly, S. (1998). Multiaccess fading channels – Part I: Polymatroid structure, optimal resource allocation and throughput capacities. *IEEE Trans. Inform. Theory*, **44**(7), 2796–2815.

Tse, D. N. C. and Viswanath, P. (2005). *Fundamentals of Wireless Communication*. Cambridge: Cambridge University Press.

Tulino, A., Lozano, A., and Verdú, S. (2005). Impact of antenna correlation on the capacity of multiantenna channels. *Trans. Inform. Theory*, **51**(7), 2491–2509.

(2006). Capacity-achieving input covariance for single-user multi-antenna channels. *IEEE Trans. Wireless Commun.*, **5**(3), 662–671.

Tulino, A. and Verdú, S. (2004). Random matrix theory and wireless communications. *Found. Trends Commun. Inform. Theory*, **1**(1).

Turin, G. (1962). On optimal diversity reception, II. *IRE Trans. Commun. Systems*, **10**(1), 22–31.

Tüchler, M. and Hagenauer, J. (2002). EXIT charts of irregular codes. In *2002 Conf. on Information Sciences and Systems*.

Tüchler, M., Koetter, R., and Singer, A. (2002). Turbo-equalization: principles and new results. *IEEE Trans. Commun.*, **50**(5), 754–767.

Tüchler, M., ten Brink, S., and Hagenauer, J. (2002). Measures for tracing convergence of iterative decoding algorithms. In *Proc. 4th IEEE/ITG Conf. on Source and Channel Coding*, Berlin, pp. 53–60.

Uddenfeldt, J. and Raith, A. (1992). *Cellular Digital Mobile Radio System and Method of Transmitting Information in a Digital Cellular Mobile Radio System*. U.S. Patent no. 5,088,108.

Uysal, M., Al-Dhahir, N., and Georghiades, C. N. (2001). A space–time block-coded OFDM scheme for unknown frequency-selective fading channels. *IEEE Commun. Lett.*, **5**(10), 393–395.

van der Meulen, E. C. (1977). A survey of multi-way channels in information theory: 1961–1976. *IEEE Trans. Inform. Theory*, **23**, 1–37.

(1994). Some reflections on the interference channel. *Communications and Cryptography: Two Sides of One Tapestry*, ed. R. E. Blahut, D. J. Costello, and T. Mittelholzer. Boston, MA: Kluwer, pp. 409–421.

Vaughan-Nichols, S. (2004). Achieving wireless broadband with WiMAX. *IEEE Computer Mag.*, **37**(6), 10–13.

Venkatesan, S., Simon, S., and Valenzuela, R. (2003). Capacity of a Gaussian MIMO channel with nonzero mean. In *Proc. IEEE Vehicular Technology Conf.*, vol. 3, pp. 1767–1771.

Verdú, S. (1998). *Multiuser Detection*. New York: Cambridge University Press.

(2002). Spectral efficiency in the wideband regime. *IEEE Trans. Inform. Theory*, **48**(6), 1319–1343.

Verdú, S. and Poor, H. V. (1987). Abstract dynamic programming models under commutativity conditions. *SIAM J. Control Optimiz.*, **25**(4), 990–1006.

Vikalo, H. and Hassibi, B. (2002). Maximum-likelihood sequence detection of multiple antenna systems over dispersive channels via sphere decoding. *EURASIP J. Appl. Signal Processing.*, **5**, 525–531.

Vikalo, H., Hassibi, B., and Kailath, T. (2004). Iterative decoding for MIMO channels via modified sphere decoding. *IEEE Trans. Wireless Commun.*, **3**(6), 2299–2311.

Vishwanath, S. and Jafar, S. (2004). On the capacity of vector interference channels. In *Proc. IEEE Inform. Theory Workshop*.

Vishwanath, S., Jafar, S., and Goldsmith, A. (2000). Optimum power and rate allocation strategies for multiple access fading channels. In *Proc. Vehicular Technology Conf.*, pp. 2888–2892.

Vishwanath, S., Jindal, N., and Goldsmith, A. (2003). Duality, achievable rates, and sum-rate capacity of Gaussian MIMO broadcast channels. *IEEE Trans. Inform. Theory*, **49**(10), 2658–2668.

Visotsky, E. and Madhow, U. (2001). Space–time transmit precoding with imperfect feedback. *IEEE Trans. Inform. Theory*, **47**, 2632–2639.

Viswanath, P. and Tse, D. N. (2003). Sum capacity of the vector Gaussian broadcast channel and uplink–downlink duality. *IEEE Trans. Inform. Theory*, **49**(8), 1912–1921.

Viswanath, P., Tse, D. N., and Anantharam, V. (2001). Asymptotically optimal water-filling in vector multiple-access channels. *IEEE Trans. Inform. Theory*, **47**(1), 241–267.

Viswanath, P., Tse, D. N., and Laroia, R. (2002). Opportunistic beamforming using dumb antennas. *IEEE Trans. Inform. Theory*, **48**(6), 1277–1294.

Viswanathan, H. and Venkatesan, S. (2003). Asymptotics of sum rate for dirty paper coding and beamforming in multiple-antenna broadcast channels. In *Proc. Allerton Conf. on Communications, Control, and Computing*.

Viswanathan, H., Venkatesan, S., and Huang, H. C. (2003). Downlink capacity evaluation of cellular networks with known interference cancellation. *IEEE J. Select. Areas Commun.*, **21**, 802–811.

Viterbi, A. J. (1998). An intuitive justification and a simplified implementation of the MAP decoder for convolutional codes. *IEEE J. Select. Areas Commun.*, **16**(2), 260–264.

Viterbo, E. and Biglieri, E. (1993). A universal lattice decoder. In *14-ème Colloque GRETSI*, Juan-les-Pins.

Viterbo, E. and Boutros, J. (1999). A universal lattice code decoder for fading channels. *IEEE Trans. Inform. Theory*, **45**(5), 1639–1642.

Vu, M. (2006). Exploiting transmit channel side information in MIMO wireless systems. Ph.D. Dissertation, Stanford University.

Vu, M. and Paulraj, A. (2003). Some asymptotic capacity results for MIMO wireless with and without channel knowledge at the transmitter. In *Proc. 37th Asilomar Conf. Signals, Systems and Computing*, vol. 1, pp. 258–262.

(2004). Optimum space–time transmission for a high K factor wireless channel with partial channel knowledge. *Wiley J. Wireless Commun. Mobile Comput.*, **4**, 807–816.

(2005a). A robust transmit CSI framework with applications in MIMO wireless precoding. In *Proc. 39th Asilomar Conf. on Signals, Systems and Computers*, pp. 623–627.

(2005b). Capacity optimization for Rician correlated MIMO wireless channels. In *Proc. 39th Asilomar Conf. Signals, Systems and Computing*, pp. 133–138.

(2005c). Characterizing the capacity for MIMO wireless channels with non-zero mean and transmit covariance. In *Proc. 43rd Allerton Conf. on Communications, Control, and Computing* pp. 623–627.

(2006). Optimal linear precoders for MIMO wireless correlated channels with non-zero mean in space–time coded systems. *IEEE Trans. Signal Processing.*, **54**, 2318–2332.

Wang, B., Zhang, J., and Host-Madsen, A. (2005). On the capacity of MIMO relay channels. *IEEE Trans. Inform. Theory*, **51**(1), 29–43.

Wang, X. and Host-Madsen, A. (1999). Group-blind multiuser detection for uplink CDMA. *IEEE J. Select. Areas Commun.*, **17**(11), 1971–1984.

Wang, X. and Poor, H. V. (1998a). Blind equalization and multiuser detection for CDMA communications in dispersive channels. *IEEE Trans. Commun.*, **46**(1), 91–103.

(1998b). Blind multiuser detection: a subspace approach. *IEEE Trans. Inform. Theory*, **44**(2), 677–691.

(1999a). Iterative (turbo) soft interference cancellation and decoding for coded CDMA. *IEEE Trans. Commun.*, **47**(7), 1046–1061.

(1999b). Space–time multiuser detection in multipath CDMA channels. *IEEE Trans. Signal Processing.*, **47**(9), 2356–2374.

(2002). *Wireless Communication Systems: Advanced Techniques for Signal Reception*. Upper Saddle River, NJ: Prentice-Hall.

Weichselberger, W., Herdin, M., Özcelik, H., and Bonek, E. (2006). A stochastic MIMO channel model with joint correlation of both link ends. *IEEE Trans. Wireless Commun.*, **5**(1), 90–100.

Weingarten, H., Steinberg, Y., and Shamai, S. (2004). Capacity region of the degraded MIMO broadcast channel. In *Proc. Intl. Symp. Information Theory*.

Winters, J. H. (1987). On the capacity of radio communication systems with diversity in a Rayleigh fading environment. *IEEE J. Select. Areas Commun.*, **5**, 871–878.

Winters, J. H., Salz, J., and Gitlin, R. D. (1994). The impact of antenna diversity on the capacity of wireless communications systems. *IEEE Trans. Commun.*, **42**(2), 1740–1751.

Wittneben, A. (1993). A new bandwidth efficient transmit antenna modulation diversity scheme for linear digital modulation. In *ICC*, pp. 1630–1634.

Wolfe, W. (1976). Amicable orthogonal designs – existence. *Can. J. Math.*, **28**, 1006–1020.

Worthen, A. P. and Stark, W. E. (2001). Unified design of iterative receivers using factor graphs. *IEEE Trans. Inform. Theory*, **47**(2), 843–849.

Wyner, A. (1994). Shannon-theoretic approach to a Gaussian cellular network. *IEEE Trans. Inform. Theory*, **40**, 1713–1727.

Yao, H. and Wornell, G. (2003a). Achieving the full MIMO diversity–multiplexing frontier with rotation based space–time codes. In *Allerton Conf. on Communication, Control, and Computing*.

(2003b). Structured space–time block codes with optimal diversity–multiplexing tradeoff and minimum delay. In *Proc. IEEE Global Telecom. Conf.*, vol. 4, pp. 1941–1945.

Yoo, T. and Goldsmith, A. (2006). On the optimality of multi-antenna broadcast scheduling using zero-forcing beamforming. *IEEE J. Select. Areas Commun.*, special issue on 4G wireless systems, **24**(3), 528–541.

Younis, W., Sayed, A., and Al-Dhahir, N. (2003). Efficient adaptive receivers for joint equalization and interference cancellation in multi-user space–time block-coded systems. *IEEE Trans. Signal Processing.*, **51**(11), 2849–2862.

Yu, K., Bengtsson, M., Ottersten, B., McNamara, D., Karlsson, P., and Beach, M. (2001). Second order statistics of NLOS indoor MIMO channels based on 5.2 GHz measurements. In *Proc. IEEE Global Telecomm. Conf.*, vol. 1, pp. 25–29.

Yu, W. (2003a). A dual decomposition approach to the sum power Gaussian vector multiple access channel sum capacity problem. In *Proc. Conf. on Information Sciences and Systems (CISS)*.

(2003b). Spatial multiplex in downlink multiuser multiple-antenna wireless environments. In *Proc. IEEE Globecom*.

Yu, W. and Cioffi, J. M. (2001). Trellis precoding for the broadcast channel. In *Proc. IEEE GLOBECOM*, San Antonio, TX, vol. 2, pp. 1338–1344.

(2004). Sum capacity of Gaussian vector broadcast channels. *IEEE Trans. Inform. Theory*, **50**(9), 1875–1892.

Yu, W., Ginis, G., and Cioffi, J. (2001). An adaptive multiuser power control algorithm for VDSL. In *Proc. Global Commun. Conf.*, pp. 394–398.

Yu, W., Rhee, W., Boyd, S., and Cioffi, J. (2004). Iterative water-filling for Gaussian vector multiple access channels. *IEEE Trans. Inform. Theory*, **50**(1), 145–152.

Yu, W., Rhee, W., and Cioffi, J. (2001). Optimal power control in multiple access fading channels with multiple antennas. In *Proc. Intl. Conf. Commun.*, pp. 575–579.

Yuksel, M. and Erkip, E. (2005). Can virtual MIMO mimic a multi-antenna system: diversity–multiplexing tradeoff for wireless relay networks. In *IEEE Communication Theory Workshop*.

Zhang, Y. and Blum, R. S. (2000). Multistage multiuser detection for CDMA with space–time coding. In *Proc. 10th IEEE Workshop on Statistical Signal and Array Processing*, Poconos, PA, pp. 1–5.

Zheng, L. and Tse, D. (2002). Communication on the Grassmann manifold: a geometric approach to the noncoherent multiple-antenna channel. *IEEE Trans. Inform. Theory*, **48**(2), 359–383.

(2003). Diversity and multiplexing: a fundamental tradeoff in multiple-antenna channels. *IEEE Trans. Inform. Theory*, **49**(5), 1073–1096.

Zhou, A. and Giannakis, G. B. (2001). Space–time coding with maximum diversity gains over frequency-selective fading channels. *IEEE Signal Processing. Lett.*, **8**(10), 269–272.

(2002). Optimal transmitter eigen-beamforming and space–time block coding based on channel mean feedback. *IEEE Trans. Signal Processing.*, **50**(10), 2599–2613.

(2003). Optimal transmitter eigen-beamforming and space–time block coding based on channel correlations. *IEEE Trans. Inform. Theory*, **49**(7), 1673–1690.

Ziemer, R. E., Peterson, R. L., and Borth, D. E. (1995). *Introduction to Spread Spectrum Communications*. Upper Saddle River, NJ: Prentice-Hall.

Index

The index covers the main text but not the various bibliographies. Figures divorced from the text coverage are indicated by bold page locators; matter in footnotes by italic locators. Where they are included in the Abbreviations List, initialisms are not expanded in the index, nor is an entry made for their expansiona.